Introduction to Materials Chemistry

Introduction to Materials Chemistry

Harry R. Allcock

A John Wiley & Sons, Inc., Publication

Library of Congress Cataloging-in-Publication Data is available.

Allcock, Harry R.
Introduction to Materials Chemistry
ISBN: 978-0-470-29333-1

Printed in the United States of America

10 9 8 7 6 5 4 3 2 1

Contents

Preface

Modern undergraduates and graduate students are keenly interested in the various ways in which chemistry impacts everyday life. This interest stems partly from career imperatives, but also from a natural interest in the relationship between fundamental science and the technology that increasingly dominates our lives. Classical chemistry courses at the university level rightly emphasize fundamental science. But courses that focus on the utilization of that chemistry are still rare.

This book is an outgrowth of a course in materials chemistry that I have taught for several years in the Chemistry Department at The Pennsylvania State University. It is a course taken by advanced undergraduates and by students who are in the first year of the graduate program in chemistry, most of whom have had little or no prior exposure to materials science. The class has also attracted undergraduates from materials science and chemical engineering programs who are seeking a general overview of the field. The subject matter is wide-ranging and, because the course extends over only one semester, the treatment aims for breadth, understanding, and perspective rather than great depth. The mathematical foundations of the field are deliberately excluded in order to emphasize chemical concepts rather than the traditional engineering or physics treatments. A challenge with this approach is that published material relevant to this subject is widely scattered in specialist books and research articles, and this presents a problem for students who seek to access background reading material. Thus, I have written this book in order to present a qualitative overview of relevant chemistry-related aspects and to provide a springboard to encourage readers to delve deeper into specific topics.

The book begins in Part I with an overview of the elementary chemistry that underlies much of modern materials science. Chapter 1 is a general introduction to the subject, Chapter 2 deals with descriptive ideas of chemical bonding, and Chapter 3 summarizes the background synthesis and reaction chemistry that play a crucial part in many aspects of materials science. Students who are familiar with elementary chemical bonding concepts may wish to move directly from Chapter 1 to Chapter

3. Special characterization techniques that are widely used in materials research are covered in Chapter 4.

Part II focuses on the core of materials chemistry—on different classes of materials, with five chapters dealing with polymers, glasses, oxide and non-oxide ceramics, metals, alloys, and composites. The emphasis in this section is on materials diversity, chemical synthesis, solid-state structure and its relationship to properties, and fabrication methods.

Part III builds on the foregoing chapters and continues with an emphasis on materials that are important in electronics (semiconductors and superconductors), energy-related applications, membranes, optics and photonics, biomedical materials, and an introduction to nanoscience and nanotechnology. Principles of device design and fabrication are increasingly emphasized in this part of the book. A glossary at the end of the book provides a ready reference to the meanings of many of the specialized terms encountered in this field.

Each chapter ends with a brief summary of future challenges in the different fields that could form the basis of class discussions and brainstorming sessions. Also included at the end of each chapter is a list of study questions and suggested reading material for more detailed study. A few of the references are to historical, groundbreaking articles that described major discoveries. Many of the study questions pose practical challenges in new materials design that students are encouraged to address either through written reports or through class discussions. Attempts by students to solve these practical problems have proved to be a popular aspect of the course. This also provides an introduction to the ways in which science and technology become integrated in the wider world.

I am grateful to several colleagues who read sections of the manuscript and made valuable suggestions. In particular, Professor John Badding at Penn State drew my attention to numerous evolving aspects of materials science. Dr. Nicholas Krogman read all the chapters and made many useful suggestions. I am also highly appreciative of the suggestions made by members of my research group and by numerous students who have taken this course and had an opportunity to see preliminary drafts of several chapters. Of course, any errors that remain are mine alone.

I hope that this book proves useful to all who have an interest in the impact of chemistry on modern technology and in the unique ways in which scientists have the knowledge, skills, and vision to bring about dramatic improvements to our way of life.

University Park, Pennsylvania   HARRY R. ALLCOCK

Introduction to Materials Chemistry

Part I

Introduction to Materials Science

1

What is Materials Chemistry?

A. DIFFERENT TYPES OF MATERIALS

Materials science is the science of *solids*, a field that encompasses most aspects of modern life. This book provides an introductory, qualitative overview of the role of chemistry in several important and expanding areas of materials science, with an emphasis on the ways in which new materials are designed, synthesized, evaluated, and used. It starts from the recognition that there are six or seven different types of solid substances that are the basis of materials science. These are polymers, ceramics, metals, element and interelement semiconductors, superconductors, optical materials, and a range of species in which small molecules are packed into ordered solids (Figure 1.1). These fields were once separate disciplines with little or no exchange of ideas across the boundaries. This is no longer true, and the central area in Figure 1.1 symbolizes research that joins and crosses the different disciplines.

For example, many of the ideas that were developed originally for polymer chemistry are found to explain some of the more puzzling aspects of ceramic science, and vice versa. Concepts that were once thought to be specific to inorganic semiconductors may explain the behavior of some polymers and small molecules in solids. Synthetic diamonds, especially diamond coatings produced by chemical vapor deposition, have characteristics that are reminiscent of both ceramics and semiconductors. Zeolites, which are inorganic materials that contain cavities that accommodate guest molecules and catalyze their reactions, have ceramic network structures but behave like the clathrates formed from small organic molecules in ordered solids. Some ceramics are superconductors, a phenomenon hitherto known only for certain metals and alloys and one inorganic polymer. Thus, a number of the most

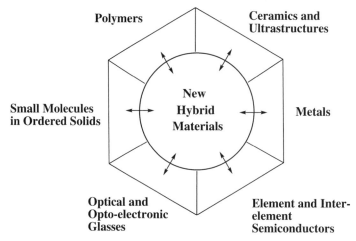

Figure 1.1. Different types of materials. Ceramics, metals, semiconductors, and optical materials are traditionally derived from inorganic sources. Polymers and small molecules in solids are normally obtained from organic or organometallic starting materials. The central area represents new types of materials that combine ideas and structures from the traditional areas in order to generate new combinations of properties. It is in this central area that some of the most important future advances can be expected.

interesting advanced developments in theory, design, and synthesis are taking place in the central area of hybrid materials shown in Figure 1.1.

This is particularly important because each of the classical materials have advantages and disadvantages. These are summarized in Figure 1.2. For instance, classical ceramics are rigid, chemically inert, and withstand high temperatures, but they are heavy, often brittle, and are difficult to fabricate into complex shapes. Most common metals are strong, tough, and good electrical conductors, but nearly all are heavy and prone to corrosion. There are also serious environmental penalties to be paid for their extraction from minerals and their refining. Inorganic semiconductors play a vital role in communications technology and computing but are difficult to purify and fabricate and are thus expensive. Classical polymers are inexpensive because they are derived from plentiful petroleum. They are easily fabricated because of their low softening temperatures, corrosion resistance, and excellent electrical insulating property. However, most polymers melt at only moderate temperatures, decompose when heated in air, and are flammable. Materials derived from small molecules packed into solids may be semiconductors, but they are brittle and melt at relatively low temperatures. These advantages and disadvantages further illustrate the importance of the central area in Figure 1.1, where cross-disciplinary research aims to produce materials that retain the advantages but minimize the disadvantages of existing materials.

All but two of the main materials areas were once based exclusively on inorganic chemistry. The exceptions were polymer science and small molecules in solids, both of which leaned heavily for their traditions on organic and organometallic chemistry. However, here, too, the boundaries are disappearing, as easily fabricated organic polymeric semiconductors replace traditional silicon semiconductors for some applications, transparent organic glasses replace the heavier silicate-based glasses

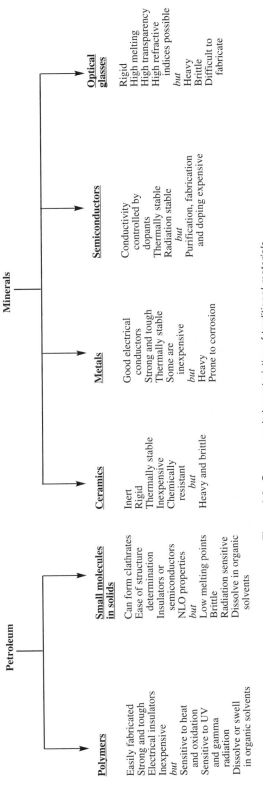

Figure 1.2. *Sources and characteristics of traditional materials.*

5

in lenses, prisms, and some optical waveguides and switches, and ultrastrong polymer–ceramic composites have grown in importance as replacements for metals. Polymers that are part organic and part inorganic have properties that cannot be duplicated by classical polymers, metals, or inorganic ceramics, and these hold considerable promise for the future.

Finally, it is essential for the reader to recognize the fundamental principle that materials science extends across all areas of chemistry, physics, engineering, biology, and medicine, and that breadth of knowledge across the physical and biological sciences is an essential requirement for understanding the materials field.

B. USES OF MATERIALS

The uses of materials are extremely diverse, and a few of them are summarized in Table 1.1 and Figure 1.3. Some fairly obvious applications are as structural materials in automobiles, aircraft, ships, trains, buildings, computer housings, nuclear reactors, and other structures (polymers, metals, and ceramics) where strength, impact resistance, and heat or radiation stability are important.

Other uses are in medical and dental applications that range from instruments, cardiovascular components (Figure 1.3a), drug delivery materials, scaffolds for tissue regeneration, and membranes (metals, ceramics, polymers). Still other applications for materials are in the electrical, computing, and communications sectors [semiconductors, electroluminescent screens, plasma displays, lasers, wiring, insulation. optoelectronic switches, and satellites (Figure 1.3b)]. Finally, a wide range of fibers, films, coatings, membranes, and elastomers are essential components of clothing, automobiles, aircraft, furniture, and manufacturing machinery. Many of these applications cannot be improved unless new materials are discovered that, for example, are

TABLE 1.1. Example Uses of Advanced Materials in Devices and Machines

Electrical	Photonics	Medical	Mechanical
Energy storage (batteries and supercapacitors)	Flat-panel displays Optical information processing	Tissue restoration and regeneration	Very strong materials (aerospace, gas turbines)
Energy generation (fuel cells, photovoltaics)	Information storage (CDs and DVDs)	Controlled drug release	High-temperature materials (jet and rocket engines, auto engines)
Superconducting magnets and train levitation	Energy conservation ("smart windows") Sensors	Artifical organs (cardiovascular. liver, pancreas,	Lightweight materials (auto, aerospace)
Information processing (semiconductors)	Fiberoptics	bone, kidney, etc.) DNA/RNA analysis	Abrasion resistance (bearings,
Sensors		Sensors for biochemical processes	machinery)
		Cell growth surfaces	Vibration damping
			Fuel and oil applications
			Seals (in rockets and hydraulic systems)
			Surface coatings

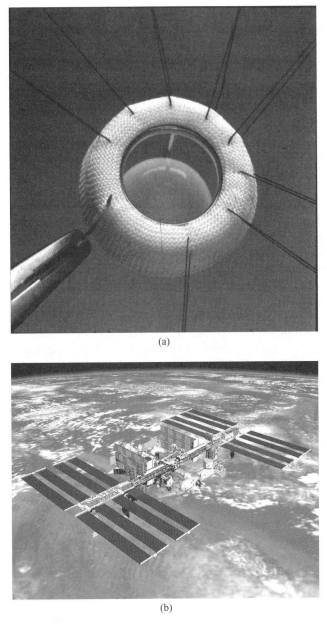

(a)

(b)

Figure 1.3. *(a) A prosthetic heart valve with a stainless-steel cage, a silicone rubber ball valve, and a polyester collar for suturing into the damaged organ. (b) The International Space Station showing the silicon semiconductor solar panels and the metal main structure (courtesy of NASA).*

fire-resistant, are unaffected by solvents or aggressive chemicals, or remain flexible at very low temperatures.

Most technology is materials limited. In other words, most advances in technology that are deemed to be desirable or even essential are blocked by the lack of materials that have the right combination of properties. For example, a certain ceramic may be stable to 2000 °C and would otherwise be used in gas turbines, but it is too brittle to withstand the shock waves that emanate from the jet engine. A promising polymer has the correct elasticity to be used in replacement heart valves, but its surface characteristics trigger a cascade of reactions that lead to the formation of a blood clot. A superconductor designed for use in levitated trains works well at −150 °C but loses its superconducting properties when warmed to room temperature.

C. APPROACHES TO PRODUCING NEW MATERIALS, NEW PROPERTIES, AND USES

There are two different approaches to research and development in materials science. The first is a traditional method used by engineers and many biomedical materials scientists. In this approach, materials are chosen from long-existing solids with a well-known, fixed set of properties that can be changed somewhat by physical manipulation (heat annealing, stretching, incorporation into composites, etc.). Typically this manipulation changes the way in which the molecules or crystallites are organized in the solid state without altering the overall molecular structure. The overwhelming credo is to improve devices or machines by new *engineering designs* using well-known inexpensive materials rather than more costly new materials. New materials often mean a lot of trouble because engineering and medical protocols have been fixed by long and arduous trial-and-error procedures. Nevertheless, this approach usually must be discarded for critical advanced applications such as aerospace developments or advanced medical devices, where no amount of manipulation of existing materials can generate the crucial combination of required properties.

The second approach is to use chemistry to produce entirely new materials. Chemists are uniquely placed to design and synthesize new materials. Chemists visualize materials down to their structure *at the molecular level*. They seek to improve materials through molecular chemistry. They ask two related questions: (1) "What is it about the molecular structure of a material that is responsible for its properties?" and (2) "How can I change the molecular structure to improve the materials properties?" This approach is more likely to generate dramatic advances in properties and performance. However, the molecular chemistry approach is successful only if it leads to a marked and often dramatic improvement in properties, because the cost of new products is always high until they can be produced and used on a large scale.

In practice, both the materials manipulation and the molecular chemistry approaches are important, and the challenge is to bring the two together. This connection of chemistry to engineering and medicine is one of the key features of materials science and biomaterials research. It becomes manifest in the sequence of events that occur along the pathway to the discovery and development of new materials. This is illustrated in Figure 1.4. The sequence starts with synthetic and mechanistic discoveries in small-molecule chemistry of the type that all chemists are familiar with. These discoveries can be in organic, inorganic, or organometallic chemistry. In addition, a few of these advances may provide access to new polymers, ceramics, semiconductors, and other materials with entirely new combinations of properties

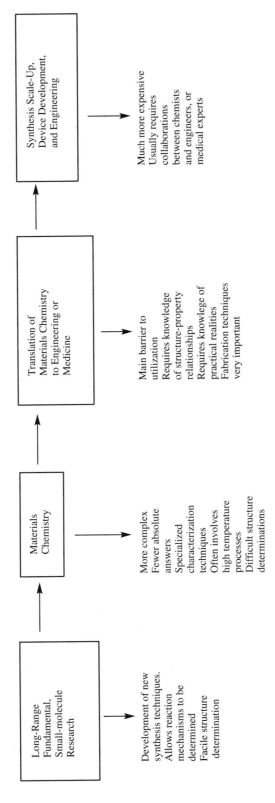

Figure 1.4. The sequence of steps from fundamental research to applications.

by processes that are described in later chapters. The characterization of these new substances by physical and analytical techniques will provide clues to their possible utility.

This second phase accounts for most of the effort that is expended by chemists in new materials research. However, it is not the end of the process, and the third step is critical. In this phase the properties of the new materials are evaluated by physicists, engineers, materials scientists, or medical experts for specific applications. This is the main barrier to utilization because the evaluation procedures are meant to mimic the behavior of the material under operational conditions in an aircraft, a space capsule, a rocket motor or jet engine, or when the material is implanted in a human being. Only if the new material survives these evaluation tests is it allowed to cross the barrier and move into the fourth stage, which is device development and engineering. Concurrent with this final phase, or perhaps preceding it, will be attempts to scale up the synthesis of the material by chemists and chemical engineers so that adequate quantities will be available. Whereas most chemists are pleased to be able to produce 100 g of a new compound, applications research normally requires kilogram quantities or more.

D. DEVICES AND MACHINES

One reason why materials science is now such an important part of modern life is because of its influence on the design and operation of devices and machines. Table 1.1 shows some of the modern technologies that are dependent for their continued progress on improvements in materials properties. Energy storage devices such as rechargeable batteries and supercapacitors have reached a plateau in their development because of the sparcity of suitable materials for solid electrolytes. The efficiency of energy generation devices such as fuel cells is seriously compromised by the limitations of the proton-conducting membranes or the ceramic electrolytes on which they depend. These two subjects are discussed in later chapters. A constant need exists for improved semiconductors, especially those that can be fabricated easily and cheaply, with future developments being driven by materials for lightweight flat-panel displays and for imaging and fabrication at the nanometer level. A related area where new materials are needed is in the field of sensors for the detection of microorganisms, toxins, biochemical products, and so on.

The field of optics has for many years depended on the use of inorganic glasses, especially silicate glasses, with various oxides employed to change the refractive index or color. However, irrespective of whether these materials are used in eyeglasses, prisms, camera lenses, or optical waveguides, they are often too heavy and brittle for many applications. For example, the replacement of the electrical circuits in aircraft or automobiles by optical circuits can realize its full potential only if lightweight materials are employed. Thus, the development of lightweight glasses is a major initiative. So, too, is the development of lightweight and easily fabricated materials for photonic switches to replace the semiconductors in electronic circuits.

The medical field has enormous potential for the use of advanced materials. New materials that can break down to harmless products in the body are needed for skin, nerve, and bone regeneration and for the controlled release of drugs. Elastomers that resist bacterial or fungal colonization are needed for artificial heart pumps, blood vessels, and catheters. Membranes for hemodialysis and blood oxygenation still need to be improved. Beyond these well-publicized devices there are many applications that do not appear in the media headlines but on which advanced technology depends.

When confronted by this broad, partial list of needed materials with new property combinations, it is easy to forget that the solutions to these problems lie in the utilization of fundamental chemistry. Without an underlying chemical approach to understanding structure–property relationships, it is unlikely that these applied developments will occur any time soon. Thus, the following chapters are designed to show the reader how chemistry can be utilized for the design and synthesis of both conventional and advanced materials.

E. THE ROLE OF CHEMISTRY IN MATERIALS SCIENCE

Modern chemistry can be divided into the study of two types of substances—small molecules and large molecules. Large-molecule science, together with ultrastructures or "extended lattices" (which are extremely large species in which a visible object consists of one giant molecule), is one of the main foundations of materials science. Small-molecule chemistry (i.e., 2–100 atoms per molecules) is the traditional basis of the chemical sciences, and it is usually the means through which students are introduced to chemistry. Small molecules provide us with most of the insights that underlie our knowledge of ceramics, polymers, and other materials and, as shown in Table 1.2, account for many of the traditional uses for chemicals in modern life. On the other hand, the more complicated molecules and ultrastructures that are used in materials science are the basis for many rapidly advancing technologies such as the other examples shown in Table 1.2. Some solids occupy an intermediate category between small and large molecules. They consist of small molecules assembled in a crystalline lattice in such a way that valuable solid-state properties are generated. These systems also fall into the category of "materials."

There are two tendencies in modern chemical research. The first is the evolution of chemistry from small molecules to macromolecules, ultrastructures, and materials. The second is the organization of chemical research around *properties* rather than

TABLE 1.2. Comparison of Uses for Small Molecules and Materials

Small Molecules	Materials
Pharmaceuticals	Structural materials
Fire retardants	Elastomers
Dyes	Textile fibers
Inks	Films and membranes
Detergents	Surface coatings
Food additives	Biomedical materials
Hydrocarbon fuels	Semiconductors
Insecticides and herbicides	Superconductors
Chemical warfare agents	Solid ionic conductors
	Optical materials (lenses, optical fibers, NLO materials)
	Photographic and lithographic materials

methodology. Chemistry has a crucial influence on nearly all aspects of materials science. The central role of chemistry stems from the unique ability of chemists to design and synthesize new materials with combinations of properties never seen before. Moreover, together with physicists and engineers, chemists search for a fundamental understanding of *why different materials are different* and how these differences can be exploited in practical applications.

There are three important aspects to modern chemistry. First, a major part of chemistry involves the development of ways to synthesize new compounds, measurement of their properties, and comparisons of their properties with predicted characteristics. It is the synthesis component that distinguishes chemists from nearly all other scientific specialists because there is little point in devising new chemical structures if they cannot be made. The importance of assembling new molecules is illustrated by the relationships shown in Figure 1.5. Thus, pioneering synthesis comes first—without it there are no new molecular structures, new properties, or uses.

The second important aspect of chemical research is to understand the relationship between molecular structure and physical or chemical properties, and to use this information to *predict* which new molecular structures might have improved properties and uses. Thus, molecular and solid-state structure determination is a key component of materials chemistry research. It follows that structure–property correlations are important as a guide to future work, for example, for molecular and materials design. In a productive research program the sequence of steps involves cycling through the *synthesis* of new molecules and materials, measurement of molecular and materials *properties*, and determination of *structure*, and then using that information as a basis for the design and synthesis of the next generation of materials.

Third, increasingly in more recent years, chemists have become involved in the initial steps to develop practical uses for new materials, often in collaboration with experts in other disciplines. But it is important to recognize that *uses are an offshoot of properties*, however elegant the method employed for the preparation of the

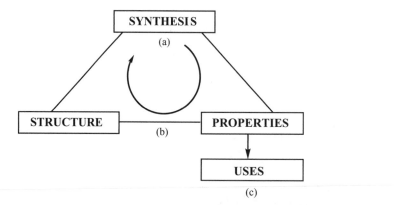

Figure 1.5. *Guiding principles of chemistry: (a) pioneering chemical synthesis comes first (without it, there are no new molecules, structures, properties, or uses); (b) structure–property correlations are important as a guide to future work, for example, in molecular or materials design; (c) uses are mainly an offshoot of properties, irrespective of structure or synthesis.*

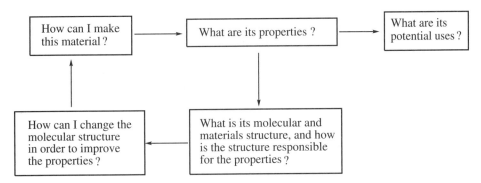

Figure 1.6. Questions that materials chemists must ask.

material or its structure determination. Many important developments in materials science occurred before the structure of a substance was known. However, a knowledge of structure is an enormous help in the understanding of both synthesis and properties and therefore for planning future work. It is in this sense that chemistry can be viewed as occupying the central area shown in Figure 1.1, an area that serves as an engine for producing novel materials and processes, from which radiate a flow of new substances that can make possible many diverse applications. The use of the general protocol shown in Figure 1.5 can be put into practice in materials science by answering a number of questions of the type shown in Figure 1.6.

Many of the following chapters are organized around the subject of *materials design*, which is the process of integrating synthesis, knowledge of structure–property relationships, and the needs of advancing technologies. It should also be noted that, in both small-molecule chemistry and materials science, attempts to study synthesis, structure, properties, or uses *on their own* (without integration with the other aspects) lowers the probability that major advances will be made.

F. A BROADER PERSPECTIVE

The scheme shown in Figure 1.7 illustrates how materials chemistry is connected to physics, engineering, biology, and medicine. Chemistry has traditionally been the crucible of new materials for these other areas, and it will continue to be so. It should also be noted that the growing connections between materials science and biology have spawned numerous advances in engineering as well as in medicine.

The further one moves from the fundamental science center shown in Figure 1.7 toward the perimeter, the more it is necessary to take into account the ways in which science and technology interface with the nonscience community. Many potential advances in technology cannot proceed because this outer interface has been neglected or presents an unsurmountable barrier. That barrier may be the high costs needed to produce a new material, it may be the traditions embedded in a particular industry, or it may arise from regulations written into law. Examples abound of new materials that could lead to significant medical, industrial, or aerospace advances, but which have been prevented from doing so by high initial costs or by regulations devised by nonscientists. So science policy, public policy, economics, and

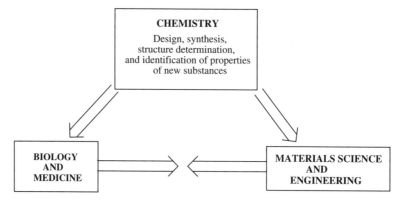

Figure 1.7. *Materials chemistry serves as an engine for the design and preparation of new compounds that help the advancement of biology, medicine, physics, and engineering, which are some of the main fields through which chemistry contributes to technology and society.*

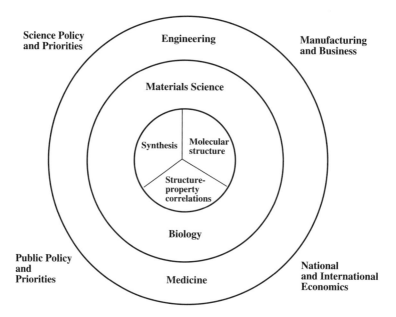

Figure 1.8. *Relationship of materials chemistry to other technical areas and to the wider world.*

manufacturing considerations must always be taken into account as aspects of chemistry move further and further from their intellectual center.

The early sections of this book deal with the fundamental (and often elementary) chemistry that underlies the main areas of materials science. Subsequent chapters review the synthesis, properties, and fabrication of a wide range of different materials, including ceramics, polymers, metals, small molecules in solids, extended lattices, semiconductors, superconductors, and optical materials. Other chapters describe how the properties of different materials are utilized to bring about improvements in medicine, batteries, fuel cells, and communications technology. Thus, the early chapters are chemistry-oriented. The later ones emphasize devices.

It is hoped that, after reading this book, and contemplating the questions at the end of each chapter, the reader will be able to delve into most of these subjects in greater detail. The references to further reading at the end of each chapter will serve as a starting point for more detailed studies.

G. TERMINOLOGY

One of the hurdles encountered by chemistry students who are entering the field of materials science for the first time is the unfamiliar terminology. This terminology is derived from many technical sources such as ceramic science, polymer science, geochemistry, metallurgy, optics, electronics, and other fields. The Glossary at the end of the book provides explanations of many of these terms. It is recommended that the reader scan this Glossary before proceeding further, and refer to it when an unfamiliar expression appears in the text.

H. EXAMPLE JOURNALS WHERE MATERIALS SCIENCE PUBLICATIONS CAN BE FOUND

ACS Nano [American Chemical Society (ACS)]

Advanced Materials (VCH)

Biomacromolecules (ACS)

Chemistry of Materials (ACS)

Journal of Inorganic and Organometallic Polymers and Materials (Springer)

Journal of Materials Chemistry [Royal Society of Chemistry, London (RSC)]

Journal of Polymer Science—Chemistry and Physics Sections (Wiley-Interscience)

Langmuir (ACS)

Macromolecules (ACS)

Materials Research Society Bulletin (MRS)

Nano Letters (ACS)

Nanomaterials (RSC)

Nature Materials

Science Materials

Soft Matter (RSC)

Solid State Ionics (Elsevier)

Physics Today

I. STUDY QUESTIONS (for class discussions or essays)

1. Wood is a material with outstanding properties, especially strength, impact resistance, and lightweight character. The earliest airplanes were made from

wood and canvas, and boats and ships were made from wood until relatively recent times. What is it about the structure of wood that gives it such desirable properties, and why is it no longer used in many common applications?

2. What are the advantages and disadvantages to the production of new materials (a) from petroleum or (b) from minerals? Are there alternatives to these two sources? If so, what are the advantages and disadvantages of these alternative sources?

3. Why is cost such an important factor that determines whether a new material is developed beyond the research stage? What applications can you think of where cost is not a significant factor, and why? Discuss how the cost of a new material might be reduced.

4. It is often said that pioneering research is relatively inexpensive but inherently inefficient, but that the development of uses is efficient but extremely expensive. Why might this be so?

5. Discuss why many engineers and most physicians prefer to use long-established materials rather than new materials that have better combinations of properties. Give examples. What would you do, as an experimental scientist, to help change this situation?

6. What might be the advantages of computers that used lasers instead of electric current, optical waveguides instead of copper or aluminum wires, and very small optical switches in place of transistors?

7. If polymers like polyethylene or polypropylene are lighter than steel, are easily fabricated into intricate shapes, and do not corrode, why do automobile manufacturers continue to use steel for most of their vehicles?

8. The rapid growth of materials science has, if anything, enhanced the need for scientists to have a deeper knowledge of small-molecule fundamental chemistry. Why is this so? Glance at the later chapters in this book before discussing this question.

2

Fundamental Principles that Underlie Materials Chemistry

A. WHY ARE DIFFERENT MATERIALS DIFFERENT?

Different substances have different properties because

- They contain different elements and different combinations of elements.
- They contain different types of chemical bonds.
- There are different sizes of the molecular units.
- Similar units within the solid-state structure may be assembled in different ways—the molecular packing arrangements may be different, or the geometry of linkage in an extended structure may vary.

These and several other factors are considered in the following sections.

B. THE ROLE OF DIFFERENT ELEMENTS

Certain elements in the periodic table play a larger role in materials science than do others. This brings about a significant simplification in the chemistry. Although most of the elements in the periodic table are used in one way or another in materials science, the 20 or so elements emphasized in Figure 2.1 are used in a disproportionate number of useful materials. These include nonmetals such as carbon, silicon, boron, nitrogen, oxygen, and phosphorus, and metals such as aluminum, magnesium, titanium, tin, iron, copper, chromium, and nickel. Other elements such as arsenic,

IA	IIA											IIIA	IVA	VA	VIA	VIIA	VIIIA
1	2	3	4	5	6	7	8	9	10	11	12	13	14	15	16	17	18
H																	He
Li	**Be**											**B**	**C**	**N**	**O**	F	Ne
Na	**Mg**											**Al**	**Si**	**P**	S	Cl	Ar
K	Ca	Sc	**Ti**	V	**Cr**	**Mn**	**Fe**	**Co**	**Ni**	**Cu**	**Zn**	Ga	**Ge**	As	Se	Br	Kr
Rb	Sr	Y	Zr	Nb	Mo	Tc	Ru	Rh	Pd	**Ag**	Cd	In	Sn	Sb	Te	I	Xe
Cs	Ba	*	Hf	Ta	W	Re	Os	Ir	Pt	**Au**	Hg	Tl	Pb	Bi	Po	At	Rn
Fr	Ra	#	Unq	Unp	Unh	Uns											

*Lanthanide series elements

La	Ce	Pr	Nd	Pm	Sm	Eu	Gd	Tb	Dy	Ho	Er	Tm	Yb	Lu

Actinide series elements

Ac	Th	Pa	U	Np	Pu	Am	Cm	Bk	Cf	Es	Fm	Md	No	Lr

Figure 2.1. Periodic table showing elements (in boldface) that play a major role in materials science.

cadmium, gallium, thallium, germanium, platinum, silver, and gold are also employed, but often in smaller quantities and for specialized uses.

The influence of different elements on properties is well known for small molecules, but in the science of solids it is particularly crucial. For example, although carbon and silicon are both in group IVA (14) of the periodic table and have many chemical similarities, the replacement of carbon by silicon in polymers, in the diamond structure, or in ceramics, brings about far-reaching changes to the properties. These differences result from the different sizes of the atoms, their electronic configuration, the ease with which the outer electrons can be lost, and the types and lengths of the bonds they form with neighboring atoms. For example, silicon has the same outer-shell electron configuration as carbon, but it is larger and forms longer covalent bonds to its neighbors. Moreover, unlike carbon, it is reluctant to form double bonds to oxygen, nitrogen, or other silicon atoms. Silicon also has the ability to expand its octet and form bonds to four, five, or six other atoms, whereas carbon is restricted to bonds to two, three, or four other atoms. Several of these differences are responsible for the fact that pure silicon (with the same extended solid-state structure as diamond) is a semiconductor whereas diamond is not. When both are doped with "impurities," silicon is a much better semiconductor than diamond.

The size of atoms increases down a group in the periodic table because additional electron shells have been added. However, atomic radii decrease in going from left to right across the periodic table because of the rising number of positive charges in the nucleus and their increased pull on the outer electrons. The ionization potential (the energy required to remove an electron from an isolated atom) decreases down a group but increases from left to right.

Thus, if we compare elements that are in different groups in the periodic table, or are in the same group but are not adjacent to each other, then the differences become profound. Some similarities exist between silicon and germanium, but far fewer between silicon and tin or lead. The introduction of transition metals changes the picture even more profoundly because of the ability of these elements to utilize their *d* orbitals, generate colored materials, enhance electrical conductivity, and provide highly reflective surfaces. Indeed, one of the most important consequences of the replacement of one element by another is the change this brings about to the chemical bonding.

C. DIFFERENT TYPES OF CHEMICAL BONDS

Theories of chemical bonding range from the purely descriptive, which are useful for visualization, to complex calculations based on quantum interpretations and approximations to the Schrödinger equation. In this book we will utilize the more descriptive approach. In the simplest interpretation of molecular and materials structure there are five types of forces that hold the components of solids together— van der Waals associations, covalent bonds, coordinate bonds, ionic assemblies, and metallic structures. These differ greatly in strength, with van der Waals forces being the weakest. The characteristics of materials with different bond types are shown in Table 2.1.

TABLE 2.1. Types of Bonds Found in Solids

Van der Waals Forces

Weak attractive forces that result from electron–electron or "dispersion" interactions; hold molecules together in the solid state, and are associated with low-melting solids and materials that are soluble in organic liquids; important in shape-fitting of molecules in the solid state.

Covalent Bonds

Strong bonds formed by overlap of atomic orbitals; exist as 2-electron (single) bonds, 4-electron (double) bonds, and 6-electron (triple) bonds, increasing in strength and decreasing in length in that order. Most covalently bonded structures are electrical insulators, except when numerous double or triple bonds are conjugated. Most covalent bonds require heat or high energy radiation before they break, although strong mechanical force may also cause bond cleavage. Thus, covalent bonds are associated with high thermal stability, especially in solids that have a 3D extended network of bonds. Polar covalent bonds are cleaved by ionic reagents.

Coordinate Bonds

Formed by donation of an electron pair from a donor atom with unused outer electrons to an electron-deficient acceptor. They are strong, thermally stable bonds, but are broken fairly easily by stronger electron donors or by polar solvents or reagents.

Ionic Bonds

Found mainly in the solid state; formed by the complete transfer of electrons from one atom to another to give cations and anions. These are strong forces in the solid state that lead to high melting temperatures. The solid-state structures are determined by the packing of spherical ions and by charge balancing. Ionic solids have low impact resistance due to cleavage down the crystal planes. They are electrical insulators in the solid state, but ionic conductors when dissolved in polar liquids.

Metallic Bonds

Metals are made up of atomic nuclei and core shell electrons organized into a lattice that is controlled by the close packing of spheres. Mobile electrons bind the structure together and are responsible for electronic conductivity and high reflectivity. This arrangement gives high strength, malleability, ductility, and impact resistance. The melting points vary according to the efficiency of binding and the degree to which metal–metal covalent bonding is also involved.

1. Van der Waals Forces and the Lennard-Jones Potential

Van der Waals forces are the weak interactions that hold neutral molecules together in the solid state. The attractions arise from the "exchange forces" between electrons in the outer regions of the interacting molecules. Van der Waals forces are responsible for the assembly of covalent molecules into crystal lattices, and the combination of attractions and molecular shape-fitting are responsible for the different packing patterns that arise in nonionic crystals. Because van der Waals forces are weak they are easily disrupted by heat, which explains why solids consisting of closely packed small molecules tend to melt at low or moderate temperatures.

In practice, the distance between atoms or groups of atoms can be defined by terms such as the Lennard-Jones potential, given in equation 1, where V is the potential energy, r is the distance between the interacting atoms, and ε is the depth of the well that corresponds to the equilibrium distance between the atoms:

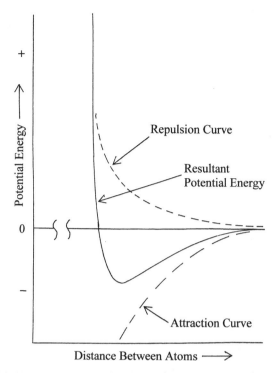

Figure 2.2. *Potential energy–atom separation curves for separate attractive and repulsive terms in the Lennard-Jones equation, and the combined interaction energies, showing how the attractions increase with decreasing separation but are counteracted by the nuclear–nuclear repulsions at short distances. The minimum in the curve represents the distance at which these two forces balance each other. The energy decrease is the force that holds covalent molecules together in a crystalline lattice, and is the driving force for shape-fitting by the molecules to allow the most efficient minimization of the overall energy. Note that the general shape of the van der Waals curve can also be used to explain the preferred conformation of a molecule caused by the attractions or repulsions of the groups connected to a rotatable bond.*

$$V(r) = 4\varepsilon \left[\left(\frac{\sigma}{r} \right)^{12} - \left(\frac{\sigma}{r} \right)^{6} \right] \tag{1}$$

Figure 2.2 shows how these forces arise by a balancing of the long-range non-bonding attractions (the 10^6 term) between atoms and molecules against the repulsions (the 10^{12} term) that result from too close an approach of the positively charged atomic nuclei.

2. Covalent Bonds

Covalent bonds are much stronger than van der Waals interactions, and considerable energy is needed to break them. For example, a Si—Si bond has an energy of 52 kcal/mol, an aliphatic carbon–carbon bond has a value of 83 kcal/mol, and a C=N bond has an energy as high as 213 kcal/mol. Covalent bonds arise between participating atoms through the pairing of outer-level electrons with opposite spin (Figure 2.3).

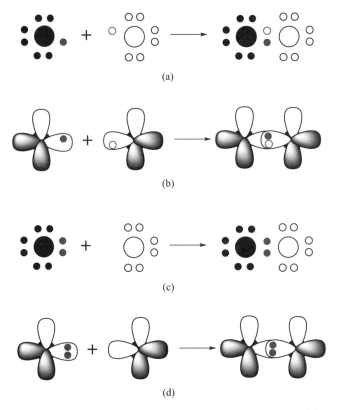

Figure 2.3. *Covalent bonding as represented by (a) the electron dot model and (b) the orbital overlap concept. Coordinate bonding, in which one atom supplies both of the electrons for the bond, is depicted in rows (c) and (d).*

One electron is provided by each participating atom. The electrons are located in orbitals, the overlap of which yields a spin-paired bond. When more than two electrons contribute to a bond (double, triple, or quadruple bonds) the bonds become shorter and stronger. Thus, the C=C bond has an energy of 146 kcal/mol, C=O has a value of 177–179 kcal/mol, and P=O has a value of 110 kcal/mol. Some of the highest covalent bond energies are found for bonds, such as C=N or P=O, which have a high polarity (see later).

a. Bond Angles. The overall shape of a molecule can be traced to its bond angles and molecular conformation. Three factors control the bond angles in an isolated small molecule—the inherent orientation of molecular orbitals, the repulsion of the electron pairs in nearby bonds, and the attractions or repulsions of groups of atoms that are linked to the central atom.

In the simplest sense the bond angles in an isolated molecule can be understood in terms of the angles between orbitals, as illustrated in Figure 2.4. Thus, in principle, a central atom with three unpaired electrons in three *p* orbitals that are oriented 90° from each other (orthogonal) will overlap the orbitals of three other atoms to give a pyramidal molecule with approximately 90° bond angles. Some representative bond angles are given in Table 2.2, and it is clear that these angles cannot be

s-orbital   **p-orbital set**   **d-orbital set**

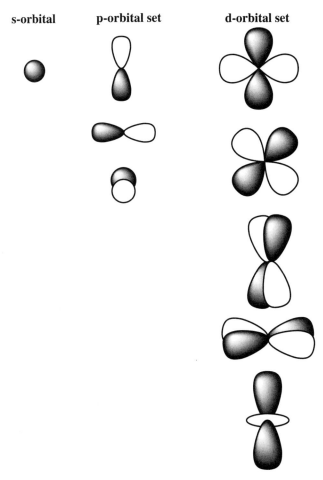

Figure 2.4. *Orientation of nonhybridized orbitals. Mixing of these generates hybrid orbitals with different bond angles, as illustrated in Figure 2.5.*

explained exclusively by the inherent orientation of simple molecular orbitals. For example, if the simple orientation of p orbitals were involved, all the H—N—H bond angles in ammonia would be 90°, and the H—O—H angle in water would also be 90°, and these are not the angles found by experiment.

In practice, the p orbitals combine with a spherical s orbital or with d orbitals to form hybrid orbitals that give rise to different bond angles. Thus, one p orbital hybridized with an s orbital gives an sp hybrid and a bond angle of 180°. The sp^2 hybrids give planar molecules with 120° bond angles, and some elements can form hybrid orbital arrangements by mixing in s orbital character to connect to four other atoms, in which case the resultant sp^3 hybrid orbitals of the central element will generate a set of 109.5° bond angles. Thus, four bonds to a central element yield a tetrahedral structure. This is the situation found in many carbon and silicon compounds and in the extended lattices of diamond and silicon. The existence of sp^3d hybrids explains the combination of 90° and 120° angles (trigonal bipyramid) when five bonds radiate from a central atom, and an sp^3d^2 arrangement is consistent with

TABLE 2.2. Experimentally Determined Bond Angles[a]

Angle	Compound	Angle (deg.)	Comments[b,c]
H—C—H	CH_4	109.5	tet
C—C—C	Saturated aliphatic	109.5	tet
C—C=C	Aromatic rings	120	
C≡C—H	Acetylenic	180	
C—O—C	::OR_2	109.5	Ether ps-tet
O—C—O	In O—CH_2—O	~109.5	
C—C(=O)—C	Ketones	~120	
H—O—H	::OH_2		ps-tet
O—Al—O	$[Al_6 O_{18}]^{18+}$	~109	In $Ca_9Al_6 O_{18}$
C—Si—C	$(CH_3)_2SiCl_2$	~109	
F—B—F	BF_3	120	tpl
H—N—H	:NH_3	107	ps-tet
H—B—H	$BH_{3\ (gas)}$	120	
B—N—B	Borazines and hexagonal boron nitride	120	Planar rings
N—B—N	Borazines and hexagonal boron nitride	120	Planar rings
O—Si—O	Silicone polymers	~109	
Si—O—Si	Silicates and α-quartz	143+	
O—P—O	PO_4^- (phosphate)	~109	tet
P—O—P	Polyphosphate		
N—P—N	Cyclic and linear phosphazenes	~120	
P—N—P	:NPR2 (cyclic and linear phosphazenes)	120–160	
Cl—P—Cl	:PCl_3	109	ps-tet
Cl—P—Cl	PCl_5 (gas)	90, 120	tbp
Cl—P—Cl	PCl_6^-	90	oct
X—Cr—X	CrX_6	90	oct
X—Pt—X	PtX_4	90	sq-pl

[a]*Symbols key* : indicates a lone pair of electrons, :: indicates two lone pairs; tet = tetrahedral, ps-tet = pseudotetrahedral, tbp = trigonal bipyramidal, oct = octahedral, sq-pl = square planar, tpl = trigonal planar.
[b]Note that bond angles in a solid-state structure may be distorted from "molecular" values by crystal-packing forces.
[c]Note also that bond angles are meaningful only for covalent or coordinate bonding systems and not for ionic crystals.

all the angles being $90°$ (octahedral) with six bonds radiating from the central element (Figure 2.5). Thus, the underlying supposition in this approach is that inherent orbital structure and hybridization control the bond angles and, thereby, the shape of a molecule.

What controls the hybridization? One factor is the mutual repulsion of the electron pairs that constitute each bond. This is known as the *valence shell electron pair repulsion* (VSEPR) concept. The basis of this approach is that electron pairs attempt to move as far away from each other as possible. Depending on the number of bonds around a central atom, these repulsions will generate specific structures with specific bond angles. Thus, two electron pairs (two bonds) will widen the bond angle to $180°$. Three electron pairs will yield a trigonal planar ($120°$) structure. Four electron pairs yield a tetrahedral structure, five will give rise to a trigonal bipyramid, and six will yield an octahedral arrangement. Moreover, if unbonded electrons are left over in "lone-pair orbitals," these, too, will repel nearby bonds and help to determine the molecular shape. For example, ammonia is pyramidal rather than trigonal planar because there is a fourth orbital—the lone pair on the nitrogen. Thus, the approximately $109°$ H—N—H angles in ammonia reflect the presence of the nonbonding

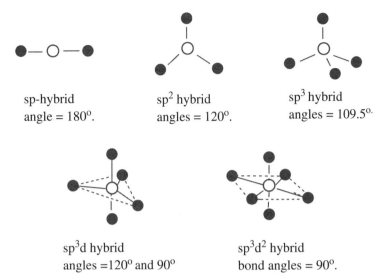

sp-hybrid
angle = 180°.

sp² hybrid
angles = 120°.

sp³ hybrid
angles = 109.5°·

sp³d hybrid
angles =120° and 90°

sp³d² hybrid
bond angles = 90°.

Figure 2.5. *Bond angles and geometries imposed by different hybrid orbital arrangements.*

lone pair, which induces a pseudotetrahedral structure (i.e., tetrahedral if the lone pair is included). In water there are two lone-pair orbitals and a pseudotetrahedral arrangement of bonds, with H—O—H angles near 109.5°. A linear bond arrangement could be the result of two large atoms or groups linked to a connecting atom to get as far away from each other as possible. Three atoms linked to a central atom can minimize their repulsions by forming a trigonal planar unit. Four atoms around a central atom require a tetrahedral structure to minimize their mutual repulsions. Five and six atoms form a trigonal bipyramidal and octahedral arrangements, respectively. Oxygen atoms in water have two lone-pair orbitals, which force the H—O—H angle to approximate to 109.5°. The same is true of the oxygens in neutral silicates or silicone polymers, and this narrows the Si—O—Si angle considerably below what might otherwise have been predicted to be 180°. In practice, the Si—O—Si bond angle is extremely variable because oxygen can readily use different hybrid orbitals in response to solid-state packing forces or physical stress. So, bond angles and molecular shapes can be understood either in terms of the orbital configurations or in terms of atomic repulsions. Although it can be argued that the inherent orientation of hybrid orbitals or VSEPR predominate, the answer probably lies in a balance of the two. Whether electron pair repulsion induces orbital hybridization or vice versa is a moot question. They have the same effect, and they work in concert.

An additional factor that must be taken into account in any discussion of bond angles is the *size* of the *groups of atoms* in the neighborhood of the bonds in question. The mutual attraction or repulsion of their electron clouds and nuclei at different distances (Figure 2.6) will also affect the bond angle. For example, an ether-type oxygen atom connected to two large groups, such as benzene rings, has a wider C—O—C angle than does the H—O—H angle in water.

The energy changes that accompany the approach of groups such as benzene rings, alkyl groups, and halogens, in a covalent molecule will be similar to the van

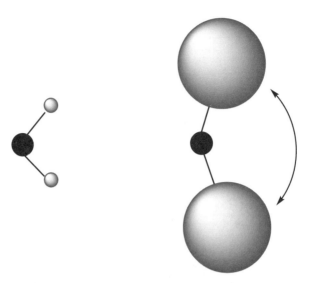

Figure 2.6. *Widening of a bond angle caused by the mutual repulsion of bulky groups attached to a central atom. Examples of such units are phenyl rings, diphenylamino, naphthyl, or neopentyl groups, and large atoms such as bromine, iodine, germanium, or transition metals.*

der Waals and Lennard-Jones interactions discussed earlier. At large distances there is a weak attraction, but strong repulsions will be encountered at close distances. These interactions are frequently simulated using calculations based on standard potentials, such as the Lennard-Jones potential, in which the (hard) repulsive term varies according to 10^{12} the distance between atoms and the (softer) attractive term varies according to a 10^6 function.

In the solid state an additional factor comes into play—the forces imposed by the packing of units into a lattice—and these frequently distort bond angles from the values found for the same molecules in solution or in the vapor state.

b. Bond Lengths. Covalent bond lengths depend on the participating elements, with the length of a nonpolar covalent single bond corresponding roughly to the sum of the covalent radii. Just as atomic radii increase down a group in the periodic table, so, too, do the covalent radii of those same elements when they are part of a bond. Typical covalent radii are shown in Table 2.3.

These values are usually estimated initially from one half the measured bond distance between atoms of the same element, although they may also be calculated from first principles. Representative bond lengths are shown in Table 2.4. As an example, the carbon–carbon bond length of ~1.54 Å suggests a covalent radius for an aliphatic carbon atom of 0.77 Å. The silicon–silicon bond length of 2.32–2.35 is one of the bases for a covalent radius for silicon of 1.11 Å. So long as the bond is relatively nonpolar, and double or triple bonding is absent, these values may be used to estimate the lengths of heteroatomic bonds. For example, the sum of the covalent radii of carbon and silicon is 1.88 Å, which corresponds closely to the experimentally measured value of 1.87 Å in methylsilane, H_3C–SiH_3. However, the sum of the covalent radii of silicon and oxygen equals 1.84 Å, whereas the experimental value of 1.5–1.64 Å is considerably shorter. Such bond shortening is probably due to the

TABLE 2.3. Covalent Radii of Selected Elements (Å)

Hydrogen	0.32	Fluorine	0.72
Boron	0.82	Chlorine	0.99
Aluminum	1.18	Bromine	1.14
Carbon	0.77	Iodine	1.33
Silicon	1.11	Magnesium	1.36
Germanium	1.22	Zinc	1.25
Tin	1.40	Titanium	1.32
Lead	1.47	Manganese	1.37
Nitrogen	0.75	Iron	1.17
Phosphorus	1.06	Cobalt	1.16
Arsenic	1.20	Nickel	1.15
Oxygen	0.73	Copper	1.17
Sulfur	1.02	Silver	1.34
Selenium	1.17	Gold	1.34

Souce: From R. T. Sanderson, *Inorganic Chemistry*, Reinhold, New York, 1967.

TABLE 2.4. Selected Examples of Bond Lengths (Å)

Bond	Source	Length	Sum of Covalent Radii[a]	Bond	Source	Length	Sum of Covalent Radii[a]
H—H	Hydrogen gas	0.74	0.64	O—H	Alcohols	0.98	1.05
C—C	Diamond	1.54	1.54	C—O	Aliphatic	1.43	1.50
C—C	Aliphatic	1.54	1.54	C=O	Aldehydes	1.22	1.50
C=C	Olefins	1.35	1.54	C—N		1.47	1.52
C≡C	Alkynes	1.20	1.54	C=N	Aromatic	1.34	1.52
C=C	Aromatic	1.39	1.54	C≡N	Nitriles	1.16	1.52
Si—Si	Silicon	2.35	2.22	Si—C	Methylsilane	1.87	1.88
Si—Si	Silanes	2.32	2.22	Si—O	Polysiloxanes	1.64	1.84
Ge—Ge	Element	2.41	2.44	Si—O	Silica	1.51	1.84
P—P	P₄	2.21	2.12	Si—N	Silicon nitride	1.57	1.80
S—S	S₈	2.05	2.04	P=N	Phosphazenes	1.56–1.60	1.81
As—As	Element	2.49	2.40	P—O	Phosphates	1.66	1.79
Al—Al[b]	Metal	2.86	2.36	P=O	Phosphates	1.52	1.79
Fe—Fe[b]	Metal	2.48	2.34	B—O	Borates	1.36	1.55
Co—Co[b]	Metal	2.51	2.32	B—N	Boron nitride	1.28	1.56
Cu—Cu[b]	Metal	2.56	2.34	B—N	Borazines	1.42	1.56
Ag—Ag[b]	Metal	2.89	2.68	Al—O	Alumina	1.76	1.91

[a]Note that the greatest discrepancies between the sum of the covalent radii and the experimentally measured bond lengths occur when the bond has a high polar or ionic component or has multiple bond character.
[b]Note also that metal–metal atom distances may reflect the packing pattern of spheres in the solid lattice rather than the influence of covalent bonds. The same applies to ions in solids.

high polarity of this bond. Note that in carbon compounds the change from single to double to triple bonds results in a decrease in bond length. Similarly, with other elemental combinations, bonds shorter than the sum of the covalent radii may indicate some form of multiple bonding. Note also that short bond lengths often lead to increased crowding of the molecular units linked to the atoms that flank that bond and cause changes to the bond angles.

c. Bond Torsion. The physical and mechanical properties of molecules and materials often depend on the ease with which the bonds can undergo torsion ("internal rotation"). Bonds with a low energy barrier to torsion allow the molecule to twist and change shape easily, and this is a crucial factor that determines whether a material is a brittle solid, an elastomer, or a gum. Materials with good impact resistance, such as rubbery polymers, are usually characterized by the presence of bonds that can absorb energy by undergoing torsion rather than by being stretched and broken. Ease of bond torsion depends on the type of bond and the size and polarity of the nearby groups (Table 2.5). For example, the barrier to internal rotation for C–C bonds can vary from a fraction of a kilocalorie per mole (kcal/mol) to ≥10 kcal/mol (40 kJ/mol) depending on the size and polarity of the atoms or groups linked to the two carbon atoms. A carbon–carbon double bond has a higher inherent torsional barrier than does the corresponding single bond and is therefore less flexible. However, the phosphorus–nitrogen double bond in phosphazene polymers has an extremely low torsional barrier (perhaps as low as 0.1 kcal) because the electron hybridization is different.

Silicon–oxygen bonds in silicone polymers have low torsional barriers, and this is one of the reasons for the high flexibility and impact resistance of these polymers. Large groups that flank a particular bond may raise the torsional barrier by non-bonded steric interference, in other words, by repulsions that become manifest as the bond in question is twisted to angles that bring the peripheral groups into a repulsive region (Figure 2.7). Hydrogen bonding or dipolar interactions may limit bond torsion via strong intramolecular *attractions*.

Torsional barriers can be measured *experimentally* by microwave, nuclear magnetic resonance (NMR), or infrared spectroscopy, but different techniques often give different values. Microwave spectroscopy is an excellent technique when applied to small molecules in the vapor state. NMR analysis can be used to estimate

**TABLE 2.5. Example Barriers to Bond Torsion
("Internal Rotation")**

Bond	Torsional Barrier (kcal/mol)
$H_3C–CH_3$	2.9
$H_3C–CF_3$	3.2
$F_3C–CF_3$	4.0
$CH_3CH_2–$ OH	3.8
$H_3C–NH_2$	2.0
$H_3CC(=O)–$ OH	41
$H_3C–OCH_3$	2.7
$H_3Si–SiH_3$	0.9
$H_3C–SiH_3$	1.7
$F_3Si–SiF_3$	0.6
$Cl_3Si–SiCl_3$	1.6
$Br_3Si–SiBr_3$	2.6
$F_2B–BF_2$	1.8
$H_2N–NH_2$	1.4
$C_6H_5–OH$	3.4
$HO–OH$	7.0

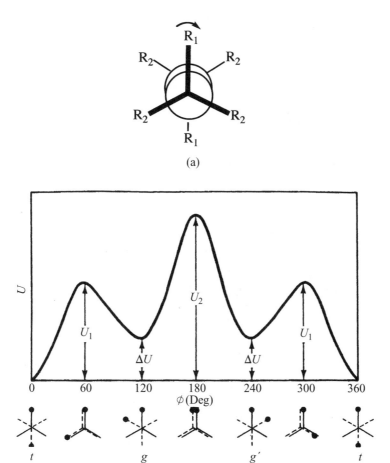

Figure 2.7. *Potential energy profile generated by torsion of a covalent bond that connects atoms with large groups that repel each other. (a) Illustration of a minimum rotational energy state that exists if the two large groups (R_1) have the maximum separation, which occurs at the trans conformation. (b) Energy changes during rotation of the bond through 360°. Note that the highest energy barrier is encountered when the two large groups are cis to each other, but that intermediate barriers exist when the large groups are in a gauche orientation.*

barriers for molecules in the liquid state. Infrared vibration-rotation spectroscopy is also used, but the results may be less accurate than those obtained by the other two methods. However, by far the largest effort is in *calculating* barriers in small molecules or biological macromolecules with the use of various force field parameters or from first principles. Unfortunately, molecules formed by most of the inorganic elements or species with more than a few atoms are difficult to study theoretically. Given the sparsity of experimental data, the calculated values may be the only information available for most bonds with which to attempt to understand molecular flexibility.

TABLE 2.6. Electronegativity Differences in Various Bonds

Bond	Electronegativity Difference	Bond	Electronegativity Difference
C—C	0	P—F	1.75
C—H	0.4	P—Cl	1.0
C—F	1.3	P—Br	0.8
C—Cl	0.55	P—O	1.35
C—Br	0.35	P—N	0.9
C—O	0.9	B—Cl	1.15
C—N	0.45	B—O	1.5
Si—C	0.7	B—N	1.05
Si—Si	0	Al—H	0.7
Si—H	0.3	Al—Cl	1.65
Si—F	2.0	Al—O	2.0
Si—Cl	1.25	Li—Cl	2.15
Si—Br	1.05	Na—Cl	2.25
Si—O	1.6	K—Cl	2.35
Si—N	0.9	Mg—Cl	1.95

d. Bond Polarity. Another factor that affects both molecular and solid-state structure and properties is the *polarity* of a covalent bond. If the two participating atoms have an equal affinity for the electron pair in the bond, the bond will be nonpolar—that is, the electrons will be shared equally between the two atoms. However, if one atom has a greater affinity for electrons than the other does, the electron pair will be statistically located closer to that atom and the bond will be polar. The C—C bond is obviously nonpolar, as is the Si—Si bond. However, the Si—N bond is polarized Si^+—N^-, and the Si^+—O^- bond is polarized similarly. The electron affinity of an atom is defined by its electronegativity. Table 2.6 gives the electronegativity differences of a number of different bonds. The *electronegativity difference* is a measure of bond polarity. Bond polarity affects many properties that range from susceptibility to hydrolysis to the way in which small molecules are assembled in a crystal lattice.

3. Coordinate Bonds

Coordinate bonding occurs when *two* electrons are donated from one atom to another (Figure 2.3). Both atoms now attain stable electron octets in the resultant compound, as illustrated for the formation of the boron trifluoride–ammonia adduct shown in Figure 2.8.

Elements such as phosphorus(III), nitrogen, sulfur, or oxygen readily donate two electrons from a lone-pair orbital to an acceptor atom such as boron, aluminum, or a transition metal. Coordinate bonds can be strong or weak depending on the elements involved. They are found in molecules in solution and in extended solid-state structures.

4. Ionic Assemblies

Ionic bonding involves the complete transfer of electrons from one atom (the cation) to another (the anion) in such a way that both participants now have a close-shell (octet) electron configuration.

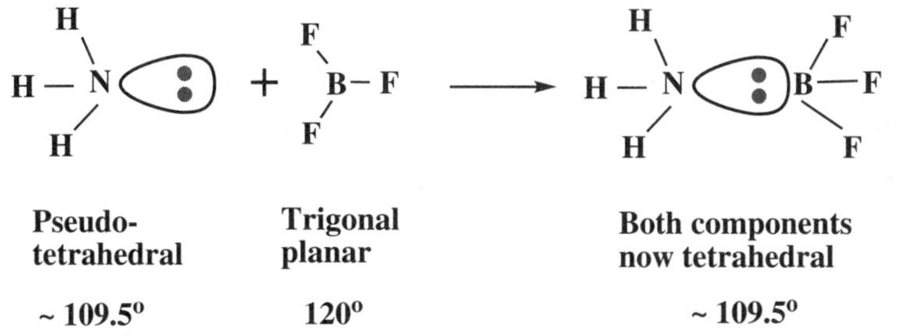

Pseudo-	**Trigonal**	**Both components**
tetrahedral	**planar**	**now tetrahedral**
~ **109.5°**	**120°**	~ **109.5°**

Figure 2.8. *Coordination bonding between ammonia and boron trofluoride. Boron has an empty bonding site (it uses only six electrons in BF₃). Thus, the ammonia lone pair is donated to boron to form a coordinate bond.*

The number of electrons transferred depends on the position of the participating elements in the periodic table. Those in group 1 lose one electron to generate a stable electron octet shell. Elements in group VIIA (17) accept one electron to form a stable octet. Elements in group 2 lose two electrons, and those in group VIA (16) accept two electrons. There are many permutations of different elements that participate in ionic bonding, and a large component of ceramic science and geochemistry is concerned with these systems. However, it is important to recognize that ionic bonding is strictly a solid-state phenomenon. Species such as sodium chloride, which are the typical ionic substances as solids, are covalently bonded in the vapor state and exist as separated solvated ions in solution. Thus, sodium chloride crystals are very poor electrical conductors because the ions cannot move, but the same ions, when solvated in aqueous media, conduct electricity. Molten salts may also be good conductors.

Because no actual bonds are formed in a purely ionic lattice, the packing of ions in a crystal is controlled by the neutralization of charges and by the size of the ions. Because different ions vary widely in size, an understanding of ionic crystal structures depends on a perception of how spheres with different diameters can be packed most efficiently into an extended structure (Figure 2.9).

The strong forces between anion and cations are responsible for the high melting points of ionic solids. However, the absence of covalent bonds in purely ionic crystals means that it is relatively easy to fracture a crystal by a sharp impact. Defects induced by the impact readily propagate between the regularly spaced ionic planes and split the crystal apart.

5. Metallic Bonding

Metals have a unique form of bonding. This bonding can be visualized as a series of spheres packed into an extended lattice with mobile electrons moving throughout the structure and serving as the "glue" that holds the solid together, as depicted in Figure 2.10.

The mobile electrons are responsible for the high electrical conductivity of many metals. Moreover, the electronic band structure of metals (see Chapter 8) underlies their high optical reflectivity. The absence of covalent bonds means that the solid-state structure can be deformed without breaking bonds. Hence many metals are

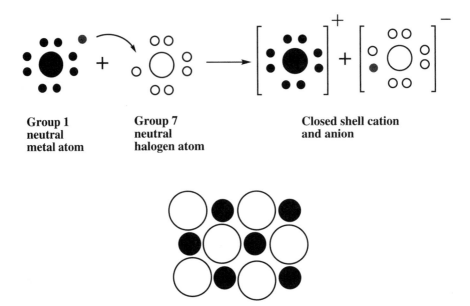

Figure 2.9. *(a) Formation of ions by transfer of a single electron from the outermost level of a metal to the incomlete outer shell of a halogen. Both species then possess a stable "octet" of electrons in their valence shell. (b) Ions pack in a crystalline lattice in such a way as to make the most efficient use of space and balance the opposing charges. The stability of ionic solids is due to the attraction of opposite charges and the efficient packing of spheres.*

also characterized by ductility and impact resistance. However, some metals such as tungsten are believed to contain covalent as well as metallic bonds, and this may be the reason for their strength and high melting points and also for their somewhat lower electrical conductivity. Note that, because the solid-state structure of metals is controlled by the packing of "spheres," the measured interatomic distances may reflect specific packing patterns rather than bond lengths.

Metallic bonding is considered in more detail in Chapter 8.

D. SIZE OF MOLECULAR UNITS

One of the most important reasons why different materials have different properties is that they contain molecular units that differ enormously in size. They range from small molecules such as naphthalene, anthracene, tetracene, or fullerenes packed in a crystal lattice, through polymer molecules with tens of thousands of atoms, to covalently bonded extended structures or ultrastructures in which the object visible to the unaided eye is one gigantic molecule. This is illustrated in Figure 2.11.

Crystals derived from small molecules tend to have low melting points and low impact strengths, and they are volatile. Polymers in solids are usually strong, flexible, impact-resistant, and nonvolatile. Ultrastructures are high-melting, strong, thermo-oxidatively stable, and nonvolatile. These differences can be attributed to the different molecular dimensions. Indeed, one feature that distinguishes most materials

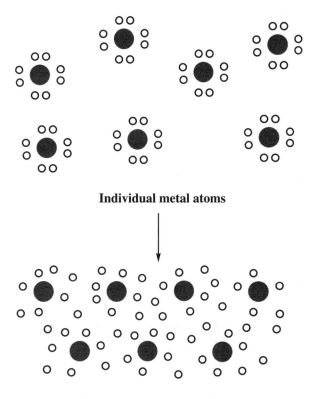

Individual metal atoms

Metal atoms in the solid state

Figure 2.10. *Metallic bonding occurs when some of the outer-shell electrons of an atom become shared with all the other atoms in the solid. These mobile electrons are sufficiently free that they are responsible for electronic conductivity. They also serve as the "glue" that holds the solid together even when it is bent, distorted, extruded, or beaten into thin sheets.*

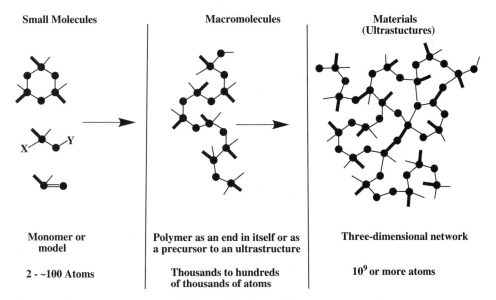

Figure 2.11. *Most of materials science revolves around large molecules or gigantic ultrastuctures. The size of these units in the solid state has a strong influence on properties.*

from other substances is the very large size of the molecules or molecular assemblies. The entanglement of long polymer chains in the solid state and the strength of the covalent bonds in the backbone are responsible for the strength, flexibility, and impact resistance of polymeric materials, and this cannot be achieved by small molecules in solids. Moreover, when an extended covalent or coordinate network exists throughout the material, it is necessary to cleave many covalent bonds to disrupt the structure, and this requires great force. Covalent bonds must also be broken before small-molecule fragments can volatilize from the material at elevated temperatures and begin a decomposition process. Hence, extended covalent solids are some of the most thermally stable materials known, often remaining intact up to temperatures above 2000 °C.

Of course, there is a continuum of properties that varies with the elements in the material, the size of the molecules, and the packing arrangement in the solid state, and it is these variables that allow new materials with improved properties to be designed and synthesized. Nevertheless, many useful materials properties depend on high molecular weight, and this factor is considered in more detail in later chapters.

E. DIFFERENT SHAPES OF COMPONENT MOLECULES AND INFLUENCE ON SOLID-STATE STRUCTURE

An understanding of solid-state structure is vital for nearly all aspects of materials science. The shape of the molecular unit plays a major role in determining the overall structure and properties of a solid (Figure 2.12).

Disk-like small molecules, such as benzene derivatives or phthalocyanines, tend to assemble into columns or tilted stacks in a crystal. Rod-like molecules may become coaligned in the solid state (see Chapter 5). Disks and rods may give rise to liquid crystalline behavior below the melting point, where the molecules can be switched from one orientation to another by the application of an electric field or by temperature changes. Slight changes to the substituents on one of these molecules will alter the liquid crystal transition temperature.

A particularly interesting situation arises when an oddly shaped molecule (the "host") cannot be fitted efficiently into a solid lattice and requires a second small molecule (a "guest") to fill the voids that would otherwise be present. These species are called *clathrates* or *molecular inclusion adducts*. Often the solid-state packing pattern may be switched from one form to another as the guest molecules enter or are removed from the lattice (Chapter 5). A variant of a clathrate structure is when an extended covalent lattice is formed with regularly spaced voids. Zeolites are ceramic-type species with this characteristic, and these materials, together with clathrates, are valuable substrates for chemical separations, as discussed in Chapter 7.

Polymer molecules, both organic and inorganic, can assume an enormous variety of conformations and overall shapes as a consequence of the twisting of the backbone bonds and the various orientations of their side chains. Two extreme situations are well known. First, if the polymer molecules form random coils in the solid state, they will interpenetrate the space occupied by neighboring molecules and form a highly disordered arrangement. What is more, if the polymer is above its glass transi-

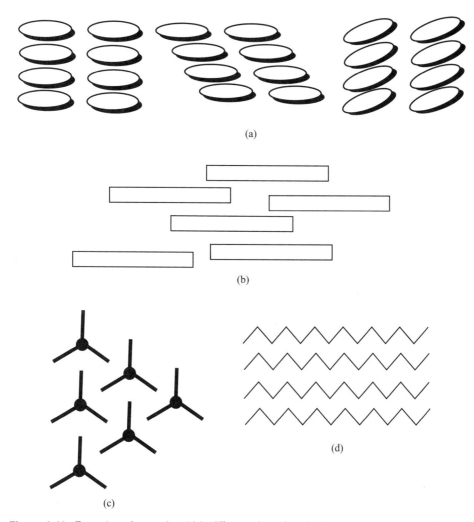

Figure 2.12. Examples of ways in which different-shaped molecules can pack in the solid state: (a) disks; (b) rod-shaped; (c) trigonal or paddle wheel molecules; (d) polymer chains.

tion temperature, the chains will be in constant thermal motion and their shapes will be changing continuously. This is the amorphous state of a polymer, and it is responsible for its ease of fabrication and resistance to impact. If the polymer is cooled below its glass transition temperature T_g, it will solidify to a glass in which the random entangled arrangement is retained. The amorphous states are favored when many different lengths of polymer molecules are present and when there is a lack of molecular symmetry and regularity along each chain. The presence of branched structures also favors amorphous character.

However, if the chains are rigid or if a regular repeating arrangement of the side groups exists along each chain, the chains will tend to become aligned with their neighbors. This coalignment will be encouraged if the material is stretched or annealed by being alternately heated and cooled. The alignment process may then cause the growth of *microcrystallites* embedded in a matrix of amorphous domains. Microcrystalline polymers are stronger than their amorphous counterparts because the microcrystallites are held together by van der Waals forces (also sometimes by ionic forces or hydrogen bonds), and they serve as thermally labile crosslinks that prevent the chains from being dragged past each other when subjected to stress or strain. Very few polymer molecules form solids that are 100% crystalline (a specialized form of polyethylene is one of the few examples) because of imperfections in the chain or side-group arrangements.

A further change in polymer architecture arises when inorganic or organic macromolecules are connected to each other by covalent or ionic crosslinks. These stiffen the structure by preventing chain slippage and, if present in supficient concentrations, will raise the thermal stability. It follows that there is a continuum of property changes as the level of crosslinking is increased. Extensive crosslinking gives stiff, strong, insoluble materials that begin to resemble ceramics. Indeed, ceramics can be viewed as extensively crosslinked inorganic polymers in which most molecular motion is inhibited and in which billions of crosslinks must be broken before the material can be induced to dissolve, melt, or decompose. As with polymers, ceramics can be amorphous, be microcrystalline, or consist of one highly ordered, crystalline giant molecule.

Thus, the importance of the solid-state structure in materials science cannot be overemphasized. If the component building units of a material can be assembled in two or more different solid-state arrangements, those different structures will generate different properties. A well-known example is the difference between graphite and diamond. Both have the same composition—elemental carbon—but have different solid state structures. Graphite consists of layers of sp^2-bonded, fused benzene rings organized into extended planes (Figure 2.13).

Each plane has a delocalized electronic structure through which electrons can move freely. Hence, graphite is a good electrical conductor, and it absorbs light strongly to give its black color. The layers are associated with their neighbors above and below mainly through weak van der Waals forces. Hence the layers can slide over each other, and this is responsible for the lubricating properties of graphite. On the other hand, diamond contains carbon atoms in the sp^3 tetrahedral state, with covalent sigma (σ) bonds connected to neighboring atoms in three dimensions to form an arrangement that extends throughout the whole crystal. There are no delocalized electrons; hence pure diamonds are transparent and colorless. Diamond powder is a hard abrasive material rather than a lubricant. Other forms of carbon such as fullerenes and nanotubes have different unit shapes and generate different solid-state structures and properties (Chapter 17).

Silicates exist in the form of a wide variety of rings, chains, and extended three-dimensional arrangements, and the solid-state structures and properties vary correspondingly. Indeed, this is one reason for the wide range of minerals found in the earth's crust. Similar situations exist with a many metal oxides, sulfides, and selenides.

A final example of the ways in which solid-state structures can alter properties is provided by composites (Chapter 9). A *composite* is a solid material that contains two or more different components. For example, two different polymers may be combined either as a homogeneous "alloy" or "blend," or as a material in which phase-separated domains are present. One polymer in a phase-separated system

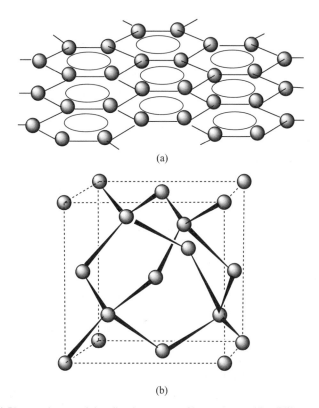

(a)

(b)

Figure 2.13. *(a) Planar, electron-delocalized structure of layers in graphite. Different degrees of offset in adjacent layers gives two forms—alpha (α) and beta (β). (b) Structure of diamond, showing puckered, saturated six-membered rings that form the building blocks of the three-dimensional architecture.*

may provide strength and rigidity; the other may provide impact resistance. Metal alloys often form phase-separated composite materials. Mineralogical composites are also well known (granite and many synthesized ceramics are examples), as are composites that contain both ceramics and polymers (ceramers). These topics are discussed in later chapters.

F. SUGGESTIONS FOR FURTHER READING

1. Cotton, F. A., *Advanced Inorganic Chemistry*, 6th ed., Wiley-Interscience, Hoboken, NJ, 1999.

2. Greenwood, N. N.; Earnshaw, A., *Chemistry of the Elements*, 2nd ed., Butterworth-Heinemann, Oxford, 1997.

3. Gillespie, R. J.; Popelier, P. L. A., *Chemical Bonding and Molecular Structure*, Oxford University Press, 2001.

4. Shriver, D.; Atkins, P., *Inorganic Chemistry* 4th ed., Freeman/Oxford, New York, 2006.

5. McMurry, J., *Organic Chemistry*, 7th ed., Brooks/Cole, 2008.

G. STUDY QUESTIONS (for class discussions or essays)

1. What help would you expect to obtain from small-molecule model studies when attempting to design a new large-molecule-based material with unique properties? What aspects of the small-molecule chemistry might be of only limited usefulness?

2. The elementary interpretation of chemical bonding presented in this chapter is mainly an aid to visualization. What additional and more useful information is likely to be obtained from more complex approaches that are based on the Schrödinger wave equation?

3. If graphite is an electronic conductor, why are crystals of benzene electrical insulators?

4. If shape-fitting is an important component of how molecules assemble in the solid state, why might polymer molecules change their shape in the solid state and even move through a solid material?

5. On the basis of the information presented in this chapter, why do some very long-chain molecules assume an extended (rigid-rod) shape even though entropy would favor random conformations?

6. Which materials properties are likely to be the most difficult to understand in terms of types of bonds, orbital arrangements, bond dipoles, and similar properties? How might the understanding of properties be simplified so that such fundamental aspects could be used effectively?

7. You are employed by a large aerospace company and have been given the task of choosing a material for the fabrication of the nose of a supersonic aircraft. What properties would you consider to be most important, and how would you design them into a new material?

8. The space shuttle is having problems with the heat-shielding tiles that protect against the heat of reentry into the atmosphere. Write a brief proposal (no more than 250 words) on how you would choose or design a material to improve the performance of the heat insulation.

9. At what point do you think that the ongoing replacement of metals in automobiles and aircraft by polymers or ceramics will end? In other words, which components of these vehicles cannot and should not be made from polymers or ceramics?

10. In some cases, the sum of the covalent radii of two atoms comes close to the measured bond distance between them, but in other cases it does not. By examining the data in Tables 2.3 and 2.4, deduce which types of bonds show the largest discrepancies, and suggest why these deviations exist.

11. Which solid has the higher density—diamond or graphite—and which allotrope can be converted to the other at high pressures?

12. Discuss circumstances in which the inherent orientation of atomic orbitals in a central element might control the molecular geometry. Why should the shape and orientation of atomic orbitals be retained in a molecule or a solid?

13. What are the essential assumptions in VSEPR theory? When do these assumptions no longer apply? To what extent can this approach be applied to solids?

14. Refer to Table 2.5. Suggest reasons for the different barriers to bond torsion that are quoted in the table.

15. Discuss the usefulness or practicality of predicting molecular shapes (bond angles, bond lengths, etc.) using a model based on primary quantum arguments versus one based on the repulsions between electron pairs.

16. Having read this chapter, summarize in your own words why different types of materials are different. Of all the contributing factors, which ones are the most important?

3

Basic Synthesis and Reaction Chemistry

A. UNDERLYING PRINCIPLES

Figure 3.1 shows the three main reaction manifolds that underlie materials synthesis. The main sources of synthesized materials are oil and minerals. As a general rule, organic compounds are obtained from oil rather than from elemental carbon. The initial steps in the synthesis of the organic compounds in oil were probably carried out by living organisms during the early history of the planet. Note, however, that considerable long-term interest lies in the use of carbon (coke) or methane as a starting point for the production of organic chemicals in preparation for the time when oil is no longer plentiful. The starting materials for the synthesis of inorganic compounds come from two sources—either directly from mineral oxides, carbonates, silicates, sulfides, sulfates, chlorides, and other compounds, or synthesized from the element rather than from a mineralogical precursor. An example of the first route is the preparation of glass and other ceramics by the reaction of silica (sand) with alkali carbonates or various oxides. For the second route, a mineral must first be reduced to the element, and the element is then used either directly as a useful material (metals, silicon) or as the starting point for further synthesis.

A third access route to materials is from plants or animals. Wood, cotton, leather, and wool are still materials of widespread importance despite the wide choice of synthetic materials.

An important principle in materials chemistry is that organic reactions are relatively slow and are controlled mainly by kinetics. The products obtained depend on time, temperature, and concentration. This is because many of the covalent bonds in organic compounds have a low polarity, and they are broken only with difficulty by polar reagents. For example, a compound such as chloroform, $CHCl_3$, or chlorobenzene is stable to water at normal temperatures and is not hydrolyzed to an

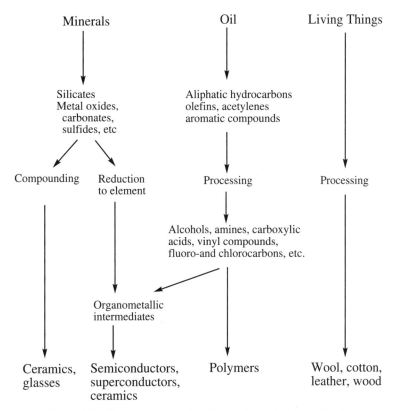

Figure 3.1. Starting points and pathways for materials synthesis.

alcohol or phenol. The slowness of many organic reactions is the reason why the mechanisms of organic reactions may be studied fairly readily because reaction intermediates can be identified or isolated, and the process may be interrupted if necessary to maximize the yield of an intermediate product.

On the other hand, the bonds in inorganic compounds tend to be more polar. They react rapidly with polar or ionic reagents. Thus, phosphorus trichloride or silicon tetrachloride, which have polar P—Cl or Si—Cl bonds, are hydrolyzed almost instantaneously to the transient species $P(OH)_3$ or $Si(OH)_4$ when brought into contact with moisture. The coordination chemistry of inorganic compounds also tends to follow very rapid reaction kinetics. Because these types of inorganic reactions are so fast, the final products may be determined by thermodynamics rather than kinetics. Often an inorganic reaction will go to completion and will establish an equilibrium concentration of products before the investigator has time to study the pathways followed.

B. STARTING POINTS FOR MATERIALS SYNTHESIS—ISOLATION OF ELEMENTS

The fields of organic and inorganic reaction chemistry are widely diversified, and the following observations introduce a few principles that are significant in materials research. The abbreviated summary given here is designed to serve as a reminder of ideas from elementary inorganic and organic chemistry and to serve as a starting point for further reading. Note that inorganic, organic, and organometallic chemistry are now very closely related fields, and that many reaction methods, mechanisms, and techniques are common to all three.

The elements that are most widely used in materials synthesis are isolated from minerals (especially oxides or sulfides) by a process of *reduction*. Reduction is accomplished in one of several ways, which are summarized in reactions 1–9:

$$Fe_2O_3 \xrightarrow[200-1200°C]{C\ (CO)} Fe + CO_2 \tag{1}$$

$$WO_3 \xrightarrow[850°C]{H_2} W + H_2O \tag{2}$$

$$B_2O_3 \xrightarrow[heat]{Mg} B + MgO \tag{3}$$

$$MgCa(CO_3)_2 \xrightarrow[1150°C]{Fe/Si} Mg + Ca_2SiO_4 + CO_2 \tag{4}$$

$$TiO_2 \xrightarrow[O_2]{C+Cl_2} TiCl_4 \xrightarrow{Mg} Ti + MgCl_2 \tag{5}$$

$$Cu_2S \xrightarrow[heat]{O_2} Cu_2O + (CuS) \xrightarrow[heat]{} Cu + SO_2 \tag{6}$$

$$\underset{\text{(molten)}}{MgCl_2} \xrightarrow[750°C]{electrolysis} \underset{\text{(cathode)}}{Mg} + \underset{\text{(anode)}}{Cl_2} \tag{7}$$

$$Al_2O_3 \xrightarrow[\substack{940-980°C \\ \text{in molten} \\ \text{Na}_3\text{AlF}_6}]{electrolysis} Al + O_2 \tag{8}$$

$$\underset{\substack{\text{impure} \\ \text{(anode)}}}{Cu} \xrightarrow[\text{aq. CuSO}_4]{electrolysis} \underset{\substack{\text{pure} \\ \text{(cathode)}}}{Cu} + \underset{\text{(anode deposits)}}{\text{other metals}} \tag{9}$$

a. Reaction of an Oxide at High Temperatures with Carbon or Hydrogen. The underlying principle is that carbon (usually in the form of coke) will react with a limited supply of oxygen to form carbon monoxide (reaction 1). Carbon monoxide is a powerful reducing agent that not only converts an oxide to the elemental form but also is itself oxidized to carbon dioxide. Alternatively, the carbon may react directly with the oxide to form carbon monoxide and dioxide.

Reduction by carbon is a widely used and inexpensive process that is employed for the isolation of iron and steel, chromium, silicon, phosphorus, tin, lead, copper, nickel, and numerous other elements. In some processes, heating is accomplished by the combustion of carbon. In others, heat is generated by striking an arc between two electrodes.

The carbon reduction process cannot be used effectively if the element reacts with carbon to form a stable carbide. For example, titanium forms a highly stable carbide, and this metal must be isolated by more complex procedures (see text below). Although iron reacts with carbon at high temperatures, the carbon content can be reduced by subsequent treatment of the molten metal with oxygen. Pure metals are seldom produced by carbon reduction, and subsequent purification steps are usually needed.

The use of hydrogen as a reducing agent is more expensive and requires more sophisticated equipment, but it can give pure material, often at lower temperatures. Thus, tungsten, which forms a stable carbide when heated with carbon, is obtained via the reduction of tungsten oxide (WO_3) by hydrogen at 850 °C (reaction 2). Germanium is isolated from GeO_2 by reduction with hydrogen at temperatures above 500 °C.

b. Reaction of an Oxide with an Element that Has a Greater Affinity for Oxygen.

b. Reaction of an Oxide with an Element that Has a Greater Affinity for Oxygen. Boron is isolated by the high-temperature reaction of boric oxide (B_2O_3) with magnesium (reaction 3). Magnesium is obtained from dolomite [$MgCa(CO_3)_2$] by heating with ferrosilicon at 1150 °C (reaction 4).

c. Isolation via Chlorination or Oxidation. Titanium is a difficult metal to extract from its ores (which explains its high cost). The ore, rutile (TiO_2) or ilmenite ($FeTiO_3$), is first heated with carbon and chlorine to produce either $TiCl_4$, or $TiCl_4$ plus $FeCl_3$, which are separated by fractional distillation (reaction 5). The high-temperature interaction of $TiCl_4$ with magnesium or sodium then yields titanium metal contaminated with $MgCl_2$ or $NaCl$. These salts must then be extracted by treatment with aqueous acid to leave a titanium sponge. Titanium has a melting point of 1667 °C, and is thus difficult to fabricate.

Copper is separated from its sulfide ores and from iron sulfide by a complex oxidation–reduction process. In this, the FeS is first oxidized by air to FeO and some of the Cu_2S forms Cu_2O. The Cu_2O then reacts with unchanged Cu_2S to form impure copper metal and SO_2 (reaction 6). The loss of gaseous SO_2 is a driving force for the reaction. Copper produced in this way is not pure enough for electrical applications, and it must be purified further by electrolysis.

d. Electrolytic Reduction. Electrolysis is an excellent method for the production of high-purity metals. The penalty that must be paid is the high cost of electricity. Hence, electrorefining plants are normally situated close to supplies of relatively inexpensive hydroelectricity. Metals such as copper, magnesium, and aluminum are produced on a very large scale by electrolysis (reactions 7–9). In the case of magnesium, molten anhydrous $MgCl_2$ or hydrated $MgCl_2$ from seawater is electrolyzed at temperatures up to 750 °C. Aluminum is produced from the mineral bauxite [$Al(O)OH$ or $Al(OH)_3$], which is first converted to Al_2O_3 and then electrolyzed at 940–980 °C in molten cryolite, Na_3AlF_6 (reaction 8).

Impure copper or nickel, produced by the carbon reduction process, is cast into anodes and is electrolyzed using an electrolyte of aqueous copper sulfate or nickel sulfate. High-purity copper is deposited on the cathode. Silver, an impurity in the copper, deposits beneath the anode during the electrorefining process. The isolation of specific metals is discussed in Chapter 8.

e. Pyrolysis and Vapor Deposition. An important access route to elemental carbon is via the pyrolysis of methane or other hydrocarbons to give carbon black or activated carbon. A similar material is soot, formed by vapor deposition in chimneys. Fullerenes and nanotubes are also produced by pyrolysis and vapor deposition. Diamond-like films are deposited on surfaces when gaseous methane or other hydrocarbons in the presence of hydrogen are heated by a plasma or by a wire at

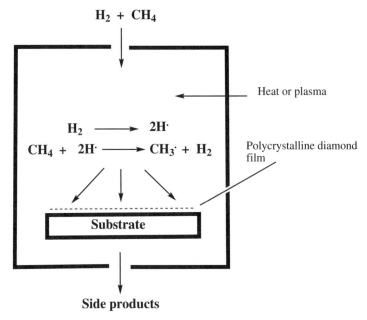

Figure 3.2. *Chemical vapor deposition method for depositing diamond-like carbon on a substrate.*

2200 °C and the products are allowed to impinge on the surfaces of metals or silicon. This chemical vapor deposition (CVD) of carbon onto metal surfaces yields thin, ultrahard films that are valuable for machine tools (Figure 3.2). Note that synthetic diamonds (i.e., discrete crystals) are produced from molten graphite by the application of heat (1500 °C) and high pressure [58,000 atmospheres (atm)].

Other sources of elemental carbon are organic polymers, pitch, or asphalt, which, when pyrolyzed in the absence of oxygen, yield carbon fiber, which is an increasingly important material in aerospace applications (see Chapter 7). Finally, one of the major sources of elemental carbon is coke—a porous residue left after the volatile components have been distilled from coal. Coke is one of the least expensive reducing agents known, and its use in steelmaking revolutionized industrial practice during the industrial revolution.

Another example of pyrolytic chemical vapor deposition is the conversion of silane, SiH_4, to films of amorphous semiconductor silicon. This is a method for the preparation of relatively inexpensive semiconductors for solar cells (see Chapter 10).

Note that some relatively pure elements are found native in mineral deposits— gold, silver, sulfur, diamond, and graphite are well-known examples.

C. PRINCIPLES THAT UNDERLIE MATERIALS SYNTHESIS

1. Importance of Halides in Materials Synthesis

The formation of chloro compounds is a key step in the conversion of nonmetallic elements to materials. This is typically accomplished by the direct reaction of the element or its oxide or of an organic compound with chlorine at elevated temperatures, as illustrated in reactions 10–13. Typical examples are the production of the chlorides of boron, silicon, aluminum, or phosphorus. However, conversion of most metals to their chlorides is more easily accomplished by reaction with hydrochloric acid (reaction 14). Organic chloro-compounds are produced not from elemental carbon but by the chlorination of hydrocarbons (reactions 15 and 16). Chlorination of methane gives CH_3Cl, CH_2Cl_2, $CHCl_3$, and CCl_4. Chlorination of benzene gives mono-, di-, and trichlorobenzenes. Brominations proceed in a similar manner.

$$Si \xrightarrow{\text{Cl}_2} SiCl_4 \tag{10}$$

$$Al \xrightarrow{\text{Cl}_2} AlCl_3 \tag{11}$$

$$P \xrightarrow{\text{Cl}_2} PCl_3 \xrightarrow{\text{Cl}_2} PCl_5 \tag{12}$$

$$B_2O_3 \xrightarrow{\text{C,Cl}_2} BCl_3 \tag{13}$$

$$Fe + HCl \rightarrow FeCl_2 + H_2 \tag{14}$$

$$C_6H_6 \xrightarrow[-\text{HCl}]{\substack{500\,°C \\ \text{Cl}_2}} C_6H_5Cl \xrightarrow[-\text{HCl}]{\text{Cl}_2} C_6H_4Cl_2, \text{etc.} \tag{15}$$

$$CH_4 \xrightarrow[-\text{HCl}]{\text{Cl}_2} CH_3Cl, CH_2Cl_2, CHCl_3, CCl_4 \tag{16}$$

Chloro and bromo compounds play two crucial roles in the synthesis of materials and their precursors:

1. They are reagents for reactions with nucleophiles such as alkoxides, aryloxides, amines, and organometallic reagents to link organic units to inorganic elements or to organic substrates, as shown in in reactions 17–22.

$$SiCl_4 + NaOR \xrightarrow[-\text{NaCl}]{} Si(OR)_4 \tag{17}$$

$$TiCl_4 + NaOR \xrightarrow[-\text{NaCl}]{} Ti(OR)_4 \tag{18}$$

$$AlCl_3 + RLi \xrightarrow[-\text{LiCl}]{} AlR_3 \tag{19}$$

$$C_6H_5Br + RLi \xrightarrow[-\text{LiBr}]{} C_6H_5R \tag{20}$$

$$BCl_3 + RNH_2 \xrightarrow[-\text{HCl}]{} B(NHR)_3 \tag{21}$$

$$PCl_5 + NH_3 \xrightarrow[-\text{HCl}]{} (NPCl_2)_3 \tag{22}$$

2. Main-group inorganic halides can be hydrolyzed readily to hydroxyl derivatives (reactions 23–26). These, in turn, condense by elimination of water to form oxide ceramics.

$$SiCl_4 + H_2O \xrightarrow[-\text{HCl}]{} Si(OH)_4 \tag{23}$$

$$BCl_3 + H_2O \xrightarrow[-\text{HCl}]{} B(OH)_3 \tag{24}$$

$$PCl_5 + H_2O \xrightarrow[-\text{HCl}]{} P(O)(OH)_3 \tag{25}$$

$$AlCl_3 + H_2O \xrightarrow[-\text{HCl}]{} Al(OH)_3 \tag{26}$$

2. Acidic Hydroxides and Condensation Reactions

Although the hydroxides of the common alkali and alkaline-earth elements (groups I and II) are well-known bases, the hydroxy derivatives of most of the non-metals and some metals are acids. Examples range from strong acids such as sulfuric $O_2S(OH)_2$, nitric O_2NOH, or perchloric O_3ClOH acids to the moderate or weak acids such as silicic $Si(OH)_4$, phosphoric $OP(OH)_3$, or boric $B(OH)_3$ acid and the amphoteric (acidic or basic depending on conditions) aluminum hydroxide $Al(OH)_3$. These species are acids because the oxygen atoms of the OH groups are linked to the central element through covalent bonds, whereas the hydrogen atom can readily dissociate from oxygen as a hydrogen cation.

The weaker polyhydroxy acids such as silicic, boric, or phosphoric acids play an important role in materials chemistry because they readily undergo *condensation reactions* to form polymers and ceramics. A condensation reaction involves the elimination of a small molecule (usually water) to bind two molecules together through an element–oxygen–element covalent linkage (reactions 27–29). If each participating molecule bears only one hydroxy group, the result is a dimer (reaction 27). Two hydroxy groups per molecule yield a cyclic molecule or a linear polymer (reaction 28), and three or four yield a crosslinked ultrastructure (reaction 29).

$$\begin{array}{c} R \\ | \\ R-Si-OH \\ | \\ R \end{array} + \begin{array}{c} R \\ | \\ HO-Si-R \\ | \\ R \end{array} \xrightarrow{-H_2O} \begin{array}{c} R \quad\quad R \\ | \quad\quad\ | \\ R-Si-O-Si-R \\ | \quad\quad\ | \\ R \quad\quad R \end{array} \tag{27}$$

$$\begin{array}{c} R \\ | \\ HO-Si-OH \\ | \\ R \end{array} + \begin{array}{c} R \\ | \\ HO-Si-R \\ | \\ R \end{array} \xrightarrow{-H_2O} \left[\begin{array}{c} R \\ | \\ O-Si- \\ | \\ R \end{array} \right]_n \tag{28}$$

$$\begin{array}{c} OH \\ | \\ B \\ HO \quad OH \end{array} \xrightarrow{-H_2O} \left[\begin{array}{c} O \\ | \\ B \\ O \quad O \end{array} \right]_n \tag{29}$$

$$\begin{array}{c} ONa \\ | \\ HO-P-OH \\ || \\ O \end{array} + \begin{array}{c} ONa \\ | \\ HO-P-OH \\ || \\ O \end{array} \xrightarrow{-H_2O} \left[\begin{array}{c} ONa \\ | \\ O-P- \\ || \\ O \end{array} \right]_n \tag{30}$$

$$\begin{array}{c} O \\ || \\ HO-S-OH \\ || \\ O \end{array} \xrightarrow{-H_2O} \left[\begin{array}{c} O \\ || \\ O-S- \\ || \\ O \end{array} \right]_n \xrightarrow{H_2O} \begin{array}{c} O \\ || \\ HO-S-OH \\ || \\ O \end{array} \tag{31}$$

$$HO-\overset{\overset{\displaystyle O}{\|}}{C}-R-\overset{\overset{\displaystyle O}{\|}}{C}-OH + HO-R'-OH \xrightarrow[-H_2O]{} \left[\overset{\overset{\displaystyle O}{\|}}{C}-R-\overset{\overset{\displaystyle O}{\|}}{C}-O-R'-O-\right]_n$$

(32)

$$HO-R-OH + HO-R-OH \underset{H_2O}{\overset{-H_2O}{\rightleftharpoons}} HO-R-O-R-OH \qquad (33)$$

The functionality of a polyhydroxy monomer may be controlled by replacement of some of the hydroxyl groups by inactive organic units such as CH_3 or C_6H_5 units or conversion to a salt linkage such as $O^- Na^+$. Thus, the functionality of silicic acid can be changed from $Si(OH)_4$ to $Si(CH_3)(OH)_3$, $Si(CH_3)_2(OH)_2$, or $Si(CH_3)_3OH$, or of phosphoric acid from $OP(OH)_3$ to $OP(ONa)(OH)_2$, or $OP(ONa)_2OH$ (reaction 30).

A significant difference between the condensation products of the weak hydroxy acids and those of the strong acids is the stability of the condensation products. Sulfuric acid loses water to form poly(sulfur trioxide) (reaction 31), but this macro-molecule is instantly hydrolyzed back to sulfuric acid when exposed to even trace amounts of water. This is not the case for the condensation products from the weak acids, which tend to be much more stable in neutral aqueous media and hydrolyze only in strong acid or base.

Organic compounds that bear hydroxyl units also participate in condensation reactions. This is true irrespective of whether the OH unit is part of a carboxylic acid structure or the functional component of an alcohol. An example is the formation of a polyester as shown in reaction 32.

Water is the molecule that is normally eliminated in materials-related condensation reactions, but the loss of ammonia or other small molecules is also possible. A characteristic of condensation reactions is that they are often controlled by equilibria rather than kinetics. This means that in a closed system the (forward) condensation reaction is in equilibrium with the reverse process in which the newly formed bond is hydrolyzed and the two original separate molecules are re-formed (reaction 33).

However, removal of the water as the reaction proceeds retards the back reaction and ensures that the condensation process goes to completion. Hence, moderate to high temperatures are usually needed to ensure that a condensation process proceeds efficiently. Condensation reactions are the basis of the synthesis of silicates, titanates, polyphosphates, polyesters such as Dacron or Mylar, polyamides such as nylons or Kevlar, polyimides, melamine resins, and many more.

3. Metathetical Exchange Reactions

The term *metathetical exchange* refers to a process whereby two molecules or ionic species exchange similar units. An example is when an aqueous solution of potassium bromide is added to a solution of sodium chloride (reaction 34).

$$K^+ + Br^- + Na^+ + Cl^- \rightleftharpoons K^+ + Cl^- + Na^+ + Br^- \qquad (34)$$

$$Na_2SO_4 + BaCl_2 \rightarrow \underset{\text{insoluble}}{BaSO_4} + 2Na^+ + 2Cl^- \tag{35}$$

$$O{=}P(OC_2H_5) + NaOC_3H_7 \rightleftharpoons O{=}P(OC_3H_7)_3 + NaOC_2H_5 \tag{36}$$

Either cation can interact with either anion. In fact, all four ions are dispersed in water, and the concept of discrete pairing does not apply. However, if one of the exchanged ion salts is now less soluble than the starting materials, it will crystallize or precipitate from solution. Partial removal of the solvent by evaporation will yield a similar result.

A classical example is the addition of aqueous sodium sulfate to a solution of barium chloride (reaction 35). Barium sulfate is so insoluble that it precipitates immediately and drives the reaction to completion. Addition of an aqueous solution of silver nitrate to a solution of sodium chloride results in the immediate precipitation of silver chloride. Because reactions of this type are so rapid, they are either controlled by equilibria or they are driven to completion by the insolubility of one of the products.

Metathetical exchange processes occur with covalent molecules also. For example, triethylphosphate can be converted partially or completely to tripropylphosphate by treatment with sodium propoxide as shown in reaction 36. The main difference from the other examples is that these reactions are slow, and they fall into a category known as *nucleophilic substitutions*.

4. Nucleophilic Substitution

The replacement of one component of a covalent molecule by another is a vital part of both organic and inorganic chemistry. A general illustration of the process is shown in reaction 37.

$$R^- + {}^{\backslash}E^{\delta+} - X^{\delta-} \longrightarrow R - E^{\backslash} + X^- \tag{37}$$

$$R \cdots E^- \cdots X \tag{38}$$

$$\begin{aligned} {}^{\backslash}E^{\delta+} - X^{\delta-} \longrightarrow E^+ + X^- \xrightarrow{R^-} R - E^{\backslash} + X^- \\ \text{or} \\ {}^{\backslash}E{-}R \end{aligned} \tag{39}$$

A nucleophilic substitution occurs when an anion (nucleophile) attacks the more positively charged atom of a polar bond. We will call this attacked atom the *central atom*. The more negatively charged component of that bond departs and its place is taken by the incoming unit. There are two extremes of this type of reaction. In the first, the incoming anion attaches itself to the central atom to give a transient intermediate that bears a negative charge (reaction 38). This is called an S_N2 reaction because it is a bimolecular process. In practice, the leaving anion may depart as the incoming anion approaches, in a synchronous process. Alternatively, the departing anion may leave before the incoming anion arrives (reaction 39)—a so-called S_N1 reaction, because the key step is a unimolecular process. S_N1 and S_N2 reactions can be identified by their kinetic characteristics—that is, whether they follow first-order, pseudo-first-order, or second-order kinetics. The more polar the original E—X bond, the more likely it is that S_N1 kinetics will prevail. A characteristic of S_N2 processes is that they are strongly influenced by the *size* of both the incoming and outgoing anions. Because the transition state in an S_N2 process is more crowded than the ground state (e.g., a pentacoordinate, trigonal bipyramidal silicon or phosphorus atom), steric factors exert a strong influence on the energy barrier of the transition state, the speed of the reaction, and the nature of the products. Moreover, because the approach of the incoming anion is likely to be from a direction furthest removed from that of the departing anion, an S_N2 reaction would be expected to lead to an inversion of configuration.

By contrast, because the leaving anion departs first in an S_N1 reaction (reaction 39), retention or scrambling of the original configuration is to be expected, and the steric size of the incoming unit is less crucial. These factors are important for understanding the nature of the products from these reactions and for predicting the course of new reactions.

5. Electrophilic Substitution

Electrophilic reactions occur when a cation attacks a compound and displaces another positively charged species, often a proton (reactions 40 and 41). These reactions are common in organic chemistry.

$$(40)$$

$$(41)$$

$$R^+ + CH_2 = \overset{\overset{\displaystyle R}{|}}{\underset{\underset{\displaystyle R}{|}}{C}} \longrightarrow R - CH_2 - \overset{\overset{\displaystyle R}{|}}{\underset{\underset{\displaystyle R}{|}}{C^+}} \tag{42}$$

For example, the nitration of toluene by a mixture of nitric and sulfuric acids proceeds by an electrophilic mechanism (reaction 40), as does the sulfonation of benzene (reaction 41). Chlorination and bromination of organic compounds involves electrophilic replacement of hydrogen by the halogen. An important ramification of electrophilic chemistry is the polymerization of olefins by carbonium ions, as illustrated in the initiation process shown in reaction 42. The carbonium ion formed in this reaction can then attack another olefin molecule and initiate a chain growth process. These reactions are discussed in more detail in Chapter 6.

6. Coordination Chemistry

Molecules that have excess valence shell electrons in one or more lone-pair orbitals can donate those electrons into the unfilled valence shells of acceptor molecules. This concept was introduced in Chapter 2. Two classical examples are the coordination compounds formed between ammonia and boranes and between phosphines and transition metals (reactions 43 and 44).

$$
\begin{array}{ccc}
H & Cl \\
H\!-\!N\!: + B\!-\!Cl & \longrightarrow & H\!-\!N\!:\!B\!-\!Cl \\
H & Cl & H \quad Cl
\end{array} \tag{43}
$$

$$
\begin{array}{ccc}
R & & R \\
R\!-\!P\!: + M(CO)_x & \longrightarrow & R\!-\!P\!:M(CO)_x \\
R & & R
\end{array} \tag{44}
$$

The synthesis of coordination compounds is normally a matter of bringing the donor and acceptor molecules together in a solvent that does not itself interfere with the process. Donor solvents such as ethers or amines are normally avoided. Crystallization or precipitation of the donor–acceptor complex then yields the solid material. Complex formation from the vapor state is also possible.

7. Branching and Crosslinking

Organic and inorganic polymer chains can be linear or branched. Branched macromolecules take several different forms as shown in Figure 3.3. Branched structures entangle more easily in the solid or liquid states. They also inhibit crystallite formation. Branching usually occurs during an inorganic or organic polymer synthesis process when tri- or tetrafunctional monomers are used along with difunctional species. Branched molecules will dissolve in suitable solvents, although the rate of dissolution might be slow.

Crosslinks are covalent or ionic connections that are formed between polymer molecules or that join rings together to form ultrastructure materials (Figure 3.3). Rubber is a useful material only because it contains crosslinked chains that prevent the polymer molecules from sliding past each other when the material is stretched. Many inorganic materials such as silicates are crosslinked through Si—O—Si linkages formed via the condensation of Si—OH groups. It is even possible to understand the properties of diamond or silicon in terms of crosslinking between every atom in the crystal. A characteristic of crosslinked species is that they are insoluble in all solvents. Light crosslinking will give a material that absorbs a solvent and swells but that will not dissolve. Heavily crosslinked materials do not even swell in potential solvents. Three methods for forming crosslinks in organic macromolecules are shown in reactions 45– 47. Reaction 45 is a common outcome for organic materials and is driven by free-radical reactions. The process shown as reaction 46 is known as a "2 plus 2 cycloaddition," and is induced by ultraviolet radiation. Ionic crosslinks exist when chains are linked through acid side groups connected to a divalent or trivalent cation such as Ca^{2+} or Al^{3+} (reaction 47). This is a common

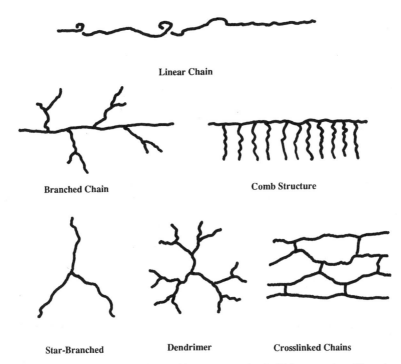

Linear Chain

Branched Chain

Comb Structure

Star-Branched

Dendrimer

Crosslinked Chains

Figure 3.3. *Different polymer molecular architectures, each of which generates different properties. (From Allcock, H. R.,* Chemistry and Applications of Polyphosphazenes, *Wiley-Interscience, Hoboken, NJ, 2003, p. 5.)*

situation in inorganic materials. Ionic crosslinks of this type can often be disrupted by replacement of the multivalent cations by monovalent cations such as Na^+. Crosslinking by condensation of hydroxyl groups is a common process in silicate chemistry.

$$CH_3 \quad CH_3 \xrightarrow[\;-H_2\;]{\substack{\gamma\,\text{rays}\\ \text{UV light}\\ \text{peroxides}}} CH_2 \mid CH_2 \tag{45}$$

$$CH=CH_2 \quad CH=CH_2 \xrightarrow{\text{UV light}} CH-CH_2 \mid\;\mid CH-CH_2 \tag{46}$$

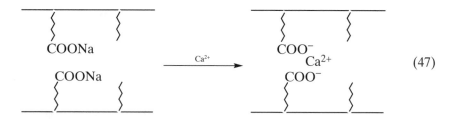

$$(47)$$

8. Polymerization–Depolymerization Equilibria

An important process involved in the synthesis and decomposition of many materials is the existence of monomer–polymer, ring–polymer, and ring–ring equilibria. Thus, some organic and inorganic polymers can be produced by the polymerization of small rings, but at high temperatures those polymers may depolymerize back to small rings and lose all their useful properties. Examples of these reactions are given in Chapter 6. Equilibria between rings and chains are also found in many ceramic synthesis processes such as in the condensation reactions of silicon diols and triols to either rings or chains. An important principle in these reactions is that they depend on a balance between energy and entropy. Ring-opening polymerizations and condensation reactions do not involve significant changes in energy. However, the breakdown of one large molecule to many small molecules involves an increase in entropy, a change that becomes amplified as the temperature is raised ($\Delta G = \Delta H - T\,\Delta S$). Thus, at high temperatures linear polymers and some lightly crosslinked materials may depolymerize to generate many small molecules. However, if the system is heavily crosslinked, this is less likely to happen because there are many more bonds to be broken. This is one reason why silicone polymers depolymerize at moderate temperatures but silicate ceramics do not.

D. ILLUSTRATIVE CHEMISTRY OF SELECTED NONMETALLIC ELEMENTS

The nonmetallic elements play a major role in all but a few of the different materials areas. The following sections introduce some of the chemistry that underlies a few well-known element systems. In general, the chemistry described here is, with minor variations, typical of many main-group elements. The behavior of some of the most important metals is covered in Chapter 8.

1. Carbon Chemistry

a. The Element. Elemental carbon is found in the earth's crust in the form of graphite and diamond. Graphite is manufactured by heating coke and silica at 2500 °C, a process that involves the intermediate formation of silicon carbide (see Chapter 7). Artificial diamonds used for jewelry and abrasives are produced by heating graphite at high temperatures and high pressures in the presence of a nickel catalyst.

b. Organic Compounds from Oil. Oil is converted to organic starting materials in an oil refinery by a combination of destructive distillation and catalytic reformation (Figure 3.4). The main distillation fractions include low-boiling C_1 to C_5 linear and branched aliphatic hydrocarbons (bp $\leq 40\,°C$), and gasoline (bp 40–180 °C), which contains C_6–C_{10} hydrocarbons, including linear, branched, and cyclic alkanes plus alkylbenzenes. Kerosene (bp 180–230 °C) contains C_{11} and C_{12} hydrocarbons. Higher-boiling fractions yield C_{13}–C_{17} diesel fuel and home heating oil (bp 230–305 °C), a range of C_{18}–C_{38} lubricating oils and waxes (bp 305–515 °C). The residue after distillation is the complex black material known as *asphalt* or "pitch." A large part of the kerosene fraction is broken down into smaller hydrocarbons by "cracking" and catalytic reformation using zeolites (Chapter 7) as acid catalysts, and this is a major source of alkenes that are one of the starting points for polymer syntheses. Figure 3.4 outlines some of the pathways that are utilized.

Thus, ethylene, propylene. and styrene are used directly for polymerization (Chapter 6). Ethylene reacts with oxygen at 300 °C over a silver oxide catalyst to give ethylene oxide, a starting material for the manufacture of poly(ethylene oxide). Ethylene oxide is acid hydrolyzed to ethylene glycol, another key starting material for several other polymers. Acetylene reacts with hydrogen chloride to form vinyl chloride and with hydrogen fluoride to yield vinyl fluoride, both of which are important monomers for polymer formation. Acetic acid reacts with acetylene over a mercury/phosphate catalyst to give vinyl acetate, and with methanol to form methyl

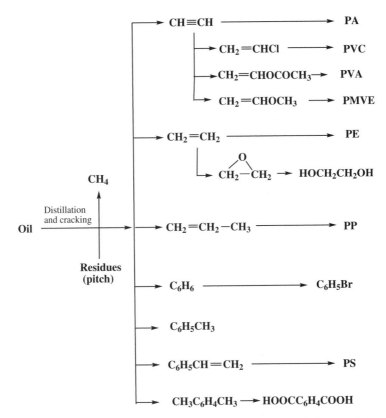

Figure 3.4. *Simplified flowchart showing pathways from oil to a few organic intermediates and materials [PA = polyacetylene, PVS = poly(vinyl chloride), PVA = poly(vinyl acetate), PVME = poly(methylvinyl ether), PE = polyethylene, PP = polypropylene, PS = polystyrene].*

vinyl ether. Thus, large segments of the polymer materials industry radiate from the small-molecule unsaturated species produced in an oil refinery.

c. Free-Radical Reactions. Free radical chemistry is mainly a characteristic of carbon compounds. Free radicals are formed when a low polarity bond is cleaved homolytically, that is, when each of the two fragments separates with one of the two electrons of the broken bond. Thus, a free radical bears an unpaired electron. Low-polarity bonds (C–C, C–H, C–S, C–Br, etc.) can be cleaved to free radicals when a molecule is exposed to heat, ultraviolet, X-ray, or γ radiation, or by mechanical stress. Free radicals are usually highly reactive species that will attack nearby molecules rapidly and sometimes indiscriminately. Aliphatic carbon compounds are especially susceptible to these reactions. The processes shown in reactions 48–54 illustrate a few of the possibilities.

Some of these reactions (48–50) are detrimental to the stability of organic materials since they lead to decomposition and uncontrolled crosslinking, but others (reactions 53 and 54) are important in polymer synthesis (see Chapter 6) because they are used to initiate free-radical polymerization reactions.

$$-\overset{|}{\underset{|}{C}}-\overset{|}{\underset{|}{C}}- \xrightarrow{\text{heat, UV,} \atop \gamma \text{ rays}} -\overset{|}{\underset{|}{C}}\cdot \;+\; \cdot\overset{|}{\underset{|}{C}}- \tag{48}$$

$$-\overset{|}{\underset{|}{C}}-H \xrightarrow{\text{heat}} -\overset{|}{\underset{|}{C}}\cdot \;+\; \cdot H \tag{49}$$

$$-\overset{|}{\underset{|}{C}}\cdot \;+\; -\overset{|}{\underset{|}{C}}=\overset{|}{\underset{|}{C}}- \longrightarrow -\overset{|}{\underset{|}{C}}-\overset{|}{\underset{|}{C}}{=}\overset{|}{\underset{|}{C}}\cdot \tag{50}$$

$$-S{-}S- \xrightarrow{\text{heat}} -S\cdot \;+\; \cdot S- \tag{51}$$

$$-\overset{|}{\underset{|}{C}}-Cl \xrightarrow{\text{heat, UV}} -\overset{|}{\underset{|}{C}}\cdot \;+\; \cdot Cl \tag{52}$$

$$R{-}N{=}N{-}R \xrightarrow{\text{heat, UV}} 2\,R\cdot + N_2 \tag{53}$$

$$R{-}\overset{O}{\overset{\|}{C}}{-}O{-}O{-}\overset{O}{\overset{\|}{C}}{-}R \xrightarrow{\text{heat, UV}} 2\,R{-}\overset{O}{\overset{\|}{C}}{-}O\cdot \tag{54}$$

Free-radical processes that take place in the presence of oxygen lead to the formation of epoxides, peroxides, and hydroperoxides (reaction 55), and this is a major reason why many polymers become brittle and decompose when exposed to strong sunlight. Fluorinated polymers and most inorganic materials are far less susceptible to free radical decomposition.

d. Oxidation Reactions. Many organic compounds are susceptible to oxidation in air at high temperatures or when exposed to high-energy radiation in the ultraviolet, X-ray, or γ-ray regions of the spectrum. This is perhaps the most serious weakness of materials that contain C–C and C–H bonds. Aliphatic C=C double bonds are especially susceptible to oxidation because oxygen can add across the double bond to form epoxides and hydroperoxides (reaction 55) which initiate the cleavage of nearby C–C single bonds. However, aromatic ring compounds are often more resistant to oxidative breakdown than are their aliphatic counterparts.

A driving force for the oxidative decomposition of organic compounds is the ease with which they can be converted to carbon dioxide. This is an exothermic process which can proceed directly and rapidly (through combustion) or indirectly through the formation of peroxides, carbon monoxide, or other species. Because carbon dioxide is a gas, it is lost readily, and this serves as a driving force for decomposition. Fire is a major hazard in the use of organic materials.

Combustion is an accelerated oxidation process fed both by the heat generated and the volatility of both carbon dioxide and the small organic molecules that are formed by the decomposition. The volatile small molecules oxidize rapidly by free-radical processes that give out more heat and lead to the emission of light. Compounds with C–Cl or C–Br bonds dissociate in a flame to form carbon and halogen free radicals, and these break the free-radical chain reactions and suppress the fire. Hence chlorinated or brominated organic compounds are added to organic polymers as fire retardants. However, chloro or bromo compounds absorb radiation and yield free radicals, and this serves as a mechanism to accelerate photodecomposition of organic polymers. Phosphorus compounds are also used as fire retardants, but mainly because they react with the combustion products to form a "char" that smothers the fire by cutting off the oxygen supply.

It is the susceptibility to oxidation reactions that places many organic materials at a disadvantage compared to inorganic alternatives.

e. Addition across Double or Triple Bonds. The addition of reagents across organic double and triple bonds was mentioned earlier. Acetylene is the starting point for the production of polymerization monomers and for the subsequent assembly of organic polymer molecules by free-radical and ionic polymerization reactions (see reaction 58), as well as for the construction of complex organic molecules. Addition across double bonds (reactions 56 and 57) are vital steps for the preparation of organic compounds. Rules and guidelines exist for the mechanisms of these addition reactions depending on the nature of the groups linked to the unsaturated center, but these are beyond the scope of this book.

From a polymeric materials viewpoint, one of the most important addition reactions is the sequential addition of olefins or vinyl compounds to double bonds, which leads to the growth of macromolecules as shown in reaction 58.

f. Formation of Organometallic Compounds. Some of the most useful building blocks in modern materials synthesis are organic derivatives of metals and metalloid elements. Organic compounds with chlorine, bromine, or iodine substituents react with metals to form organometallic compounds. This is illustrated by the interactions shown in reactions 59– 62. Thus, organolithium species are formed from the interaction of metallic lithium with 1-chlorobutane to give *n*-butyllithium. Magnesium reacts with bromobenzene to yield the Grignard reagent, phenylmagnesium bromide. These and related reagents are vital species for synthesis and as initiators for polymerization reactions. Ferrocene is prepared by the reaction of cyclopentadiene with sodium, followed by treatment of the cyclopentadienide with ferrous chloride (reaction 63). Ferrocene itself can be lithiated, and the products are intermediates for inorganic polymer formation. Triethylaluminum is produced by the reaction of ethylene with hydrogen and aluminum (reaction 61). This compound is also a key reagent for polymer synthesis (Chapter 6).

$$RCl \xrightarrow{\text{Li}} RLi + LiCl \tag{59}$$

$$RBr \xrightarrow{\text{Mg}} RMgBr + R_2Mg + MgBr_2 \tag{60}$$

$$CH_2{=}CH_2 \xrightarrow{\text{Al}+H_2} Al(CH_2CH_3)_3 \tag{61}$$

$$ECl_x \xrightarrow[-\text{LiCl or MgCl}_2]{\text{RLI or RMgBr}} ER_x \tag{62}$$

(E = a metallic or nonmetallic element)

$$\tag{63}$$

2. Silicon Chemistry

A comparison of four typical main-group elements—boron, carbon, silicon, and phosphorus—is given in Table 3.1. It will be obvious that, despite their proximity within the periodic table, significant differences exist between carbon and silicon.

a. The Element. Silicon is the second most abundant element in the earth's crust (oxygen is the first). The isolation of silicon from sand was mentioned earlier. The initial step involves a high-temperature reduction of SiO_2 with carbon in an electric furnace, as shown in reactions 64 and 65. The silicon produced by this method is not pure enough for semiconductor use. Hence, it is purified further by reaction of the silicon and silicon carbide impurity with chlorine to give $SiCl_4$ and CCl_4, which are separated by fractional distillation. Reduction of the pure $SiCl_4$ to silicon is accomplished by treatment with zinc or magnesium (reactions 66 and 67). Other methods for the isolation of elemental silicon include the chemical vapor deposition of SiH_4,

TABLE 3.1. Comparison of Four Typical Main-Group Elements

Property	Boron	Carbon	Silicon	Phosphorus
Method of isolation	B_2O_3 reduced with magnesium	Native and combustion of oil or coal	Reduction with carbon	Reduction with carbon
Connectivity in compounds	3, 4	2. 3. 4	4, 5, 6	3, 4, 5, 6
Orbital hybridization	sp^2, sp^3	sp, sp^2, sp^3	sp^3, sp^3d, sp^3d^2	sp^3, sp^3d, sp^3d^2
Double bonds	Yes	Yes	No (except in rare cases)	Yes
Halides plus water	Hydrolyzed	Stable (CCl_4)	Hydrolyzed rapidly	Hydrolyzed rapidly
Polarity of E—Cl bond	High	Low	High	High
Physical state of oxide at room temperature	Solid	Gas	Solid	Solid
Ketones/aldehdes or analogs	No	Yes	No	Yes (e.g., phosphine oxides)
Polymers by hydrolysis of halides or alkoxides	Yes	No	Yes	Yes
Electrical conductivity via E—E bonding	No	Yes (graphite); no (diamond)	Yes (semiconductor)	No
Stability of hydrides to oxygen	No	Yes	No	No
Oxide ceramics	Yes	No	Yes	No
Ease of photocleavage of E—E bond	Probably low	Moderate	High	Low
Absorption in the visible region by —E—E—E— bonds	No	Yes (graphite); no (diamond)	Yes	No

the reaction of SiI_4 with hydrogen, or the reaction of Na_2SiF_6 with sodium. The fabrication of ultrapure silicon (10^{-9}–10^{-12}% impurities) into semiconductor components is described in Chapter 10.

$$SiO_2 + C \rightarrow Si + SiC + CO \qquad (64)$$

$$SiC + SiO_2 \rightarrow Si + CO_2 \qquad (65)$$

$$Si + SiC \xrightarrow{Cl_2} SiCl_4 + CCl_4 \qquad (66)$$

$$SiCl_4 \xrightarrow[-MCl_2]{Mg\ or\ Zn} Si\,(pure) \qquad (67)$$

b. Silicon Reaction Chemistry. Figure 3.5 illustrates some of the primary reaction pathways used for the assembly of silicon compounds. The figure shows three reaction manifolds. The first starts from silica and leads into ceramic silicon chemistry. The second begins with elemental silicon or silica and yields semiconductor silicon. These two represent the inorganic side of silicon behavior. The third leads via $SiCl_4$ into part of the *organo*silicon platform.

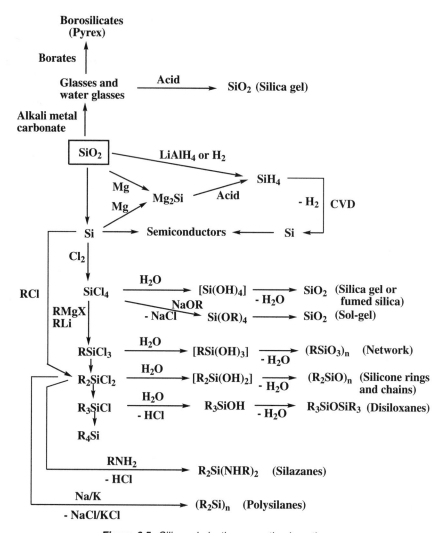

Figure 3.5. *Silicon derivatives—synthesis pathways.*

Totally inorganic materials can be accessed from sand by treatment with alkaline or alkaline-earth oxides or carbonates at temperature high enough to generate a molten reaction mixture. These reagents cleave some of the bonds in silica to generate linear, branched, or ladder chains or various-sized rings. This lowers the melting point and is the ancient process for the production of "soda glass." The more alkali used, the more Si—O—Si bonds are converted to Si—O$^-$ M$^+$ units or (for divalent cations, Si—O$^-$ $\cdots$ M^{2+} $\cdots$ O$^-$—Si ionic crosslinks). Thus, the alkali serves as a reagent for breaking covalent crosslinks. Ultimately, silica may be broken down to small-molecule water-soluble silicates such as sodium silicate, Si(ONa)$_4$, (water glass).

The enormous variety of different known silicates results from the permutations of accessible chains, ladders, and rings with different cations such as Na$^+$, K$^+$, Ca^{2+},

Mg^{2+}. Al^{3+}, or $Fe^{2+ \text{ or } 3+}$. Various chain and ladder structures that can be formed are described in Chapter 7.

Silicon tetrachloride provides a route to synthetic silica. Hydrolysis of $SiCl_4$ vapor yields finely divided "fumed silica" (reaction 68). Silica gel is produced by the acidification of aqueous solutions of sodium silicate (reaction 69). Another process for the production of synthetic silica is via the "sol–gel" process, in which silicon alkoxides (prepared either from $SiCl_4$ and alcohols or from silicates and sodium alkoxides) are hydrolyzed (reaction 70) (see Chapter 7). The utilization of silicon as a semiconductor is covered in Chapter 10.

$$SiCl_4 \xrightarrow[\substack{-HCl \\ \text{(fast, uncontrolled)}}]{H_2O} \underset{\text{ultrastructure}}{SiO_2} \tag{68}$$

$$SiO_2 + Na_2CO_3 \xrightarrow[-CO_2]{} Si(O^-Na^+)_4 \xrightarrow{HX} \underset{\text{silica gel}}{SiO_2(OH)} \tag{69}$$

$$SiCl_4 \xrightarrow[-NaCl]{NaOEt} Si(OEt)_4 \xrightarrow[\text{controlled}]{H_2O/EtOH} \underset{\text{"sol–gel" silica}}{SiO_2(OH)(OEt)} \tag{70}$$

$$SiCl_4 \xrightarrow{RLi \text{ or } RMgX} \underset{\text{(R=Me, Et, Pr, Ph, vinyl, etc.)}}{RSiCl_3, R_2SiCl_2, R_3SiCl, R_4Si} \tag{71}$$

$$Me_3SiCl \xrightarrow[-HCl]{H_2O} Me_3SiOH \xrightarrow[-H_2O]{} Me_3SiOSiMe_3 \tag{72}$$

$$Me_2SiCl_2 \xrightarrow[-HCl]{H_2O} [Me_2Si(OH)_2] \xrightarrow[-H_2O]{} (Me_2Si-O)_{3,4,5\ldots} \tag{73}$$

$$MeSiCl_3 \xrightarrow[-HCl]{H_2O} [MeSi(OH)_3] \xrightarrow[-H_2O]{} \text{ultrastructure network} \tag{74}$$

$$Si + 2MeCl \xrightarrow{\text{Rochow process)}} Me_2SiCl_2 \tag{75}$$

Organosilicon chemistry is a very broad field, in which silicon derivatives of most of the known organic units are accessible. From a materials perspective, some of the most important reaction pathways radiate from silicon tetrachloride (reaction 71).

Silicon tetrachloride reacts with Grignard reagents or organolithium reagents to replace chlorine atoms by organic units. For example, the reaction of $SiCl_4$ with methylmagnesium bromide or methyllithium provides access to a series of increasingly substituted silicon derivatives such as $MeSiCl_3$, Me_2SiCl_2, Me_3SiCl, and Me_4Si. Phenyllithium gives a similar series of phenyl derivatives. These species are important because hydrolysis of the remaining Si—Cl bonds yields organo*silanols*, which may condense to eliminate water and form Si—O—Si linkages. Depending on the number of hydroxyl groups attached to silicon, the condensation reactions may form network ultrastructures, linear polymers or small rings, or simple dimers called *disiloxanes* (reactions 72– 74). Silicone polymers are formed from dimethyldichlorosilane, Me_2SiCl_2, by hydrolysis and condensation to rings, followed by ring-opening polymerization of these to high polymers (Chapter 6). However, the large-scale starting materials are not produced via the Grignard or organolithium route (which is too expensive), but by the Rochow process—the reaction of methyl chloride with hot silicon (reaction 75).

Alternative pathways to organosilicon compounds are through fluorosilanes, produced from silicon tetrachloride and fluorine, or through aminosilanes, which are the products formed when $SiCl_4$ or organochlorosilanes react with ammonia or organic amines. This is a brief glimpse of silicon reaction chemistry, but more detailed descriptions are available and some are listed in Section E at the end of this chapter.

c. Differences from Carbon Compounds. A number of striking differences exist between the chemistry of carbon and silicon (Table 3.1):

1. As mentioned earlier, silicon–silicon double bonds are rare and very labile unless stabilized by the presence of bulky organic units on each silicon atom.
2. Although alcohols and phenols are stable species in organic chemistry, their counterparts in silicon chemistry tend to condense rapidly to form Si—O—Si linkages.
3. Organic diols may be dehydrated in an intramolecular process to form ketones, but, as seen above, the silicon counterparts tend to undergo intermolecular condensation to give rings and chains. Carbon–silicon bonds may be cleaved by acids or bases, but carbon–carbon bonds are more resistant to cleavage under these conditions. This difference is a consequence of the polarity of the carbon–silicon bond.

One of the most striking differences between the two elements is the behavior of the element–hydrogen bonds. Methane is a fairly stable molecule in air, but silane (SiH_4) is spontaneously flammable in the atmosphere. This can be traced to the higher polarity of the $Si^{\delta+}$—$H^{\delta-}$ bond compared to the relatively nonpolar and oppositely polarized $C^{\delta-}$—$H^{\delta+}$ bond. Thus, Ph_3SiH reacts with n-butyllithium to give Ph_3SiBu plus LiH while Ph_3CH reacts with the same reagent to give Ph_3CLi plus BuH. The Si—H bond is stable to neutral water but is hydrolyzed by aqueous acid or base. The C—H bond is stable to water, acid, and base.

3. Boron Chemistry

Some of the principles described for silicon chemistry also apply to boron chemistry (Figure 3.6), and the following summary only covers the main similarities and differences.

a. The Element. Elemental boron is isolated from boric oxide (B_2O_3) by a reaction with magnesium to give a product that is 95–98% pure. Other methods include the dechlorination of BCl_3 by zinc, the reduction of BBr_3 by hydrogen, the thermal decomposition of BI_3, and the electrolysis of molten borates. Boron hydrides (see later) decompose to pure boron and hydrogen when heated at high temperatures. Boron forms a large number of different structures in the crystalline solid state that range from B_{12} icosahedra to icosahedra with an additional boron framework radiating out from the central unit. Boron is an important dopant for use in the production of silicon semiconductor wafers.

b. Borides. Boron forms a large number of compounds with other elements such as carbon, silicon, and many metals. These species cover a wide range of

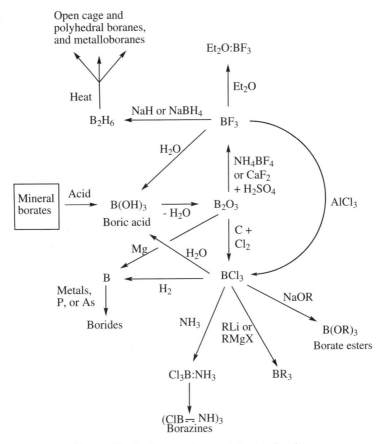

Figure 3.6. Synthesis pathways in boron chemistry.

stoichiometries. Boron carbide powder is used as an abrasive, and many metal borides are used as high-temperature, chemically resistant ceramic materials. Borides are synthesized by a variety of processes. Direct, high-temperature interaction of the elements is one method. The use of boron to reduce the oxide of another element is another, or the coreduction of boric oxide and a second element oxide with carbon. The solid-state structures of borides are exceedingly diverse and complex, but when boron is present in greater amounts than the other element (say, carbon), boron–boron bonding to form B_{12} icosahedra connected by atoms of the other element may dominate the structure. Borides are used in nuclear reactors as neutron shields and control rods because of the high absorptivity of ^{10}B for neutrons.

c. Borates. Boron and its compounds are obtained from mineral borates, which are water-soluble salts of boric acid, $B(OH)_3$. Typically the sodium and calcium salts of boric acid are found in large quantities, apparently after crystallizing from solution in ancient, evaporating lakes. Acidification of the salts yields boric acid, the thermal dehydration of which gives boric oxide (B_2O_3). Because the three hydroxyl groups in boric acid can condense randomly to form B—O—B bonds, the structure of B_2O_3 consists of a mixture of extended crosslinked chains and six-membered rings that crystallizes only with difficulty. This is the basis of borate glasses. As with silica, the introduction of alkali metal ions cleaves the chains and lowers the crosslink density. The addition of borates or boric oxide to silica turns soda glass into borosilicate glass, a much tougher material with a greater resistance to thermal shock.

d. Boron Halides. As with silicon chemistry, the halides of boron play a crucial role in providing access to a wide range of other derivatives and to rings and chains. The two most important halides are BF_3 and BCl_3, which are obtained by the pathways shown in Figure 3.5. Boron trifluoride is formed when B_2O_3 is heated with NH_4BF_4 (ammonium fluoroborate) or when B_2O_3 or sodium borate ($Na_3B_4O_7$) is treated with CaF_2 and concentrated sulfuric acid. Boron trichloride is produced either by treatment of BF_3 with Al_2Cl_6 or by the combined carbon reduction and chlorination of B_2O_3. Both BF_3 and BCl_3 are gases at room temperature.

Like the reactions of $SiCl_4$, both BCl_3 and BF_3 react with organometallic reagents to give organoboron compounds and with sodium alkoxides or aryloxides to yield borate esters. Boron trifluoride is used as an initiator in some vinyl addition polymerizations. One of the more important uses for BCl_3 is in its reactions with ammonia or primary amines to give borazines, which are six-membered ring compounds with a structural resemblances to benzene. Borazine rings are a starting point for the pyrolytic synthesis of boron nitride ceramics (see Chapter 7).

e. Boron Hydrides. One of the remarkable features of boron chemistry is the formation of a range of boron hydrides with quite unexpected structures. Their relationship to materials science is mainly through their use as precursors to ceramics, but from a scientific viewpoint they represent one of the most important developments in inorganic chemistry since the mid-1950s. Their synthesis starts with the reduction of BF_3 by NaH to give diborane (B_2H_6). This molecule is spontaneously flammable in the atmosphere. Note that this is a dimer of the expected BH_3. However, unlike ethane with its C—C bond, the structure of B_2H_6 involves *a hydrogen-bridged structure* as shown in Figure 3.7 held together by so-called banana bonds.

The three-center hydrogen bridge is also characteristic of a wide range of more complex boranes. Thus, pyrolysis of diborane at temperatures above 100°C brings about the progressive loss of hydrogen and the formation of higher fused boranes, including pyramid-like molecules, folded net-like and closed-cage B_{10} structures, open-face B_{11} cages, and remarkably stable B_{12} icosahedral compounds. Moreover, many of these form carbon derivatives (carboranes) and metallo compounds (metalloboranes and metallocarboranes). Some of these molecules have two open cages

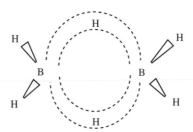

Figure 3.7. *Structure of diborane illustrating 3-electron "banana" bonds.*

connected by a metal in a manner reminiscent of ferrocene. The closed cage systems are much more stable to the atmosphere than is diborane. Note that in these structures the bonding includes both B—B and B—H—B linkages, and that the boron atoms usually bear additional hydrogen atoms at the perimeter of the molecule. Altogether, these structures form a wide interlinked chemistry that is still being explored, and that offers many future prospects for materials science.

4. Phosphorus Chemistry

The chemistry of phosphorus is exceedingly diverse and, like that of silicon and boron, has both its inorganic and organic-related aspects (Figure 3.8). The inorganic part revolves around phosphates, phosphites, and other salts together with interelement materials that contain phosphorus. The organic aspects are even broader than those of silicon. Like silicon, phosphorus plays an important role in polymer chemistry—specifically through a broad class of polymers known as *polyphosphazenes*. The nomenclature of phosphorus compounds is complex, and Figures 3.8 (here) and 3.9 (later) provide a starting point for understanding the naming system.

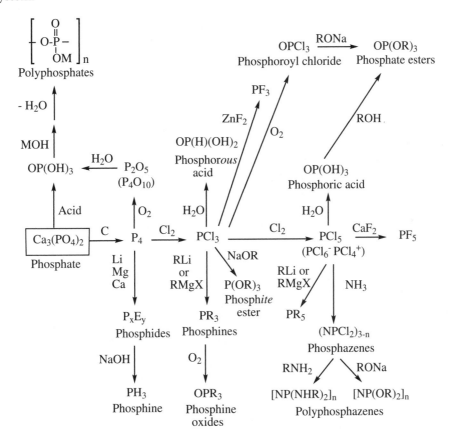

M = a cation, R = an alkyl or aryl group

Figure 3.8. *Phosphorus reaction chemistry.*

a. The Element. Phosphorus is produced from calcium phosphate mineral by heating with carbon and silica at 1400–1500 °C. The overall process is summarized in reaction 76, with the silica serving to remove the calcium as an insoluble "slag." The element distills from the reactor as P_4 tetrahedra and is condensed to the solid or liquid states. The solid is stored under water. Elemental phosphorus exists in several different solid state forms, known as *allotropes*. The most reactive of these is "white phosphorus," which oxidizes rapidly and reacts readily with halogens and other reagents. This form of the element ignites spontaneously in air at temperatures just above room temperature. Phosphorus is characterized by a number of different oxidation states that range from P(III) to P(VI). The penta- and hexavalent states are believed by some to make use of the $3d$ orbitals, a characteristic that occurs only with the second- and later-row elements. The hexavalent state exists in the form of R_6P^- anions.

$$Ca_3(PO_4)_2 \xrightarrow[\text{heat}]{C+SiO_2} P_4 + CaSiO_3 + CO \tag{76}$$

b. Phosphides. A large number of binary compounds exist, formed by the direct thermal reactions of phosphorus with various metals or their oxides. For each metal phosphide system, several stoichiometries of the two elements are usually found, all of which lead to the existence of a very wide range of different materials. Phosphides that contain group III main-group metals, such as GaP, are valuable semiconductors.

c. Phosphorus Halides. A number of hydrolytically unstable chlorides are known which include PCl_3, PCl_5, and $P(O)Cl_3$. The first two are produced directly from phosphorus and chlorine (Figure 3.8). The third is a product from the partial hydrolysis of PCl_5. PCl_3 is a highly reactive liquid. PCl_5 is normally produced from PCl_3 by further chlorination in an organic solvent such as monochloro- or dichlorobenzene. Note that PCl_5 in the solid state exists as the salt $PCl_4^+PCl_6^-$. Phosphoroyl chloride [$P(O)Cl_3$] is a reactive liquid at room temperature. The importance of these chlorides is that they are intermediates for the synthesis of a very wide range of organophosphorus compounds.

d. Phosphorus Acids, Phosphates, and Phosphites. Numerous hydroxy acids of phosphorus and their complex condensation products are known. These are derived mainly from phosphorus(III) and phosphorus(V) compounds. The two most important hydroxy acids are phosphor*ic* acid [$P(O)(OH)_3$] and phosphor*ous* acid, ostensibly $P(OH)_3$, but in practice, $HP(O)(OH)_2$. Phosphoric acid is the basis of inorganic phosphates, in which one to three protons are replaced by metal or ammonium cations. Phosphorous acid gives rise to phosphites, $P(OM)_3$ or $P(OR)_3$. Phosphoric acid is produced when P_2O_5 (really P_4O_{10}) reacts with water or when PCl_5 is hydrolyzed. Phosphorous acid is formed by the hydrolysis of PCl_3. These species can undergo condensation to produce chains and rings (metaphosphates or polyphosphates) consisting of P—O—P linkages. The complexity of these condensation processes is beyond the scope of this book, but the reader can explore this subject in greater detail through the references at the end of this chapter.

e. Organophosphorus Chemistry (Phosphines, Phosphine Oxides, Phosphites, and Phosphate Esters).

A large number of organic derivatives of phosphorus are known. The five main classes from a materials viewpoint are phosphines (PR_3), phosphine oxides, $[P(O)R_3]$, phosphoranes (PR_5), and phosphate or phosphite esters, $P(O)(OR)_3$ or $P(OR)_3$ (Figures 3.8 and 3.9). Organophosphines are accessible by the reactions of PCl_3 with organolithium or Grignard reagents.

Organophosphines are widely used as coordination ligands in transition metal complexes, including polymerization catalysts. Phosphine oxides (OPR_3) are stable species formed by the oxidation of phosphines. Phosphoranes arise through the reactions of PCl_5 with organometallic reagents or other nucleophiles. They are trigonal bipyramidal molecules that are often exceedingly stable. Phosphate esters are formed either by the reactions of PCl_5 with alcohols or phenols or their sodium salts, or by the direct reactions of P_2O_5 or phosphoric acid with alcohols or phenols. Phosphite esters arise from the reactions of PCl_3 with alcohols or phenols. Phosphorus also forms a broad range of amino derivatives when PCl_3 or PCl_5 interact with ammonia or amines.

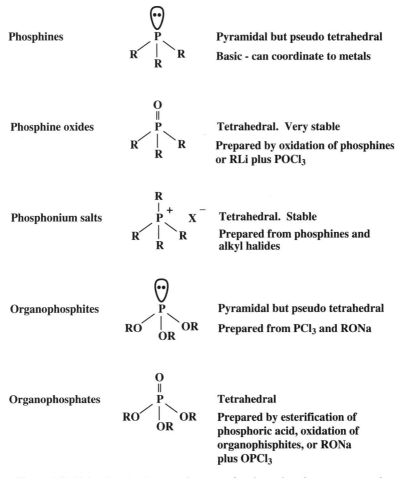

Phosphines	Pyramidal but pseudo tetrahedral Basic - can coordinate to metals
Phosphine oxides	Tetrahedral. Very stable Prepared by oxidation of phosphines or RLi plus $POCl_3$
Phosphonium salts	Tetrahedral. Stable Prepared from phosphines and alkyl halides
Organophosphites	Pyramidal but pseudo tetrahedral Prepared from PCl_3 and RONa
Organophosphates	Tetrahedral Prepared by esterification of phosphoric acid, oxidation of organophisphites, or RONa plus $OPCl_3$

Figure 3.9. *Molecular structures and names of various phosphorus compounds.*

f. Phosphorus in Polymers. Polyphosphates are the products of condensation reactions between metal or ammonium phosphates that bear two hydroxyl groups. Some of these polymers are found in mineral formations, but others are produced in the laboratory. However, the broadest class of phosphorus-containing polymers are the polyphosphazenes. One of the main methods for their preparation involves the reaction of PCl_5 with ammonia or ammonium chloride to give the cyclic species, $(NPCl_2)_3$. This is then induced to undergo a ring-opening polymerization to high-molecular-weight poly(dichlorophosphazene), which is then subjected to halogen replacement reaction using amines, alcohols, or phenols. This topic is reviewed in more detail in Chapter 6.

5. Interelement Compounds

A significant component of modern materials science involves the chemistry and properties of solids derived from two or more elements. Examples include semiconductors such as GaAs, GaP, AlP, ZnSe, and InP. Other binary solids, such as BN, AlN, Si_3N_4, or SiC, are high-temperature ceramics. We have already seen how certain interelement materials, such as SiC, are produced inadvertently during the reduction of silicon oxide by carbon. In general, binary solids are prepared by the high-temperature reactions of the pure elements, usually in a sealed system. Some binary species are similar to metal alloys in the sense that the ratio of the different elements in the material can vary over a wide range to generate a continuous series of properties. However, others exist in definite stoichiometries that suggest specific solid-state structures, such as CdS or GaAs, with a fairly narrow range of acceptable compositions.

6. Small Rings, Cages, and Short Chains

Phosphates, phosphazenes, organosiloxanes, organosilazanes, borazines, trioxane, cyclic esters, cyclic amides, carboranes, and fullerenes all exist in the form of rings or cages. Some of these have been mentioned in earlier sections of this chapter. Nearly all form components of materials, as is discussed in later sections of this book.

E. SUGGESTIONS FOR FURTHER READING

1. Cotton, F. A., *Advanced Inorganic Chemistry*, 6th ed., Wiley-Interscience, Hoboken, NJ, 1999.
2. Greenwood, N. N.; Earnshaw, A., *Chemistry of the Elements*, 2nd ed., Butterworth-Heinemann, Oxford, 1997.
3. Ellis, A. B.; Geselbracht, M. J.; Johnson, B. J.; Lisensky, G. C.; Robinson, W. R., *Teaching General Chemistry: A Materials Science Companion*, American Chemical Society, Washington, DC, 1993.
4. Carey, F. A.; Sundberg, R. J., *Advanced Organic Chemistry*, 5th ed., Wiley-Interscience, Hoboken, NJ, 2007.
5. Smith, M. B.; March, J., *Advanced Organic Chemistry: Reactions, Mechanisms, and Structure*, 5th ed., Wiley-Interscience, Hoboken, NJ, 2000.
6. Jones, M., *Organic Chemistry*, 3rd ed., Norton, 2004.

F. STUDY QUESTIONS (for class discussions or essays)

1. The extraction of pure elements from their ores requires a major expenditure of energy in the form of heat or electrical power. Thus, the recycling of elements is an important objective. Select one widely used element and trace the methods used for its extraction from minerals and its purification, followed by processes that could be used for its recovery and recycling. For each step, estimate the relative amount of energy that would be required for the processing of each kilogram.

2. Some of the differences between the chemistry of silicon and carbon are described in this chapter. Through background reading, develop this subject in more detail and identify those differences that are significant in materials science.

3. Hydroxides can be acidic, basic, or amphoteric (either acidic or basic). Discuss the factors that are involved in controlling this behavior.

4. The condensation reactions of weak hydroxy acids are important processes in materials science. Discuss the meaning of this statement.

5. The chemical reactions of silicon, phosphorus, and boron have certain features in common that are utilized in materials synthesis. Discuss them.

6. Review the various ways in which the elements used in materials chemistry are obtained from their minerals.

7. Identify the compounds in the following list that have a molecular structure at room temperature that differs from the one implied by the formula, and explain the reasons for the difference: BF_3, PCl_3, PCl_5, B_2H_6, $Al(C_2H_5)_3$.

8. Explain why entropy is an important factor in determining the thermal stability of a material.

9. If oil becomes too scarce to use for the large-scale synthesis of carbon-based materials, which reaction pathways would you choose to synthesize them from coal?

10. Are there viable sources of silicon compounds other than sand? Suggest some alternatives, and describe how they would be used as starting materials.

11. If white elemental phosphorus is flammable when heated in air, why are phosphorus compounds widely used as fire retardants?

12. Trace the changes in reaction mechanisms that accompany the transition from the chemistry of carbon compounds to those of silicon, boron, and phosphorus. How can you explain these differences?

4

Structure Determination and Special Techniques for Materials Characterization

A. PURPOSE

The driving force for the expansion of materials science is the design and synthesis of new forms of solid-state matter. However, the synthesis of new materials is largely meaningless unless the structures and properties of those new materials are also investigated. Only by understanding the relationship between structure and properties can new materials be designed in a rational way. Thus, in this chapter we examine techniques that play a major role in materials characterization. Although some of these methods will probably be familiar to the reader because of their widespread use in small-molecule chemistry, others are more specialized and will be given greater attention. The following techniques are divided into those that give information about the interior structure of solids and those that probe the nature of the surface. The relevance of solution-based analytical methods is considered briefly at the end of this chapter.

B. ANALYSIS OF BULK MATERIALS

1. Elemental Microanalysis

Despite the existence of more sophisticated techniques, elemental microanalysis to determine the ratios of different elements continues to be the first step required for the characterization of a new material. This can be accomplished by classical "wet chemistry" or combustion analytical techniques of the type that are sometimes taught in inorganic or analytical chemistry courses, or it may involve physical analyses such as X-ray photoelectron spectroscopy. Whatever the techniques employed, the primary purposes are to determine which elements are present and their ratios. Elements such as C, H, N, O, Si, P, S, and the halogens are analyzed routinely by wet chemistry and combustion methods. Some elements such as P or Si are difficult to analyze in the presence of oxygen.

2. Infrared–Raman Spectroscopy

Infrared and Raman spectroscopy (vibrational–rotational spectroscopy) provide indications of the types of bonds that are present in a solid. Thus, bonds with different lengths, strengths, bending, and torsional characteristics absorb different wavelengths of infrared radiation, and the absorption maxima may be characteristic of the types of linkages present. These absorptions generally occur within the 4000–400 cm^{-1} region of the spectrum. Only those vibrations that result in a rhythmic change in dipole moment are detected in the infared. Thus, bonds such as O—H and N—H bonds absorb infrared radiation between 3000 and 3700 cm^{-1}. Ionic sulfate absorbs near 1100 and near 500 cm^{-1}, and the ammonium ion is detected from absorbances near 3100 and 1400 cm^{-1}. Characteristic frequencies can be used to distinguish between aliphatic and aromatic C—C bonds, different types of C—O bonds, and a wide range of different inorganic structures. The actual absorbances may result from bond-stretching motions or vibration–rotation modes, both of which depend on bond strength and length.

Vibrations that involve no dipole moment change may be detected by Raman spectroscopy, which is a technique that analyzes the visible light scattered by the material and detects the wavelength shifts that occur during scattering.

A particular advantage of infrared and Raman spectroscopy for materials analysis is that the substance may be studied as a powdered solid or thin film. A powder can be mixed to a paste (a mull) in mineral oil or ground with potassium bromide and pressed into a pellet. It does not have to be soluble in a suitable solvent. Soluble, film-forming materials (particularly polymers) are cast on a sodium chloride plate by allowing the solvent to evaporate from a solution.

3. Solid-State Nuclear Magnetic Resonance Spectroscopy

Solution-state NMR is perhaps the most widely used characterization technique in small-molecule chemistry. It is employed extensively for studies of the structure of compounds that contain C, H, D, F, Si, P, and many other elements. Materials that dissolve in suitable solvents (particularly polymers) can be examined readily by this technique. However, the analysis of *solids* by NMR techniques is a more challenging problem. The NMR method depends on the relaxation of nuclear spin states that are induced by radiofrequencies. When a small molecule is dissolved in a solvent, these spin relaxations depend on the atoms to which the target atom is attached— with different intramolecular environments affecting the relaxation process. The different environments are denoted by the "chemical shifts." Thus, the environments of hydrogen or carbon atoms in an organic compound will be recognizable from the chemical shift.

However, the atoms in solid materials are connected to a more extended framework of neighbors, and the nuclear spin relaxations can become coupled to those of a large number of other atoms. This leads to NMR line broadening, a phenomenon that disguises many of the near-neighbor interactions that are of interest to the scientist. The magic-angle spinning technique is used to help resolve some of these problems. Two situations are commonly encountered. In the first, if the material has a random network amorphous structure, there may be so many different local magnetic field environments around an atom, that the lines remain broad, though not so broad as those without magic angle spinning. Moreover, if the atoms of interest are part of a regular crystalline structure, in which the nuclei are surrounded by a uniform regular environment of other atoms, then the resolution and narrowness of the lines can be impressive. Solid-state NMR analysis is now feasible for a wide range of nuclei, including H, D, C, Si, B, N, P, F, Li, Na, Cs, Al, and V. Oxides can also be studied in special cases.

4. Thermal Analysis

a. Differential Scanning Calorimetry (DSC). DSC analysis is a method that detects thermal transitions in a material. It is particularly effective for the identification of glass transition temperatures (T_g), crystalline melting temperatures (T_m), or liquid crystalline transitions (T_{lc}). If the material decomposes as the temperature is raised, DSC analysis can also identify the temperature at which this process begins (T_d) and the various stages in the thermal decomposition. The technique depends on a measurement of the heat evolved or absorbed when a material undergoes a phase transition. The apparatus is shown schematically in Figure 4.1.

The sample to be examined is placed in one of the two small pans. The second pan is either empty or contains a control material that has no transitions in the temperature range of interest. Both samples are then heated by individual electric heaters, with the temperature being raised typically at a rate of 10 °C per minute. The current flow to the heaters is monitored continuously. When an endothermic transition occurs in the test material, more electrical current will be needed to maintain the temperature rise, and the increased current is recorded and plotted on a graph. An exothermic transition will require less background heat and current,

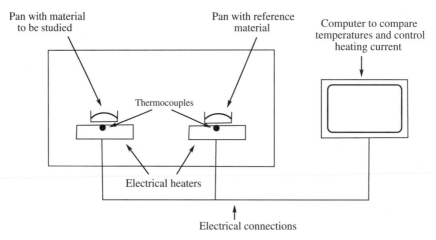

Figure 4.1. Schematic diagram of equipment for differential scanning calorimetry (DSC).

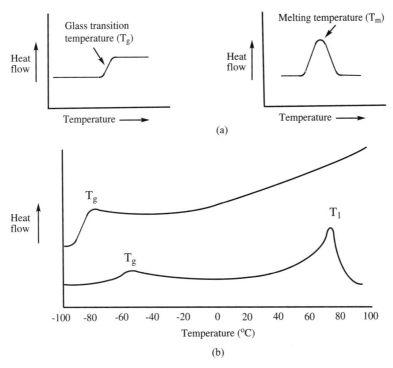

Figure 4.2. Top: *DSC scans showing idealized profiles for glass transitions and melting temperatures.* Bottom: *Experimental DSC curves for (a) an amorphous polymer and (b) a polymer with both a glass transition and a melting transition.*

and this, too, will be recorded and translated onto the graph. Glass transitions are exothermic. Conversely, the melting of a crystalline material is endothermic because energy must be supplied to separate the component molecules. Figure 4.2 illustrates the profile generated by a differential scanning calorimetry (DSC) experiment. Decomposition temperatures may be exothermic or endothermic depending on the chemistry of decomposition.

A related technique is *differential thermal analysis* (DTA), which uses similar principles except that, instead of current flow, the instrument measures the difference in temperature between the two samples as the temperature is raised. DTA gives information similar to that from DSC, but is perhaps less sensitive to hard-to-detect transitions.

b. Thermogravimetric Analysis (TGA). Thermogravimetric analysis measures the weight lost by a sample as the temperature is raised. Decomposition usually results in the formation of volatile small molecules; hence the onset of decomposition is detected by the onset of weight loss. TGA is commonly used to assess the thermal stability of polymers and the progress of reactions that turn preceramic materials into nonoxide ceramics.

The apparatus is quite simple in principle. A small sample pan is weighed continuously on a microbalance inside an oven. The temperature is raised automatically according to a predetermined program (typically 10 °C per minute), and the weight

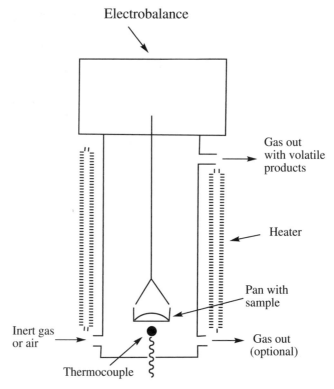

Figure 4.3. *Schematic diagram of thermogravimetric analysis (TGA) apparatus. The weight of the sample is measured as the temperature of the sample is raised. The upper gas outlet allows volatile decomposition products to be collected and analyzed.*

of the sample is monitored continuously (Figure 4.3). A plot of weight versus temperature is then obtained.

The dramatic weight loss shown in the curve in Figure 4.4a is caused by the polymer depolymerizing catastrophically to the monomer, which volatilizes from the system. The curve in Figure 4.4b illustrates the less precipitous weight decline that occurs when a preceramic polymer is heated to convert it to a ceramic.

TGA results yield values such as the initial decomposition temperature (T_d) or the temperature at which 10% or 50% of the weight has been lost $(T_{10}$ or $T_{50})$. In the case of a preceramic pyrolysis, an important result is the percentage of residue material left after heating is complete (the "char yield"). However, if the decomposition products have a sufficiently high molecular weight that they are nonvolatile, TGA experiments may seriously over-estimate the thermal stability of a material.

c. Thermomechanical Analysis (TMA). The evaluation of a structural material from an engineering viewpoint requires that its mechanical properties be studied as a function of temperature. This is accomplished in several different ways. First, measurements of the coefficient of thermal expansion are important for a variety of applications, and specialized equipment is available to simplify these studies. Also,

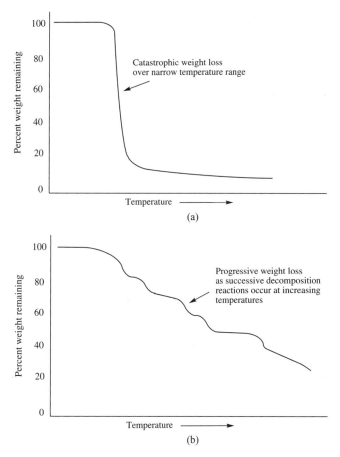

Figure 4.4. *Weight loss profiles for (a) a material that decomposes or depolymerizes catastrophically at one specific temperature and (b) a substance that decomposes progressively as the temperature is raised. Note that profile (a) may also be characteristic of a material that decomposes or depolymerizes at a temperature below the weight falloff, but the decomposition products do not volatilize until a higher temperature is reached.*

the hardness or softness of the material as the temperature is raised or lowered can be assessed by an indentation technique. A probe (vaguely resembling the ones used to play old phonograph records) is placed under a load, and its penetration into the surface of the sample is measured as the temperature is raised. At the T_g the material changes from a glass to a more flexible material that is indented more easily. This provides a measurement of the T_g.

Another device for thermomechanical analysis uses disks that rotate and vibrate to measure the flexibility of a material at different temperatures. Here again, glass transition temperatures may be detected. Elastic modulus and compliance are other properties that are measured. Variable-temperature stress–strain analysis is also used to assess engineering properties, as described in the following section.

5. Stress–Strain and Impact Analysis

The stress–strain behavior of a material can be measured by means of an Instron machine. The sample, in the shape of a "dogbone" has its ends clamped between the jaws of a machine that attempts to tear the material apart (Figure 4.5).

Brittle materials do not elongate when stress is applied, but break suddenly (Figure 4.6). Viscoelastic materials stretch under the same conditions and eventually undergo viscous flow. If these experiments can be carried out with the samples at different temperatures, the investigator can assess how the engineering properties change with temperature.

Impact resistance is measured by allowing a pendulum weight to swing down to hit a sheet or block of notched material to measure the force required to induce fracture. The initial height of the weight above the sample can be varied to change the force of impact.

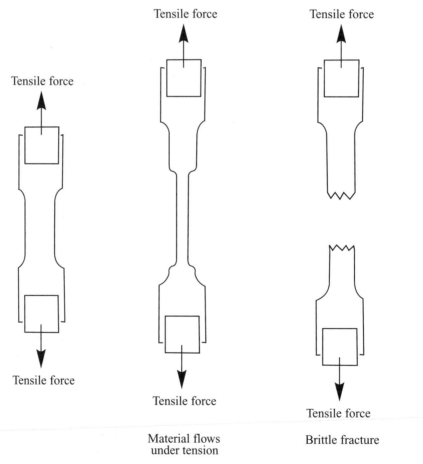

Figure 4.5. "Dogbone"—type evaluation of a material in an Instron machine. Two jaws connected to a robust mechanic device attempt to stretch the material. Depending on its physical properties, the material may elongate before failing, or it may snap in a brittle fracture.

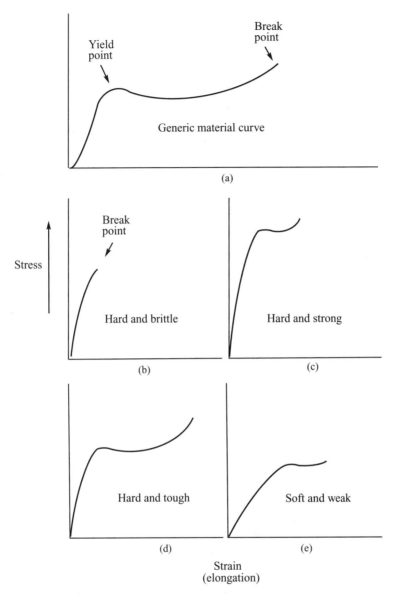

Figure 4.6. *Different profiles in the stress–strain curves are characteristic of various materials. These curves are a vital component in decisions about the engineering uses of materials.*

Engineering-type testing also involves measurements of compression strength and torsional stress and strain.

6. X-Ray Diffraction

a. Powder X-Ray Diffraction. Powder X-ray diffraction is a characterization technique that is widely used in inorganic materials science. The powdered material is packed into a borate glass capilliary tube and a narrow beam of X rays is passed through it. The main X-ray beam passes straight through the sample, but diffracted beams emerge as widening cones of diffraction around the path of the main beam. The theory of X-ray diffraction is beyond the scope of this text, but, in summary, crystalline substances are characterized by layers of atoms or molecules arrayed on a three-dimensional lattice. A large number of different planes can be drawn through the lattice, each set of planes having a different spacing from its neighbors. Diffraction occurs as the incident X-ray beam is "reflected" (actually diffracted) from all of these layer planes. Because the crystallites in the powder are randomly oriented, all possible planes will be sampled by the beam. Hence, the pattern emerging from the crystal will be a series of cones of X rays around the emergent main beam. The number of cones will in principle equal the number of different layer spacings in the sample. The diffraction pattern is sampled in one of three ways: (1) a flat piece of photographic film may be placed in the path of the emergent rays, and the film developed reveals the circles that represent the intersection of the film with the diffraction cones (Figure 4.7); (2) an electronic area detector may replace the piece of film, to provide a quicker means for recording the data; or (3) a narrow strip of photographic film may be curved into a cylinder that encircles the sample. The advantage of this arrangement is that it removes distortions that are inherent in the use of a flat detector.

The next step is to use the recorded data to work backward to calculate the diffraction angles θ in the crystals that gave rise to the recorded pattern, and from those angles to determine the layer spacings d that were responsible for the diffraction. This is accomplished with use of the Bragg equation, $n\lambda = 2d \sin\theta$, where λ is the radiation wavelength. Thus, the dataset consists of a set of layer spacings, and

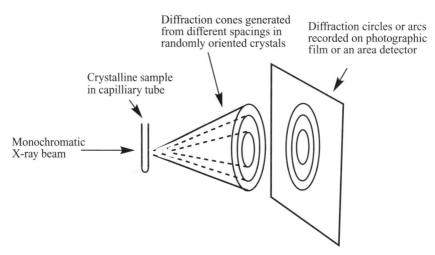

Figure 4.7. *Powder X-ray diffraction involves passing a thin beam of monochromatic X rays through a powdered crystalline sample contained in a capilliary tube. Each set of identical layer spacings gives rise to one diffraction angle. Because the crystals are randomly oriented, all angles of incidence and diffraction are sampled, generating a cone of X rays. Concentric cones (from different layer spacings) intersect a piece of photographic film or an electronic detector and are recorded as concentric circles. Measurement of the diffraction angles from the recorded image and the distance from the sample yields the layer spacings from the Bragg equation,* $n\lambda = 2d \sin\theta$.

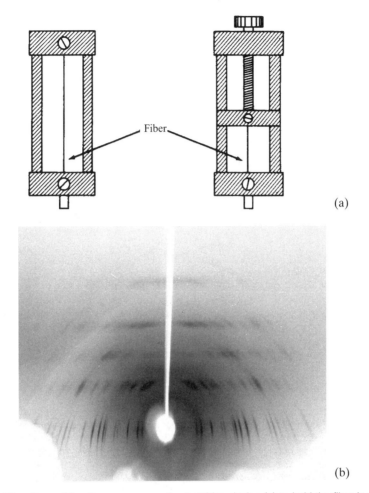

(a)

(b)

Figure 4.8. *Fiber X-ray diffraction uses a sample stretching device (a) to hold the fiber in the X-ray beam. A typical polymer fiber X-ray pattern is shown in (b). Each layer of short arcs gives information about the repeating distance along the polymer chain, while individual arcs on a particular layer give clues to the separation between adjacent parallel chains and the groups attached to those chains.*

the investigator then attempts to match that set to the values for known materials. A very large compilation of these powder datasets is available. Although powder diffraction data can occasionally be used to determine new molecular and materials structures, their main use is to identify substances in terms of previously studied materials.

A variation of the powder diffraction technique is used for polymers. If a polymer fiber is stretched so that the chains are aligned along the fiber axis, the fiber can replace the capilliary tube of powdered crystals described above (Figure 4.8a). The diffraction pattern from an oriented crystalline fiber will then consist of concentric cones, but cones that coalesce into spots rather than arcs as the material is stretched. The pattern can be used to calculate the repeating distance along the polymer chains, the chain conformation, and the separation between chains. However, the

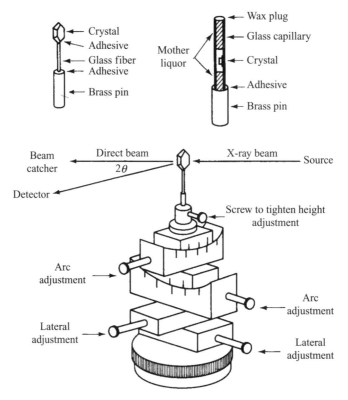

Figure 4.9. *A single-crystal X-ray diffractometer allows a narrow beam of monochromatic X rays to strike a crystal that is mounted on a device (a goniometer head) that allows it to be oriented in any direction. As the crystal is moved, diffracted X rays flash out as each set of planes within the crystal come into the diffracting position (determined by the Bragg angle). The intensity of each diffracted beam depends on destructive interference between interlayer features. Both the diffraction angle and the intensity of each reflection are captured by an array detector (a flat-panel electronic device) that transmits the data to a computer where the overall crystal structure and the location of individual atoms in the structure are calculated.*

resolution of the structural features is hardly ever as good as can be obtained for small molecules in crystals studied by single-crystal methods.

b. Single-Crystal X-Ray Diffraction. The organization of molecules, atoms, or ions in a crystal allows the location of those units to be identified with precision using single-crystal X-ray diffraction. Moreover, this technique permits the positions of the atoms within a molecule to be established, and this information, in turn, leads directly to the determination of bond lengths and bond angles.

The equipment for single crystal X-ray diffraction is shown schematically in Figure 4.9. A collimated beam of monochromatic X rays is allowed to impinge on a single crystal of the material under investigation. The crystal is mounted on a device (a goniometer head) that allows specific crystal faces and layer planes to be aligned with the instrument. The goniometer head in turn is mounted on an apparatus that (sometimes by means of a wrist movement) can move the crystal so that any set of layer planes can be brought into the diffracting position relative to the incident X-ray beam. This is under computer control, with the movements orchestrated in terms of the space group (packing arrangment) within the crystal. Diffracted X-rays flash out as the crystal is rotated and tilted, and these are detected using a piece of photographic film (the original historical method), a scintillation counter, or an electronic flat-panel detector (the most modern method). This last method allows very rapid collection of data, which is important when thousands of individual reflections must be recorded. Computer programs are then used to analyze both the positions of the reflections (and therefore the diffraction angles, and the layer spacings) and the intensities of each reflection. The intensities are employed to determine the phase interference interactions generated by different atoms in a molecule, and this, in turn, gives precise data on the location of atoms within each molecule.

7. Refractive Index and Chromatic Dispersion

a. Refractive Index. Optical glasses used in lenses, prisms, optical waveguides, and switches have two important characteristics—their refractive index and the degree to which they split white light into the component colors (chromatic dispersion). *Refractive index* is a measure of the speed of light through the glass or crystal compared to its (faster) speed in a vacuum. Thus, the refractive index of a solid or liquid is always higher than that of a vacuum (or air). Typical values for solids are in the range of 1.3–2.4. Refractive index is measured in several different ways. Although the measurement of the refractive index of a liquid is a routine procedure (just place a drop of the liquid on a refractometer plate and measure the angle), obtaining the same information for a transparent glass or crystal is another matter. Two of the commonest techniques are refractive index matching of the material to that of a liquid in which it is immersed, and measurement of the critical angle.

In the liquid immersion method, the investigator assembles a series of different organic liquids to provide a range of refractive indices, and the material is successively immersed in each of these until the sample can no longer be seen as a separate phase. The refractive index (RI) of the material is then the same as the refractive index of that liquid. If the material is swelled by or dissolves in organic media, aqueous solutions of salts can be used instead. A second method uses the apparatus shown schematically in Figure 4.10.

Figure 4.10. *Method for measuring the refractive index of a solid. A film of the material is solvent cast onto one face of a prism that has a known refractive index. A beam of light of known wavelength is then passed through the prism to sample the film from the inner side. The prism is then rotated until the critical angle is reached where total internal reflection causes the beam to be diverted back into the prism. A knowledge of the critical angle and the refractive index of the prism then enables calculation of the refractive index of the test material.*

A film of the material under investigation is cast onto one face of a prism that is mounted on a calibrated rotation stage. A monochromatic beam of light from a laser is allowed to pass through the prism and strike the prism–material interface. The prism is rotated until the angle of the incident beam to the prism–material interface reaches the critical angle θ_c, at which the beam is reflected back into the prism. Because $\theta_c = \sin^{-1}(1/n)$, the refractive index n can be calculated if the refractive index of the prism is known. In practice, the apparatus should be calibrated using materials with known refractive indices. Note that RI changes with the wavelength of the light; the RI to blue light is higher than the RI to red light. Hence the wavelength of light used for the measurement must always be specified.

b. Chromatic Dispersion. Chromatic dispersion arises because different wavelengths of light are bent at different angles when passing through an interface that separates two substances with different refractive indices. Blue light is slowed more than red light when passing from a low-RI phase (say, air) to a higher-RI phase. Hence the blue light is bent more. Chromatic dispersion leads to color fringing when white light is focused by a single lens and is, of course, responsible for the separation of white light into its component colors as it passes through a prism or through raindrops to form a rainbow. It is also responsible for the distortion of pulsed signals transmitted down an optical fiber waveguide, since the light source may not be absolutely monochromatic, or the fiber material may not be uniform. Therefore, closely pulsed signals propagated at different frequencies will move at different speeds and will eventually overlap. These issues are considered further in Chapter 14. Chromatic dispersion can be measured directly by timing the transmission of a monochromatic light pulse as it passes down an optical fiber and by comparing the speed for different frequencies. Typically, several different lasers (generating different frequencies) are connected to one end of an optical fiber, and the arrival time of each pulse at the other end is measured for each frequency. Other methods for the measurement of optical dispersion include phase shift techniques, which are beyond the scope of this book.

8. Magnetic Susceptibility

A paramagnetic material is attracted into a magnetic field, while a diamagnetic substance is weakly repelled. Paramagnetism arises from materials that contain unpaired but uncoupled electrons. Diamagnetism results from the circulation of spin-paired electrons induced by the magnetic field, and the weak repulsion is due to the induced magnetic field generated by that circulation. Because paired electrons are found in all molecules, all solid materials have a diamagnetic component, which tends to offset and weaken any paramagnetic properties.

Magnetic susceptibility is a measure of para- or diamagnetism. It is determined experimentally in several ways. The simplest approach is by the use of a Gouy balance in which the sample is suspended between the poles of an electromagnet, and its weight is determined with the magnetic field on and off (Figure 4.11).

If the material is a powder, it can be packed into a sample tube divided into two halves. The other half is filled with a material of known susceptibility. The compartment division is placed at the midpoint of the magnet. If the upper sample is more paramagnetic than the reference material, the tube will appear to be heavier with

Balance

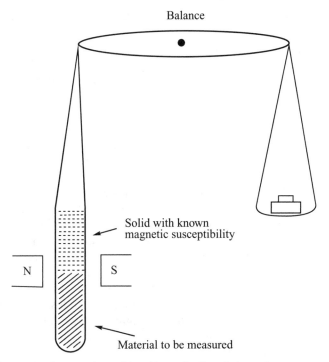

N   S

Solid with known
magnetic susceptibility

Material to be measured

Figure 4.11. Schematic diagram of a traditional form of a Gouy balance for measurement of magnetic susceptibilities. The sample (usually contained in a tube of nonparamagnetic material) is suspended between the poles of a permanent magnet. In the simplest method, half of the tube is filled with the sample of interest and the other half, with a material of known magnetic susceptibility. The material with the highest magnetic susceptibility is drawn into the field, and the apparent weights with the field on and off allow the magnetic susceptibility of the sample to be estimated. An alternative approach eliminates the control material and uses a powerful electromagnet. The weight of the sample with the magnetic field on and off is related to the magnetic susceptibility of the sample material, and if the strength of the magnetic field is known, the magnetic susceptibility can be calculated directly. The most modern devices use superconducting magnets and measure the influence of the paramagnetism of the sample on the behavior of the current in the magnet.

the magnetic field on than when it is off. The weight difference is then used in the equation $\Delta W \times g = [(K - K_r)H^2A]/2$, where K is the magnetic volume susceptibility of the material, K_r is that of the reference, H is the field strength in gauss, g is the gravitational constant, and A is the cross–sectional area of the tube. The magnetic field should be in the range of 5000–15,000 G. The level of paramagnetism detected can be used as an indicator of the number of unpaired electrons.

However, materials that have unpaired but *coupled* electrons (i.e., where the electron spins are aligned within a solid-state lattice) generate what is known as cooperative magnetism, and these are characterized by very large magnetic suseptibility values. These materials are called *ferrimagnetic*, *antiferrimagnetic*, and *ferromagnetic*, and their behavior depends on the strength of the applied magnetic field. Ferromagnetism exists below a certain critical temperature and is unusual in the sense that the susceptibility increases rapidly with increases in field strength but then saturates. Ferromagnetic materials have a parallel alignment of the spins of the atomic magnets and they are magnetic without the influence of an external field.

Examples include iron, cobalt, nickel, and their alloys and certain rare-earth elements.

The uses of magnetic susceptibility include (1) identification of unpaired electrons in the material, (2) detection of aromaticity in organic solids, (3) detection of free radicals in solids, and (4) study of transition metals and their compounds.

9. Electrical Conductivity

The measurement of electrical conductivity is a crucial aspect of research in metals, semiconductors, superconductors, solid battery electrolytes, and fuel cell membranes. The measurements may be as unsophisticated as determining the resistance between two electrodes separated by a known distance on the surface of a solid, or they may involve elaborate experiments using a four-probe device or alternating current impedance measurements. In general, conductivities determined for compressed powders are less useful than values obtained from films or monolithic solids. It is also necessary to use separate techniques for electronic and ionic conductors.

The four-probe technique is appropriate for materials that have electronic conductivities in the range of 10^{-4}–10^{4} S/cm ($\Omega^{-1}\,cm^{-1}$). The device consists of four electrodes inline, with current applied to the two outer electrodes, but with the voltage measured across the inner two electrodes (Figure 4.12). The probes are tips of osmium or tungsten carbide, which either rest on the surface or penetrate the surface to sample the bulk material.

The purpose of this arrangement is to ensure that the results are not distorted by the contact and spreading resistance associated with the outer probes. It also provides a constant current flow through the material, which simplifies interpretation of the voltage detected between the two inner electrodes.

The two-probe technique does not provide this simplification, but it is used for materials with a low electronic conductivity (10^{-8}–1 S/cm). A known voltage is applied across the two electrodes and the current passing through the material is

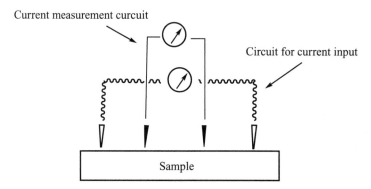

Current measurement curcuit

Circuit for current input

Sample

Figure 4.12. A four-probe electrical conductivity- or resistivity-measuring device. The indentation probes are oriented inline and are separated by ~1 mm. A known current is passed through the sample via the two outer indentation probes, but the resistivity is measured between the two inner probes by means of a high-impedance electrometer. This arrangement is used because the high impedance of the measuring electrometer ensures that negligible current passes between the two inner probes compared to the current passing between the two outer probes. Hence, this method works well for semiconductors, polymers, and other materials with conductivities ranging from 10^{-4} to 10^{4} S/cm.

measured. Low-conductivity materials can also be studied as thin films, with two electrodes applied to opposite faces of the film.

Measurements of *ionic* conduction in a solid require the use of another technique, known as *impedance analysis*. This is appropriate for ionic conductivities in the range of say 10^{-8}–1 S/cm. Ionic conductors suffer from polarization of the electrodes under the influence of a direct current because ions collect at the electrodes and the conductivity falls as the experiment proceeds. For this reason, measurements must be made with only minimal perturbation of the system and with the use of alternating current (AC).

Impedance (Z) is a measure of the opposition offered by a material to the flow of an alternating current at a given frequency. Unlike a measurement of resistance using direct current, AC resistance depends on two parameters. The first is the resistance to the flow of charge, which depends on the ratio of V_{max} to I_{max}, and the second is the difference in phase angle by which the current lags behind the voltage. Thus, Z is a complex quantity with real and imaginary components that correspond to the resistance and the reactance, respectively. The experiment requires the application of a small direct current across a film of the sample, together with a superimposed alternating current. The experimental setup is illustrated in Figure 4.13. The electrochemical potential and the current are monitored as a function of frequency. The complex impedance is then split into its real (resistive) and imaginary parts and the real component is plotted against the imaginary part (Figure 4.14).

10. Transmission Electron Microscopy

A transmission electron microscope (TEM) resembles an optical microscope except that the lenses of the microscope are replaced by coils that focus an electron beam that has passed through the sample. The resolution attainable by a TEM may be as small as 0.2 nm (2 Å) compared to the theoretical limit for an optical microscope, which is in the range of 0.2 μm (microns). Figure 4.15 shows the general layout of a TEM unit.

Because electrons are absorbed strongly by materials, the sample can be no more than 50 nm thick. When heavier inorganic elements are present, the sample must be even thinner. For polymer samples, the magnification may be in the range of 50–100 times.

C. SURFACE AND THIN-FILM ANALYSIS TECHNIQUES

1. Scanning Electron Microscopy

Scanning electron microscopy (SEM) is one of the most widely used techniques in materials science. The surface of the material is sputter-coated with a thin layer of gold or another conductive material and is scanned by a thin beam of electrons. The operation of an SEM apparatus is shown in Figure 4.16. The sample must be maintained under vacuum to avoid the scattering of electrons by air.

Electrons produced from a heated filament in an electron gun are accelerated toward a hollow anode and are then focused on a point at the specimen surface and

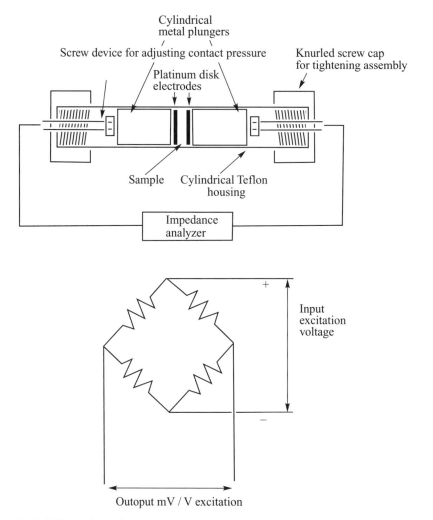

Figure 4.13. Cell and circuit diagram for measurements of ionic conductivities of solids using electrochemical impedance spectroscopy. The circuit consists of a resistor and a capacitor connected in parallel, with the electrical behavior monitored with an AC impedance analyzer.

rastered across it by means of one or more magnetic lenses. As it pauses at each point on the surface, the beam causes the ejection of slow electrons from the atoms at the surface, and these secondary electrons are detected by a positively charged grid placed above the sample. In a typical experiment, the rastering process samples 1000 points along each line for 1000 different lines to give a total of one million pieces of information. This information is processed electronically and displayed on a screen as a map of the surface, with a resolution of typically 4 nm. The types of elements present on the surface can be identified by a second type of radiation emitted by the sample—the backscattered fast electrons—the density of which depends on the size of the atomic nuclei. Thus, the location of different elements on the surface as well as the topographical features can be mapped by use of the two techniques.

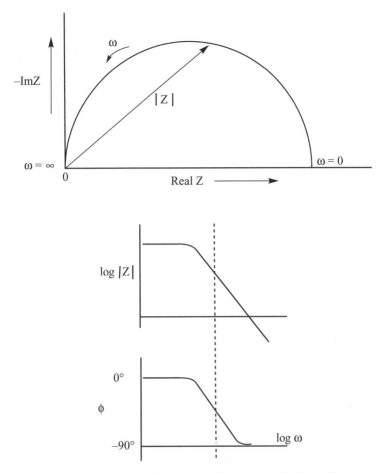

Figure 4.14. *A Nyquist plot with the inpedance vector **Z** shown as a function of the imaginary versus the real impedance, with the frequency increasing from right to left.*

2. Scanning Tunneling Microscopy (STM) and Atomic Force Microscopy (AFM)

a. Scanning Tunneling Microscopy. This is a technique for surveying the surface of a material down to the size of molecular or atomic features. A minute tip with a radius between 10 and 100 nm is moved across the surface of interest such that an electric current is able to tunnel through the small vacuum gap between the tip and the material. The current flowing across the gap depends on the distance to the surface and the electronic structure in that region of the surface (Figure 4.17). In STIM work the electronics are configured in such a way that separation between tip and surface is kept constant by variations in the current. Thus, a record of the current variations provides a profile of the surface.

b. Atomic Force Microscopy. In AFM the tip is attached to a cantilever device that allows the tip to move vertically as it covers the surface in a so-called contact

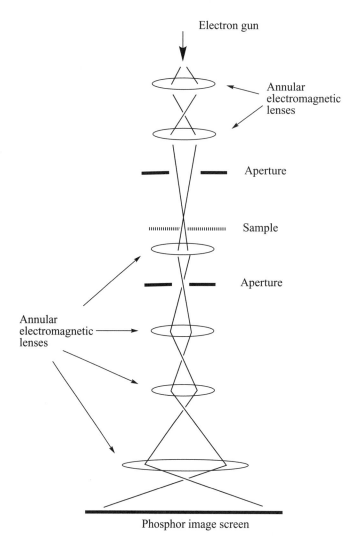

Figure 4.15. *Schematic of a transmission electron microscope. The ring-shaped electromagnetic lenses focus the monochromatic electron beam through the thin sample, enlarging the image until it impinges on a phosphor screen, where the object can be viewed or photographed. The equipment must be maintained under vacuum because of the ease of electrons scattering in air.*

mode (Figure 4.18). This vertical movement is in response to the attractive or repulsive forces generated between the tip and atoms on the surface. The vertical movement of the tip is amplified optically by means of a laser beam reflected off the cantilever, and this provides a profile of the topography of the surface. In practice, the tip is sometimes used in a "tapping mode" in which the tip moves up and down to tap and sample the surface. The advantage of the tapping technique is that it reduces the influence of lateral forces that cause the tip to twist as it is moved over surface features.

The effectiveness of the AFM method depends on the size of the probe tip. Probes are typically fabricated from silicon or silicon nitride or, increasingly, from

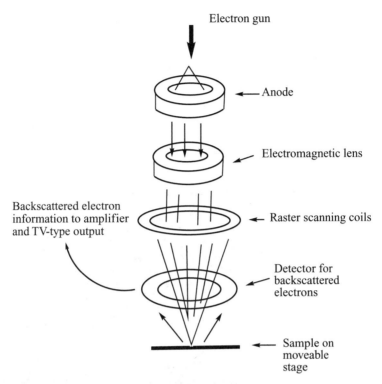

Figure 4.16. *A scanning electron microscope rasters a focused beam of electrons over the surface of a sample. Electrons scattered back from the surface are detected and analyzed, and the data are converted to a TV-type image of the surface.*

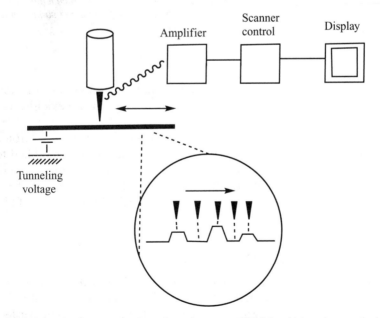

Figure 4.17. *Schematic of a scanning tunneling microscope (STM) in which a sharp probe is scanned across a surface maintaining a constant distance from that surface. Changes in the distance between the probe tip and the atom-scale surface features translate into changes in the electric current that tunnels between the surface and the probe. Thus, after amplification of the current, a profile of the surface can be created with features detectable down to sizes of 0.2 nm.*

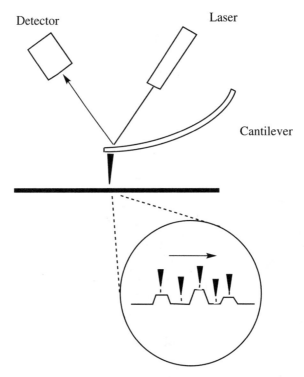

Figure 4.18. *Atomic force microscopy (AFM) resembles STM in the sense that a probe is rastered across a surface, but differs because the probe is free to move vertically on the end of a cantilever. The vertical movement is detected and amplified by means of a laser beam reflected off the cantilever. The equipment can be configured in two main ways: (a) the tip skims the surface features at a separation distance controlled by surface repulsive forces, and (b) the probe taps the surface at a high frequency in a manner that avoids damage to soft surfaces.*

single-wall nanotubes (see Chapter 17). Single-wall nanotubes have the advantage of small diameter, which gives better resolution, and physical flexibility. However, they are expensive.

Atomic force microscopy has several advantages over scanning electron microscopy for surface investigations. The surface does not need to be modified to make it electronically conductive. Moreover, it gives a three-dimensional rather than a two-dimensional map of the surface. In addition, AFM can be used with the surface exposed to air or under water, which is especially useful for the study of biological molecules or cells on surfaces.

3. X-Ray Photoelectron Spectroscopy (XPS)

X-ray photoelectron spectroscopy spectroscopy [also known as *electron spectroscopy for chemical analysis* (ESCA)] is a method for analyzing the outer 5–50 Å of a surface. The techniques allows identification of the elements in the surface layer and also provides information about the environment surrounding a particular element. The method involves use of a monochromatic beam of soft X rays (typically Mg K_α or Al K_α radiation) directed at the surface in a vacuum chamber (Figure 4.19).

These X rays displace electrons from the inner shells of the surface atoms, and the energies of the ejected electrons are analyzed by means of a detector. Because the energies of the ejected electrons reflect the energy levels of the inner orbitals from which they originated, a measurement of the kinetic energy of the detected electrons is diagnostic for the type of element. Moreover, the same element in a different environment will give a different peak in the spectrum. Actual spectra may contain multiple peaks that arise not only from different atomic environments but also from the ejection of electrons from different orbitals. Electron spin interactions may also yield multiple peaks. All elements in the periodic table, except for hydrogen and helium, can be studied in this way.

Auger spectroscopy is a related technique that generally uses an electron beam instead of X rays. Ejection of the first electron from an inner shell causes a higher-energy electron to take its place, thereby emitting energy that can lead to the ejection of an outer electron. This secondary electron (the Auger electron) has an energy that can be traced to a specific element and electronic configuration. In practice several different Auger electrons may be emitted, each resulting from different transitions within the atomic shell. The overall result is an alternative way to identify different elements in a surface and to ascertain their environment.

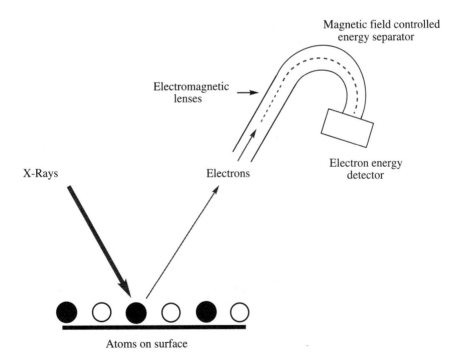

Figure 4.19. *Principle behind X-ray photoelectron spectroscopy. A narrow X-ray beam ejects inner-core electrons from surface atoms. The electrons have different energies depending on their source within the shells of each atom. These electrons are guided into a hemispherical path separator, and their original binding energies are measured. The energy profile spectrum is characteristic of (a) the type of atom and (b) the environment of the atom. The relative amounts of different atoms in the surface region can also be determined.*

4. Total Internal Reflection Infrared Spectroscopy

Although infrared spectroscopy is typically used for the analysis of the bulk phase, one variant of the technique, *total internal reflection infrared spectroscopy* (TIR-IR), is particularly important for surface analysis. In this method, a beam of infrared light is induced to bounce back and forth between the materials surface and the face of a totally internally reflecting glass plate or crystal, such as germanium or thallium bromide iodide. The surface is sampled at each bounce of the beam. This technique is useful for the analysis of thin films cast or deposited on a surface, and also provides information about how the surface composition differs from that of the bulk material.

5. Ellipsometry

Ellipsometry is a method that is widely used to measure the thickness of thin films (angstoms to micrometers thick), the roughness of surfaces, and/or the refractive index of thin films. It is routinely employed in the semiconductor industry to measure the thickness of polymer films spun cast on silicon wafers (Chapter 10).

The method involves reflection of a monochromatic, linearly polarized light beam (see Chapter 14) from the surface of a thin, transparent film. The reflected beam is elliptically polarized, with a degree of ellipticity that depends on the film thickness (Figure 4.20). The degree of ellipticity is measured from the reflected light intensity transmitted through a second linear polarizer as it is rotated through different angles. Although any convenient wavelength of light can be employed, a helium–neon laser ($\lambda = 632.8\,\text{nm}$) is commonly used.

Ellipsometry works well for a single thin film on a surface or for multiple-layer stacks, since the reflection from each layer can be identified and analyzed. Interpretation of the data requires not only knowledge of the ellipticity of the reflected light but also the refractive index of the material. Thus, a knowledge of one (determined by some other technique) can give the other. Surface roughness estimates are possible through the measurement of film thickness at different points across the

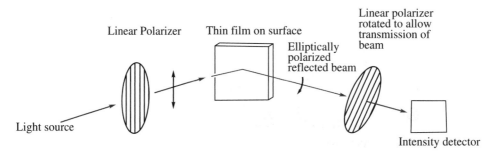

Figure 4.20. *Ellipsometry makes use of the behavior of a linearly polarized light beam when it reflects from a surface. Linearly polarized light consists of two wave vectors at 90° (perpendicular) to each other and in phase. Reflection from both surfaces of a thin film causes the two wave components (one in the p plane and the other in the s plane) to move out of phase, converting the linear polarization to elliptical polarization. The change in polarization is a function of film thickness and can be measured by analyzing the change in angle through a second linear polarizer.*

surface, specifically, by moving the film incrementally so that the point of contact of the incident beam samples different areas.

6. Contact Angles

The chemical structure of a surface often differs from the composition of the interior of the material. This may be the result of reactions of the surface with air or moisture, or it may reflect the preference of certain components of the solid to concentrate at the surface. A simple method to investigate surface character is to measure the angle made by a drop of water or other liquid in contact with the surface. Hydrophobic surfaces will repel a water droplet and generate a high contact angle (Figure 4.21). Conversely, a hydrophilic surface will induce the droplet to spread on the surface and give rise to a low contact angle. These values are measured by means of a telescope mounted parallel to the surface

Contact angles can range from close to zero degrees to values in excess of 100°. For example, glass has a contact angle near 50°, gold a value of 77°, polyethylene 94°, and fluorinated polymers near 100°. "Superhydrophobic" surfaces have contact angles as high as 150° or 160°. Because the hydrophobicity or hydrophilicity of different chemical groupings is known, the contact angle should provide clues about which groups populate the surface. In many cases, the character of the surface will depend on its prior treatment. For example, a polymer surface that has been in contact with water may have reorganized to place the most hydrophilic units close to the surface. Air is generally considered to be a hydrophobic medium. Hence the same polymer may show evidence of a more hydrophobic interface if it has been stored in dry air. Moreover, the measured contact angle may depend on whether

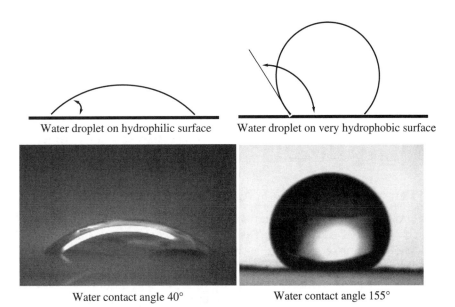

Water droplet on hydrophilic surface   Water droplet on very hydrophobic surface

Water contact angle 40°   Water contact angle 155°

Figure 4.21. *Water contact angles are measured by observation through a small telescopic image of a water droplet placed on a surface. Contact angles of hydrophobic liquids on a surface will be very different and will provide a secondary measure of hydrophobicity or hydrophilicity.*

the droplet is static, advancing, or retreating on the surface. Contact angles are extremely useful for understanding both adhesion and biomedical compatibility.

D. SOLUTION ANALYSIS TECHNIQUES

1. General Comments

Although the analysis of materials in the solid state is the most direct approach to characterization, it is often necessary to couple that information to data that can be obtained only from solution data. Thus, information about composition, molecular architecture, molecular weights, and other properties is often crucial to understanding solid-state structure and yet cannot be obtained directly from the techniques described earlier in this chapter. A discussion of these solution-based techniques is beyond the scope of this book, but the following brief comments will provide a basis for further study.

2. Solution NMR Spectroscopy

Solution-state ^1H, ^{13}C, ^{31}P, ^{19}F, ^{28}Si, and other. NMR spectra can provide valuable information about the molecular structure of small molecules and uncrosslinked polymers that are soluble in suitable solvents. The method can even be used for lightly crosslinked polymers if they can be swollen in suitable media. However, heavily crosslinked polymers and ceramics must be studied by solid-state NMR techniques, which, as discussed earlier, provide less detailed information. One area where solution-state NMR may be useful for insoluble materials is when the products of thermal or chemical decomposition need to be identified. Many of these products are soluble and can be studied by conventional solution-state NMR techniques.

3. Solution-State Light Scattering

Light-scattering methods yield information about the molecular size of molecules in solution, specifically molecular weights of macromolecules and the size of colloidal particles. This is a method that has been used mainly to study polymers rather than other types of materials.

4. Gel Permeation Chromatography

This is another method for the study of polymer molecular weights. The principle on which it is based is as follows. The equipment consists of a tube packed with porous microspheres of a material such as polystyrene. A solution of a polymer is pumped through the column, and the dissolved macromolecules have an opportunity to enter the pores in the stationary substrate. Because of the pore size, smaller molecules have a greater probability of entering the pores and being retained or delayed compared to larger molecules, which may be swept through the column without serious delay. Thus, the solution that emerges first from the column contains the larger molecules and the smallest solute molecules emerge last. In practice, if the polymer sample contains a distribution of molecular weights, this distribution will be reflected in the elution time. The eluting solute molecules are detected by their influence on the refractive index of the emerging solution or by some other method such as infrared absorption. Because most detectors give a plot of elution time versus an indirect estimate of the number of molecules emerging at any time during the experiment, it is possible to estimate both the molecular weight distribution and the maximum molecular weights among the molecules present. This information can then be used to understand the influence of molecular weight on solid-state properties such as rigidity, viscous flow, and materials strength.

E. SUGGESTIONS FOR FURTHER READING

1. Ellis, A. B.; Geselbracht, M. J.; Johnson, B. J.; Lisensky, G. C.; Robinson, W. R., *Teaching General Chemistry: A Materials Science Companion*, American Chemical Society, Washington, DC, 1993.
2. *Chem. Rev.* (Special Issue on Force and Tunneling Microscopy), **97**(4) (1997).
3. Wang, Z. L., "New developments in transmission electron microscopy for nanotechnology," *Adv. Mater.* **15**:1497–1514 (2003).
4. Hohne, G. W. H.; Hemminger, W. F.; Flammerheim, H.-J., *Differential Scanning Calorimetry, Analytical and Bioanalytical Chemistry*, Vol. 380, Springer, 2004.
5. Allcock, H. R.; Lampe, F. W.; Mark, J. E., *Contemporary Polymer Chemistry*, 3rd ed., Pearson, Prentice-Hall, 2003, Chapters 14–21.
6. Ladd, M. F. C.; Palmer, R. A., *Structure Determination by X-Ray Crystallography*, 3rd ed., Plenum Press, 1993.
7. Silverstein, R. M.; Webster, F. X.; Kiemle, D. J., *Spectrometric Identification of Organic Compounds*, 7th ed., Wiley, 2005.
8. Fyfe, C., *Solid State NMR for Chemists*, C.F.C. Press, Guelph, 1983.
9. Sanders, K. M.; Hunter, B. K., *Modern NMR Spectroscopy: A Guide for Chemists*, 2nd ed., Oxford 1993.
10. Becker, E., *High Resolution NMR: Theory and Chemical Applications*, Academic Press 1999.

F. STUDY QUESTIONS (for essays or class discussions)

1. The term "magic-angle spinning" is used in the section on solid-state NMR spectroscopy. What does it mean, and what is its significance?
2. You have been given a sample of an unknown solid material. What is the sequence of tests that you would recommend for finding out what it is?
3. What are "elastic modulus" and "compliance"? Do some research in the library to find out what these terms mean. In what types of applications might they prove to be critical properties?
4. A narrow beam of X rays is passed through a small sample of a solid material that is rotated around the vertical axis. The X-ray scattering or diffraction is detected by means of a piece of photographic film curved around the vertical axis of the sample. The pattern detected after development of the film consists

of a series of sharp spots arrayed along parallel lines. What does this tell you about the material?

5. In X-ray diffraction experiments, why is a borate glass capillary tube used to contain the sample instead of an ordinary glass capillary?

6. Study a book on single-crystal X-ray diffraction until you understand the process of molecular structure determination. Pay special attention to the concept by which the intensity of a reflection and its phase are correlated, and why this is important.

7. Explain why direct current is used to measure electronic conductivity, but alternating current must be employed to measure ionic conductivity.

8. Suggest situations in which SEM techniques might be more useful for studying a surface than STM, AFM, or XPS methods, and vice versa.

9. TGA, DSC, and TMA are three techniques that are widely used in materials science. Explain the principles involved and the advantages and disadvantages of each.

10. Why is the glass transition temperature of a solid a crucial property that controls many or most of its uses? How is this property measured?

11. X-ray crystallography, NMR spectroscopy, Fourier transform infrared spectroscopy, and elemental analyses all give useful and complementary information about a material. Pick one material and show how a combination of these methods might provide evidence for its structure.

12. Why is the structure determination of an insoluble solid more difficult than the corresponding determination for a soluble small molecule? Give examples.

13. How much of the internal structure of a solid can be inferred from the surface structure as determined by SEM, STM, or XPS methods? What additional information would be needed to make an unambiguous assessment of the internal structure?

14. Thermal analysis methods, such as DSC, TGA, and TMA, are widely used to characterize materials. Under what circumstances may these techniques give misleading information? Suggest specific examples.

Part II

Different Types of Materials

5

Small Molecules in Solids

A. IMPORTANCE OF SMALL-MOLECULE MATERIALS

Small molecules in solids play an important role in electronics, photonics, and molecular separations. Small molecules can become assembled in the solid state in several different ways, all of which provide a useful introduction to the behavior of the more complex materials discussed in later chapters.

First, small molecules often become packed into a highly ordered crystalline three-dimensional lattice. Such crystalline solids are among the simplest and easiest materials to understand, and they can be studied with great accuracy by single-crystal X-ray diffraction techniques (see Chapter 4). This self-assembly of small molecules into an ordered three-dimensional array may occur not only when the molecules crystallize from solution or from the melt but also when they are condensed directly into the solid state from the vapor.

Second, the formation of crystals from solution or from molten mixtures may yield crystalline solids that contain more than one small-molecule species. These can involve anion/cation or charge transfer associations, or they may be solids in which neutral guest molecules occupy space between the host molecules. These last systems are formed from molecules that, because of their shape, crystallize with significant empty space (free volume) in the lattice, into which small "guest molecules" can fit. They are called *clathrates* or "molecular inclusion compounds," although they are not "compounds" in the usual sense of the word.

Third, small molecules in solids may exist in the form of an amorphous glass, in which the molecules are oriented randomly. This situation is less common for small molecules than it is for polymers or ceramics (see Chapters 6 and 7), but it can occur with small molecules if a molten liquid is cooled rapidly, and if the molecules are highly flexible or have irregular shapes.

Introduction to Materials Chemistry, by Harry R. Allock
Copyright © 2008 by John Wiley & Sons, Inc.

97

Finally, within a discrete temperature range, small molecules sometimes occupy a phase that is intermediate between a crystalline or amorphous solid and a liquid—the so-called liquid crystalline state.

B. PACKING OF SMALL MOLECULES IN THE SOLID STATE

The packing arrangements of small molecules in a crystal are determined mainly by three factors—shape-fitting, molecular dipoles or charges, and hydrogen bonding. These are considered in turn.

1. Shape-Fitting

Figure 5.1 illustrates the principles involved in shape-fitting. Neutral molecules attempt to assume a minimum energy orientation in terms of the attractions and repulsions between them and their near neighbors. Thus, van der Waals–type forces (see Chapter 2) are the main influence on the shape-fitting process, and the separation between adjacent molecules and their orientation relative to each other are a result of this energy minimization. Occasionally more than one packing arrangement will be acceptable, and this is responsible for crystal structure changes as the solid is heated, cooled, or subjected to pressure.

It is possible to calculate the minimum–energy packing arrangement using known or predicted interaction parameters such as a 6:12 Lennard-Jones function (again see Chapter 2), but in practice it is usually quicker to carry out a single-crystal X-ray diffraction study and examine the experimentally determined packing arrangement.

Small molecules come in so many different shapes that it is not possible to generalize about shape-fitting. Nevertheless, certain well-known examples exist that

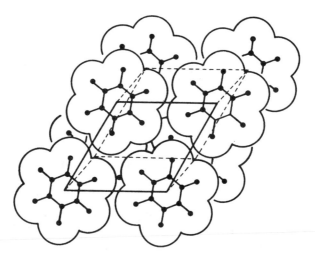

Figure 5.1. A simple classical example of shape-fitting, showing how a motif in the form of hexamethylbenzene can be assembled on a lattice to generate a two-dimensional array that minimizes the free volume in the crystal. (From Bunn, C. W., Chemical Crystallography, *Oxford University Press, 1963.*)

provide a starting point for understanding the relationship between molecular shape and solid-state structure. As shown in Chapter 2 (Figure 2.12), disk-shaped molecules frequently stack in the solid state like plates or saucers. Fused aromatic ring systems in the solid state often assume a tilted stack arrangement. The same is true of phthalocyanine molecules.

Short-chain linear aliphatic organic molecules assume a zigzag conformation and pack closely as shown in Figure 2.12. Some paddle-wheel-shaped molecules may pack in a way that interleafs their "paddles" or in a more open structure that generates cavities or tunnels. Examples of these possibilities are described later. Spherical molecules such as fullerenes assemble in the solid state much like metal atoms since there are relatively few alternative packing arrangements for spheres.

2. Dipolar or Charged Molecules

Small molecules with dipoles or separated charges attempt to achieve minimum-energy packing arrangements by bringing the positively charged regions close to the negatively charged parts of neighboring molecules (Figure 5.2). In practice, the crystal structure is the result of some compromise between charge neutralization and shape fitting, and is difficult to predict by modeling techniques. Here again, an X-ray structure is usually the fastest way to understand how these factors achieve a balance.

Note that there is an important difference between the packing of *uniatomic ions* in solids and the packing of charged molecules. The ions are spherical, and the variations in packing opportunities are far fewer than in nonspherical molecules.

3. Hydrogen Bonding

The crystal structures of molecules such as amino acids, hydroxycarboxylic acids, dicarboxylic acids, phenols, amines, or water may be strongly influenced by the opportunities for hydrogen bonding. Hydrogen bonds, although weak by comparison with covalent or coordinate bonds, nevertheless exert a strong influence on packing because of the proximity of functional groups in the solid state and the limited molecular motions that are possible in a solid. Water ice is a well-known example of a hydrogen bonded solid, and this system is described later.

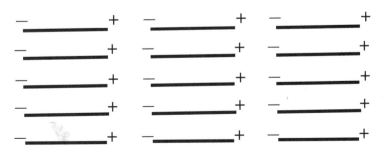

Figure 5.2. Molecules with dipoles or charged ends may become organized in the solid state in a manner that brings opposing charges into close proximity.

C. SELF-ASSEMBLY BY CRYSTALLIZATION

One of the extraordinary aspects of solid-state chemistry and physics is the way in which small molecules in solution or in the molten state assemble at the surface of a growing crystal and select a minimum-energy orientation. Considerable sampling of different orientations must take place at the interface, augmented by the need to conform to the crystal structure on the existing surface. This is why seed crystals are sometimes needed to initiate crystal growth, and why crystallization of a new compound may be difficult until nano- or microsized crystals are present in the laboratory to initiate the crystallization process. It is also interesting that impurity molecules are usually excluded from the crystal because of the forces that favor the deposition of molecules in an orientation that matches those that already exist at the surface. Shape-fitting or polar forces undoubtedy explain this behavior. The exclusion of impurity molecules underlies the common laboratory practice of purification of reaction products by recrystallization. The purification of silicon for semiconductor applications is critically dependent on this phenomenon (see Chapter 10).

Another aspect of template-induced crystallization is the deposition of molecules from the vapor state on to a crystalline surface. The crystal structure of the deposited material may be controlled by the structure of the substrate.

D. SPHERICAL MOLECULES SUCH AS FULLERENES IN THE SOLID STATE

Perhaps the simplest mode of molecular packing in the solid state is provided by C_{60} fullerene ("buckyball") molecules. These spherical molecules pack into a crystal lattice in the manner described elsewhere for ions or metal atoms. In other words, because of the spherical shape, the assembly pattern is determined by the most efficient way to assemble spheres in three dimensions, like a collection of ping-pong balls.

Figure 5.3a shows the crystal packing arrangement of C_{60} at room temperature. It is a face-centered cubic arrangement, which means that it can be visualized by assembling a first layer of spheres, with each one in contact with six others. The second layer is constructed from spheres placed in the depressions in the first layer, and the third layer is formed from spheres placed in the depressions in the second layer. Successive layers then repeat this sequence. At ambient temperature the spheres undergo rotation or tumbling in place within the lattice. However, the presence of metal atoms such as potassium trapped in the remaining space in the lattice, as in M_3C_{60}, yields a solid that is a room-temperature electronic conductor, but becomes a superconductor when cooled below $18\,K$ (see Chapter 11). The electrical conductivity is a consequence of electrons being transferred from the metal atoms to the π orbitals of fullerene molecules, and, because the spheres are in close contact, the electrons are free to move from molecule to molecule. The face-centered cubic (FCC) structure can accommodate no more than three metal atoms per C_{60} molecule (Figure 5.3b), and the presence of six metal atoms, as in M_6C_{60}, results in a reorganization of the lattice and a loss of electrical conduction.

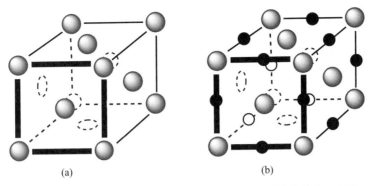

(a)   (b)

Figure 5.3. *(a) Packing of C_{60} molecules in a face-centered cubic (FCC) lattice. (b) Incorporation of alkali metal cations (small closed circles) into the octahedral spaces in the C_{60} lattice. A ratio of three cations to each C_{60} allows the original FCC lattice to be retained and gives electrical conductivity. A higher ratio of cations causes the additional ions to occupy tetrahedral holes in the lattice, and that structure is nonconductive.*

E. DISK-SHAPED MOLECULES AND OTHER FLAT STRUCTURES

Many disk-, ribbon- or wafer-shaped organic molecules exist. Examples include benzene and symmetric trisubstituted benzenes, triphenylene, coronene, and phthalocyanines (see Figure 5.4). Ribbon- or wafer-shaped structures such as tetracene and pentacene are related species.

A characteristic of many disk- or ribbon-shaped molecules is that they tend to self-assemble in a way that generates columns of stacked disks or short ribbons. The molecules in each stack may be oriented at right angles to the axis of the stack like a pile of plates or saucers, or they may be tilted (see Figure 2.12). A stacked packing arrangement provides opportunities for two phenomena to appear—discoid liquid crystalline behavior, or the possibility of electron movement down a stack. These are considered in turn.

1. Liquid Crystallinity from Disk- or Wafer-Shaped Molecules

Liquid crystallinity is a phenomenon that occurs when a solid is heated but does not undergo a clean transition from solid to liquid at the melting point. Instead, an intermediate phase (a mesophase) is formed that retains some of the order associated with the solid state but has fluidity more like that in the molten state. Similarly, if the molten material is cooled, it may not solidify at the melting point but will form a liquid-like phase with some crystalline-type association between individual molecules. The liquid crystalline phase is important because it allows molecules to be switched from one orientation to another by means of an electric field (Chapter 14). The true crystalline solid state does not allow this to happen, and the isotropic molten state permits too much random reorientation to be useful. Several quasi-disk-shaped molecules, such as cholesteryl benzoate (Figure 5.5), form liquid crystalline phases. However, the most heavily studied liquid crystalline systems are based on rod-shaped molecules (see later).

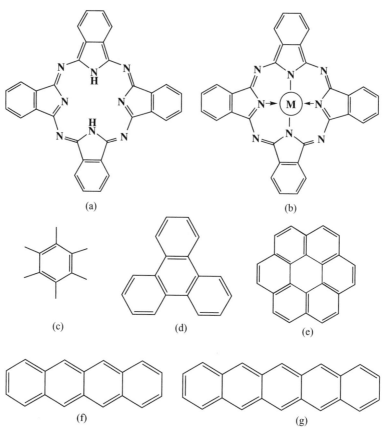

Figure 5.4. *Examples of disk- or ribbon-shaped molecules that self-assemble in the solid state to form stacked structures: (a) phthalocyanine; (b) metallophthalocyanines; (c) substituted benzene molecules; (d) triphenylene; (e) coronene; (f) tetracene; (g) pentacene.*

   The stacking of disk- or wafer-like molecules encourages liquid crystalline behavior because the van der Waals cohesive forces that bind molecules in a stack may be stronger than those that bind adjacent columns. Thus, columns may separate from each other at a certain temperature, but the columns themselves or parts of columns remain intact. Even if the columns break up, the coorientation of the molecules may remain (see Figure 5.6a).

2. Electronic Phenomena from Disk-Shaped Molecules in the Solid State

These phenomena are associated with delocalized (aromatic) molecules. When such molecules are stacked in columns in the solid state, the π orbitals of one molecule overlap their counterparts in the molecules above and below in the stack. This generates an extended π system with electron delocalization extending down the whole column. Under favorable circumstances, this yields a semiconductor- or conductor-type band structure rather than discrete molecular orbitals (for details, see Chapter

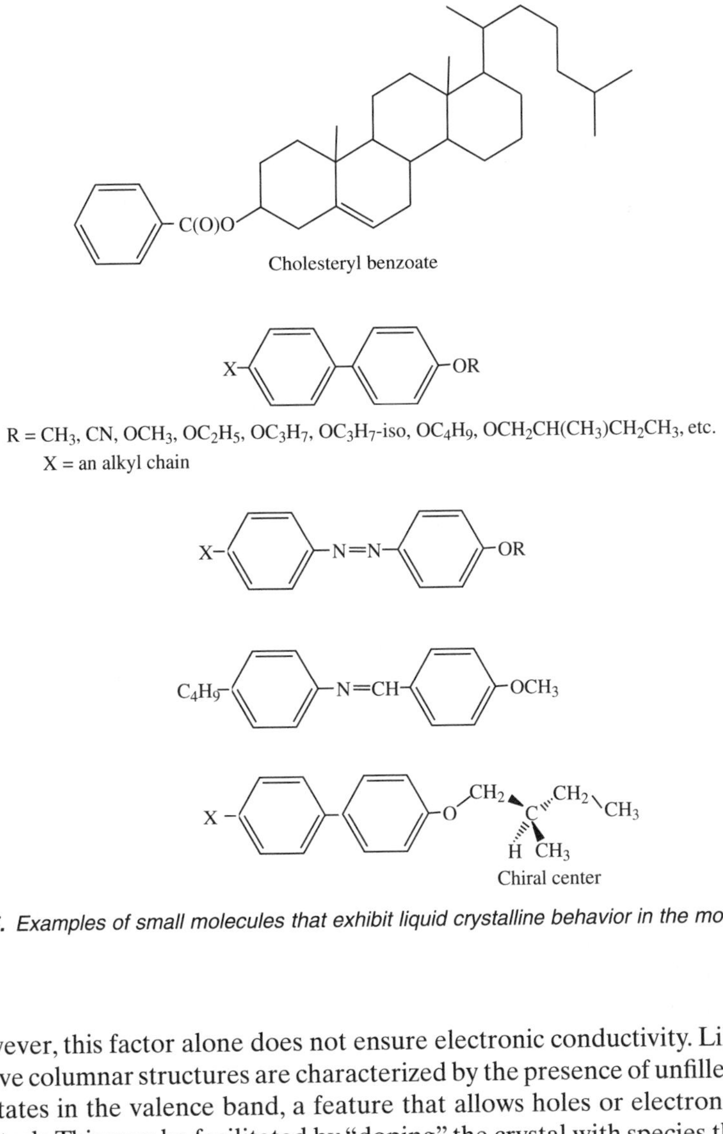

Figure 5.5. *Examples of small molecules that exhibit liquid crystalline behavior in the molten or solution state.*

10). However, this factor alone does not ensure electronic conductivity. Like metals, conductive columnar structures are characterized by the presence of unfilled highest-energy states in the valence band, a feature that allows holes or electrons to move down a stack. This may be facilitated by "doping" the crystal with species that accept or donate electrons from or to each column. For example, iodine molecules accommodated between the stacks can accept electrons to form, for example, I_3^-. The unfilled energy levels in the host column at the top of the valence band (i.e., unpaired electrons) correspond to radical cations. This permits metal-level conductivity via fast migration of the resultant holes (positive charges) down each column. Alternatively, doping with species that readily release an electron to the stack (such as metal atoms) will yield unpaired electrons (radical anions) that may be charge carriers. Hole conduction is more common. If the dopants occupy sites between the molecules in a stack, the disks will become separated beyond the distance required for molecular orbital overlap—hence the need for dopants to be intercalated *between the stacks* rather than between the molecules in each stack. Doping can also be accomplished by adding or removing electrons by contact of the solid with the

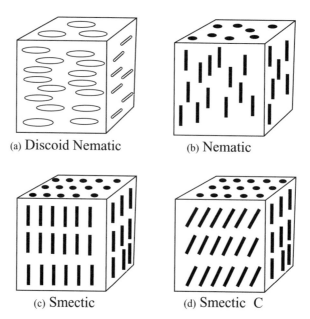

(a) Discoid Nematic   (b) Nematic

(c) Smectic   (d) Smectic C

Figure 5.6. *Four of the many different ways in which small molecules in solution or in the semimolten state can generate liquid crystalline arrays. Discoid nematic (a) systems arise when plate-like molecules become organized in a coplanar orientation but without the formation of the separate stacks that are typical of the fully crystalline state. Rod-shaped molecules can also form nematic (b) liquid crystals, or, if the molecules assemble in layered arrangements, they generate smectic (c) systems. The tilted arrangement in the smectic C (d) systems facilitate reorientation in liquid crystalline devices, especially if the individual molecules have a chiral architecture.*

surface of electrodes in an electrochemical cell. With these facts in mind, we can now examine a few specific examples.

Phthalocyanines are large, electron-delocalized, disk-shaped molecules that may or may not have a metal at the center. Copper phthalocyanine (Figure 5.4b, where M = copper) is a typical metal-containing example, which crystallizes in tilted coplanar stacks. Electrical conduction occurs when the crystals are doped with electron acceptors such as iodine. The ratio of dopant to phthalocyanine has a crucial influence on the conductivity. The arrangement of phthalocyanine and dopant molecules in a crystal is shown schematically in Figure 5.7.

Coronene is a disk-shaped aromatic organic molecule, shown in Figure 5.4e. It crystallizes from the vapor state in tilted stacks of the type depicted in Figure 5.8. Pure coronene does not form conductive crystals, but doped crystals show semiconductivity.

Tetracene and pentacene molecules [structures (f) and (g) in Figure 5.4], despite their wafer rather than disk shape, pack in the solid state in a manner reminiscent of discoid molecules, and crystals derived from these compounds can be made electrically conductive. The tilted packing scheme that forms when the molecules are deposited from the vapor state is illustrated schematically in Figure 5.9. Such monolayer films are cited as being suitable for field-effect transistors.

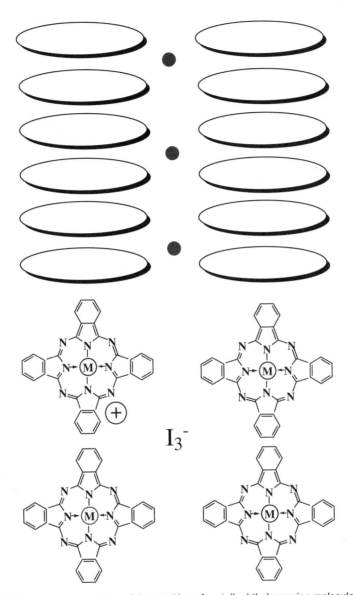

Figure 5.7. Schematic representation of the stacking of metallophthalocyanine molecules (elipses) in the solid state showing the location of dopant molecules (closed circles) in the interstack regions. Electrical conductivity occurs when electrons are removed from the phthalocyanine stack by a dopant such as iodine to give I_3^- ions. The radial positioning of the metallophthalocyanine molecules in the lower part of the diagram (view down the stacks) is schematic only and does not represent the actual radial orientation. Moreover, the molecules may exist in tilted stacks (Figure 2.12) to facilitate π-orbital overlap.

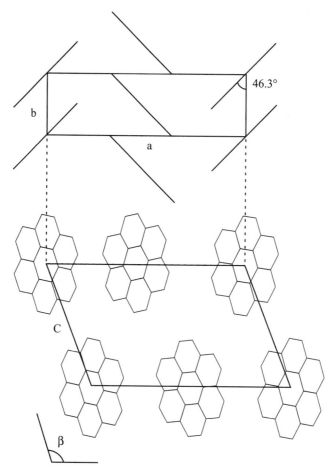

Figure 5.8. Self-assembly of coronene molecules in the solid state following vapor deposition, showing the tilted stack arrangement that allows overlap of the π-orbitals on adjacent molecules. [From Matsui, A. H.; Mizuno, K.-I., J. Phys D: Appl Phys. **28**:B242–B244 (1993).]

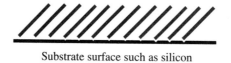

Substrate surface such as silicon

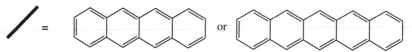

Figure 5.9. Ribbon-shaped molecules such as pentacene self-assemble into a tilted stacked arrangement when deposited from the vapor onto the surface of a semiconductor. Doping with electron acceptors or donors, together with patterning of the layer, allows devices to be constructed.

F. ROD-SHAPED MOLECULES

Rod-shaped small molecules can self-assemble in the liquid crystalline state. Examples of typical rod-shaped compounds that give rise to liquid crystalline behavior are shown in Figure 5.5. Molecules that fall into this category include derivatives of biphenyl, aromatic azo compounds, or nanotubes. In a truly ordered crystalline system the rods would be highly organized in three dimensions. In the liquid crystalline state they may have alignment along one direction only and be laterally disordered like closely packed logs floating down a river. These are called *nematic liquid crystalline systems*. Alternatively, the orientation may be along two axes (smectic) as shown in Figure 5.6. Chiral smectic C systems contain rod-like molecules that have a chiral group at one end. This situation gives a tilted–twisted orientation of the molecules in the liquid crystalline state. Spontaneous ordering in the liquid crystalline state underlies the use of rod-shaped molecules in liquid crystalline displays (Chapter 14).

G. CHARGE TRANSFER COMPLEXES

Tetrathiofulvalene (TTF) and tetracyanoquinodimethane (TCNQ) (Figure 5.10) are two planar molecules that form stacks when assembled in the solid state. The stacked structure allows overlap of the π orbitals in neigboring molecules within each stack, and this provides a pathway for electrical conduction. Pure TTF is not an electrical conductor, but it can *relinquish* an electron to form a radical cation, and the partly oxidized stacks are semiconducting. TCNQ is a molecule that readily *accepts* an electron to form a radical anion, and partly reduced stacks of these molecules are also semiconducting.

    One of the most important solids derived from these two species is formed by the cocrystallization of TTF and TCNQ to give adjacent stacks of the two molecules.

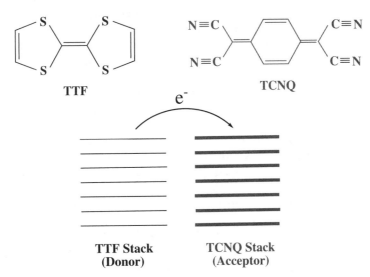

Figure 5.10. Formation of electrically conductive crystals by the cocrystallization of TTF and TCNQ.

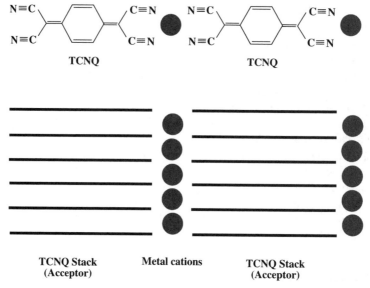

Figure 5.11. *TCNQ stacks become electrically conductive when doped with metals (solid circles).*

This allows electrons to be transferred from a TTF stack to a TCNQ stack to generate a useful.semiconductor (Figure 5.10). The conduction behavior can be visualized in terms of electons moving in one direction through the TCNQ stacks and positively charged "holes" moving in the opposite direction along the TTF stacks.

Semiconduction does not necessarily require the presence of both TTF and TCNQ in the same solid—other electron donors or acceptors will serve the same purpose. For example, the salt [TTF$^+$] Cl$^-$ is a semiconductor, with the semiconductivity resulting from the movement of holes along the TTF stacks. TCNQ becomes semiconducting in the form of salts of the type [TCNQ$^-$] M$^+$, where M is a metal such as potassium, rubidium, or thallium (Figure 5.11). It should also be noted that TCNQ is a building block in the fabrication of semiconducting nanowires (see Chapter 17). TCNQ also forms conductive charge transfer complexes with coronene in which the two components alternate down the tilted stacks.

H. CLATHRATES—MOLECULAR INCLUSION ADDUCTS

Clathrates are crystalline solids that contain guest molecules trapped within the host lattice. Clathration often occurs with host molecules that have an unusual shape that prevents them from packing efficiently to minimize their energy. In such cases the guest molecules occupy the free volume in the lattice and help to stabilize the solid-state structure. It sometimes happens that two or more different host crystal structures exist—one without the guest molecules in which the host molecules attempt to make the best use of the available volume, and another arrangement when the guest molecules are present. Because the clathration phenomenon depends on the solid-state structure, dissolving the crystals in a solvent or volatilization of the host breaks up the adduct.

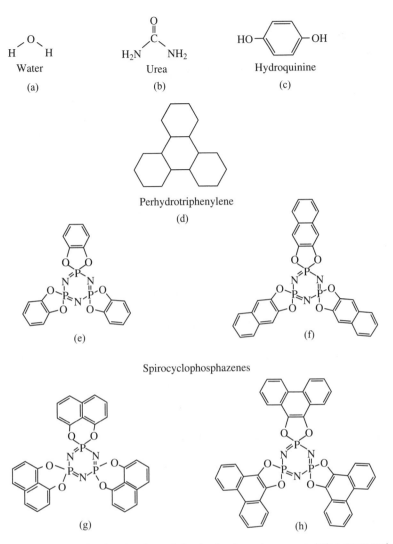

Figure 5.12. *Example molecules that form clathrate structures by encapsulating guest molecules in the solid state.*

A wide variety of small molecules form crystalline clathrates, and a few examples are shown in Figure 5.12. They range from organic molecules such as hydroquinone, urea and thiourea, cyclic ethers, perhydrotriphenylene, and cyclodextrins, through inorganic–organic molecules such as cyclophosphazenes, to totally inorganic species such as water or the coordination complexes of nickel, palladium, and platinum, and other inorganic elements. Comprehensive accounts of these different systems are beyond the scope of this book but can be traced through references given at the end of this chapter. In a broader sense, crown ethers and cryptands, zeolites, and layered aluminosilicates (Chapter 7) can also be included in this category.

Guests become incorporated into a clathrate crystal lattice in several different ways. Solvent molecules may be trapped in the host lattice as the host crystals are growing from solution. Alternatively, the guest molecules in the form of a gas or liquid may trigger a host crystal structure change in order to penetrate the existing crystals. Removal of the guest molecules by heating in vacuum may cause the host lattice to revert to its original form. These phenomena have been used to separate guest molecules according to size and shape, to provide cavities or tunnels for the safe storage of unstable guest molecules, and to provide a template for the polymerization of monomers trapped in host tunnels. The following examples illustrate several representative systems.

1. Clathrates of Water Ice

The ability of water to crystallize under pressure and trap hydrocarbon molecules such as methane is well known. Indeed, the presence of large amounts of methane trapped on the ocean floor is due to the formation of a clathrate under pressure. Methane–water clathrates are considered to be a potentially valuable source of energy for future development. Chunks of the white clathrate are occasionally found floating on the ocean surface, and these samples ignite when a flame is applied. Moreover, the water ice–methane clathrate is responsible for the blockage of natural-gas pipelines.

The clathration phenomenon depends on the ability of water to crystallize around guest molecules through the formation of hydrogen bonds between hydrogen and oxygen. Thus, clathrate formation is favored when guest molecules serve as templates around which the ice structure can form, even though the forces between host and guest are weak van der Waals attractions. Three different crystal structures, each with different cavity sizes, have been identified. All three are based on a tetrahedral arrangement of hydrogen atoms around each oxygen. The different structures apparently arise because of the different dimensions of the guest molecules. The three cavity sizes are shown in Figure 5.13. The type 1 structure, with a cavity diameter of 4.4 Å, accommodates tumbling methane molecules (diameter 4.3 Å). The type 2 structure is found when propane is clathrated. Here the cavity has a diameter

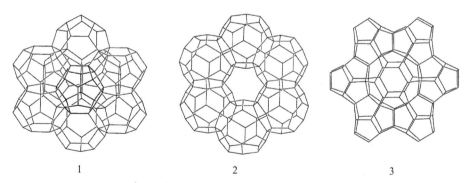

1   2   3

Figure 5.13. *Three arrangements of hydrogen-bonded water molecules in ice–hydrocarbon clathrates showing the location of the central cage in each structure. Each corner in the polyhedra represents an oxygen atom bonded either covalently or through hydrogen bonds to four hydrogen atoms. Structure **1** exists with methane as the guest, structure **2** is formed with propane, and structure **3** is formed with adamantane. [From Kirchner, M. T.; Boese, R.; Billups, W. E.; Norman, L. R., J. Am Chem. Soc. **126**:9407–9412 (2004). Reproduced with permission.]*

of 6.8 Å, which is appropriate for the presence of the 6.3 Å maximum dimension of propane. The third structure appears when adamantane (diameter 6.5–7.1 Å) is the guest. This cavity size varies from 5.8 to 7.2 Å at different locations in the cage. Because of this steric restriction, the adamantane molecules do not tumble in place.

2. Urea and Thiourea

Both urea and thiourea form honeycomb-like hydrogen-bonded arrays in the solid state. Hexagonal columnar prisms or plates are formed when these compounds are crystallized at atmospheric pressure from organic solvents, or when a solvent is added to the pure solid crystalline material. The crystals contain solvent molecules as guests. The guest molecules are accommodated in pseudohexagonal tunnels (also called "channels") that penetrate the crystals (Figure 5.14). The tunnel diameter for urea clathrates is 5.25 Å, whereas the corresponding value for thiourea is 6.1 Å. Thus, the thiourea system can accommodate larger guest molecules. Guest molecules that

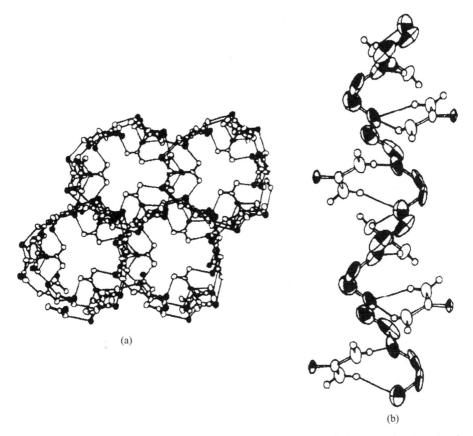

(a)

(b)

Figure 5.14. (a) View down the tunnel axis formed by hydrogen-bonded urea molecules, showing how, in the absence of other guest molecules, urea molecules line the tunnels. (b) View normal to the tunnel axis illustrating how helical poly(ethylene oxide) chains occupy the tunnels in the urea solid-state structure. [From Chenite, A.; Brisse, F., Macromolecules **24**:2221–2223 (1991), reproduced with permission.]

have been trapped in urea crystals include linear alkanes and alkenes, aliphatic alcohols, carboxylic acids, esters, and ethers, as well as polymers such as poly(ethylene oxide) (Figure 5.14). Moreover, the separation of guest molecules from a mixture of solvent molecules is also possible.

3. Perhydrotriphenylene

Two small-molecule systems with trigonal symmetry are well known for their ability to form clathrates. These are based on perhydrotriphenylene (PHTP) (Figure 5.12d) and a series of spirocyclophosphazenes (Figure 5.12e–h).

Perhydrotriphenylene is a saturated tetracyclic molecule with the trigonal shape shown in Figure 5.12d. The puckered conformation of one stereoisomer of this molecule is depicted in Figure 5.15. Thus, this is a disk-like molecule with the capacity to form stacks within a solid-state structure, similar in some respects to the behavior of phthalocyanines and coronene. Recrystallization of this compound from numerous organic solvents leads to the incorporation of solvent molecules into the lattice. With butadiene as the guest, the crystal structure takes the form shown in Figure 5.15.

The tunnels that form in this structure are especially appropriate for the clathration of linear hydrocarbons such as *n*-heptane or *n*-nonane. They also provide an environment for the polymerization of unsaturated molecules such as 1,4-butadiene, isoprene, or pentadiene to linear polymers after exposure of the clathrate to X rays.

4. Cyclophosphazenes

Spirocyclophosphazenes are paddle-wheel-shaped molecules of the types shown in Figure 5.12e–h. Like most clathrate structures, their intercalation behavior was discovered through discrepancies between the expected and the experimentally determined composition. The five phosphazene molecules shown in Figure 5.12 have differing clathration properties that can be traced to the nuances of their structure. Compound (e), tris(*o*-phenylenedioxy)cyclotriphosphazene, crystallizes in the structure shown in Figure 5.16a, which illustrates the presence of 5-Å-diameter tunnels. These tunnels penetrate the hexagonal crystals down the *c* axis and provide the space in which guest molecules are found. In the absence of guest molecules (i.e., when the compound is sublimed or heated above the melting point and cooled) the crystal structure is monoclinic, with no evidence of tunnels but with indications of inefficient packing of the molecules. Exposure of the monoclinic form to the liquid or vapor of the guest, such as benzene, toluene, or alkanes, causes a change in the crystal structure to hexagonal. When observed under a microscope, the crystals are seen to swell as guest molecules penetrate into the lattice from the ends.

Compound (f) in Figure 5.12 behaves in a similar way but has longer "arms." This leads to the formation of tunnels with a wider diameter (~10 Å), which can accommodate larger quantities of guest molecules (Figure 5.16b). Compounds (g) and (h) in Figure 5.12 also form clathrates, but with the guest molecules trapped in cavities rather than tunnels.

Clathration by the tunnel-forming systems (e) and (f) (in Figure 5.12) allows the separation of guest molecules on the basis of size and shape ("molecular recogni-

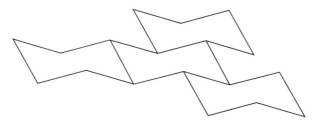

Trans-anti-trans-anti-trans-perhydrotriphenylene

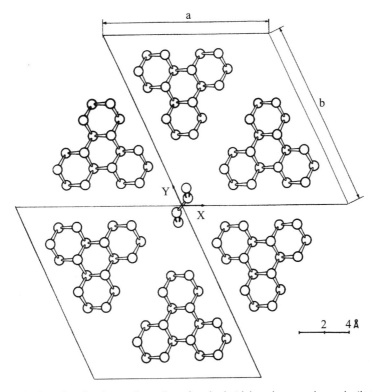

Figure 5.15. *Puckered molecular conformation of perhydrotriphenylene, and a projection of the unit cell found in its inclusion adduct with butadiene. The butadiene molecules occupy a tunnel between the host molecules. (From Farina, M., in Atwood, J. L., Davies, J. E. D., MacNicol, D. D., eds.,* Inclusion Compounds, *Vol. 2, Academic Press, 1984, Chapter 3, p. 82.)*

tion"). This occurs either through selective uptake of vapor state guest molecules or by recrystallization from mixtures of guest solvents. Thus, species (e) is able to preferentially clathrate *n*-heptane with the total exlusion of cyclohexane. *n*-Hexane is selected over benzene from a liquid mixture of the two. *m*-Xylene with 96% purity is separated from a mixture that also contains *p*-xylene and ethylbenzene. The potential use of this behavior in petrochemical separations is obvious.

Species (e) also absorbs unsaturated molecules such as methyl methacrylate, styrene, butadiene, isobutylene, and *p*-divinylbenzene into its tunnels, and these monomers have been polymerized by exposure of the clathrate to gamma radiation.

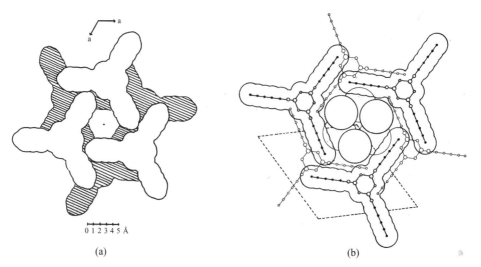

(a)   (b)

Figure 5.16. (a) View down the c-axis of hexagonal crystals of tris(o-phenylenedioxy)cyclotriphospha zene showing one of the 5-Å-diameter tunnels in which benzene or smaller molecules can be retained. (b) Representation of the tunnel structure of tris(2,3-dioxynaphthalene)cyclotriphosphazene showing the wider (10-Å) tunnel that can accommodate larger amounts of benzene (spheres) or bulkier mole-cules than in structure (a). (From Allcock et al. in Atwood et al. eds. Inclusion Compounds, Vol. 1, Academic Press 1984, Chapter 8.)

Some of the polymers formed in this way are difficult or impossible to isolate from normal nontemplated polymerization reactions.

5. Hofmann and Werner-Type Complexes

These are some of the oldest clathrate systems known. Hofmann complexes are inorganic coordination species formed by nickel cyanide in association with amino derivatives of manganese, nickel, copper, and cadmium. An example is $Ni(NH_3)2Ni(CN)_4 2C_6H_6$. The complexes are formed by mixing aqueous solutions of the metal ammine ion with the metal cyanide and an organic guest. The clathrates have a sheet structure in the solid state, with organic molecules such as benzene, biphenyl, or dioxane trapped between the sheets as shown in Figure 5.17a.

Werner-type systems are formed from propellor-shaped coordination complexes that involve metals such as nickel or cobalt associated with ligands such as isothio-cyano (NCS) and an amine such as pyridine or para-substituted pyridines. Organic guest molecules such as naphthalene or 2-methylnaphthalene are trapped in the lattice (Figure 5.17b).

6. Cyclodextrins, Cryptates, and Crown Ethers

These ring systems are related to clathrates, but form stable complexes with metal ions in solution as well as in the solid state. Thus, different principles are involved that do not depend solely on solid-state molecular packing arrangements.

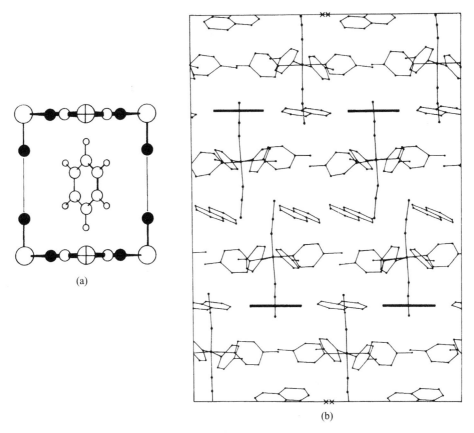

Figure 5.17. *Clathrates formed by transition metal coordination complexes. (a) Hofmann's benzene clathrate, Ni(NH₃)₂Ni(CN)₄.2C₆H₆, illustrating the position of the benzene molecules within the inorganic structure. The large open circles are 6-coordinate Ni, and the crossed large circles are square planar nickel. The solid circles are nitrogen atoms, the open circles are carbon, and the small open circles are nonammonia hydrogen atoms. (b) Molecular packing in a Werner complex, Ni(NCS)₂(4-MePy)₄.2(naphthalene) showing the layer structure in which the naphthalene molecules are trapped. (From illustrations by Iwamoto, T.; Lipowski, J., in Atwood, J. L.; Davies, J. E. D.; MacNicol, D. D., eds.,* Inclusion Compounds, *Vol. 1, Academic Press, New York, 1984.)*

I. SUGGESTIONS FOR FURTHER READING

1. Marks, T. J., "Interfaces between Molecular and Polymeric Metals: Electrically conductive, Structure-Enforced Assemblies of Metallomacrocycles", *Angew. Chem. (Int. Ed. Engl.)* **29**:857–879 (1990).

2. Puigdollers, J.; Voz, C.; Fonrodona, M.; Cheylan, S.; Stella, M.; Andreu, J.; Vetter, M.; Alcubilla, R., "Copper phthalocyanine thin film transistors with polymeric gate dielectric," *J. Noncryst. Solids* **352**:1778–1782 (2006).

3. Matsui, A. H.; Mizuno, K.-I., "Crystallization and exiton luminescence of coronene crystals," *J. Phys. D: Appl. Phys.* **26**:242–244 (1993).

4. Wudle, F.; Wobschall, D.; Hufnagle, E. J., "Electrical conductivity by the bis(1,3-dithiol)-bis(1,3-ditholium) system," *J. Am. Chem. Soc.* **94**:670–672 (1972).

5. Batail, P., "Molecular conductors," *Chem. Rev.* **104**:4887–4890 (2004).

6. Bendikov, M.; Wudl, F.; Perepichka. D. F., "Tetrathiofulvalenes, oligocenenes, and their buckminsterfullerene derivatives: The brick and mortar of organic electrics," *Chem. Rev.* **104**:4891–4946 (2004).

7. Atwood, J. L.; Davies, J. E. D.; MacNichol, D. D., eds., *Inclusion Compounds*, Vols 1–3, Academic Press, New York, 1984.

8. Allcock, H. R.; Siegel, L. A., "Molecular inclusion compounds of tris(o-phenylenedioxy)-phosphonitrilic trimer," *J. Am. Chem. Soc.* **86**:5140–5144 (1964).

9. Allcock, H. R.; Sunderland, N. J., "Separation of polymers and small molecules by crystalline host systems," *Macromolecules*, **34**:3069–3076 (2001).

10. Farina, M., "Inclusion polymerization," in Atwood, J. L.; Davies, J. E. D.; MacNichol, D. D., eds., *Inclusion Compounds*, Vol. 3, Academic Press, New York, 1984, pp. 297–329.

11. Kirchner, M. T.; Boese, R.; Billups, W. E.; Norman, L. R., "Gas hydrate single-crystal structure analysis," *J. Am. Chem. Soc.* **126**:9407–9412 (2004).

J. STUDY QUESTIONS (for class discussions or essays)

1. Explain why some small organic molecules become semiconductors when their crystals are doped. Give specific examples and describe why the doping process increases the conductivity.

2. Review the process of clathration and discuss its implications for separations science and for guest molecule isolation and reactions. Be specific in describing both the hosts and the guest molecules. Why might potentially explosive guest molecules be stored safely inside a clathrate host?

3. Describe the molecular structural features of some small molecules that predispose them toward forming liquid crystals. Explain why small changes in the structure of these molecules either change the temperature range of liquid crystallinity or eliminate the formation of a liquid crystalline phase completely.

4. Suggest molecules, other than TTF and TCNQ, that might form electronically conductive stack structures in the solid state.

5. Can single-crystal X-ray crystallography be used to study guest molecules trapped in a clathrate lattice? When might this not be possible? Which other techniques described in Chapter 4 might be useful in studying guest molecules trapped in clathrates?

6. What influence might organic substituents attached to the C_{60} cage have on the molecular packing arrangement and on the electrical behavior? Would you expect the larger fullerenes such as C_{70} or C_{84} to pack like C_{60} and have the same electrical behavior when doped?

7. Benzene crystals at −30 °C contain molecules that are *not* stacked in close-packed columns but form puckered layers (see the definitive paper by E. G. Cox in Reviews of Modern Physics, **1958**, *30(1)*, pp. 159–182). Speculate why benzene molecules do not form saucer-like stacks in the solid state comparable to phthalocyanines or some other disk-like molecules.

8. Examine the molecules shown in Figure 5.5. What features do these molecules have in common that would favor liquid crystalline behavior. Which

type of liquid crystalline arrangement (Figure 5.6) might each of these molecules adopt? Draw other molecular structures that might also generate liquid crystallinity.

9. Check in the literature to find out how phthalocyanines and their metalloderivatives are synthesized and how crystals are grown.

10. If tetracene or pentacene is vapor-deposited on a flat surface to give a tilted array (Figure 5.9), suggest reasons for the tilting. Would the deposition of a second layer mimic the tilting of the first layer?

11. What uses, in addition to molecular separations, can you think of for the solid formed from the molecules in Figure 5.12, 5.15, and 5.16?

12. The co-crystallization of two different molecules, as shown in Figure 5.10, is relatively rare. Discuss molecular structural factors that might favor this phenomenon.

6

Polymers

A. OVERVIEW

Two of the largest areas of materials chemistry involve polymers and ceramics, and both of these areas share similarities at the fundamental science level. A consideration of polymers as a prelude to the study of other materials has some important advantages. For example, many of the principles that govern the chemistry of ceramics can be understood in terms of polymer science. Thus, the underlying concepts of condensation reactions, molecular and materials flexibility, crosslinking, composites, and optical properties follow naturally from a knowledge of macromolecules and their behavior. Moreover, polymers are now widespread in technology, where they are used to enhance the properties of ceramics, metals, semiconductors, and optical materials. The differences between classical polymers and ceramics arise from the behavior of carbon compounds compared to those of the other main-group elements. Hybrid inorganic–organic polymers represent a halfway stage between classical polymers and ceramics, with some of the property advantages of both. Here we focus on the organic and inorganic–organic systems, as a prelude to a consideration of ceramics in Chapter 7.

Polymers are important materials because they are the basis of a large part of modern science and technology. Applications as diverse as lenses, housings for computers and telephones, photographic film bases, paint, thermal insulation, textile fibers, membranes, automobile and aircraft interiors and structural components, and a host of biomedical devices are based on polymers. The main reasons why polymers represent such an important part of materials science is that they can be fabricated easily, are corrosion-resistant and lightweight, and are generally inexpensive. They are designed and synthesized to give properties that range from elastomers to glasses, from soluble to insoluble materials, and from totally inert substances to reactive solids. They can also be colored by the introduction of dyes, and reinforced by the addition of other polymers or ceramics. Polymers are also key materials in the fabrication of semiconductor integrated circuits and in the development of optical devices (see Chapters 9 and 12).

In the following sections we consider first the main methods used for the synthesis of polymers, and then review structure–property relationships in these materials. Once the relationship between structure and properties is understood, it becomes possible to design new polymers with better combinations of properties. Finally, the characteristics of some example polymers are summarized at the end of this chapter. The ways in which polymers are utilized in important applications are considered in later sections of this book.

B. SYNTHESIS OF POLYMERS

1. General Principles

Polymerization processes constitute one of the most remarkable aspects of modern chemistry. For many years, prior to the 1930s, the existence of macromolecules was believed to be impossible. Yet, today an enormous range of different polymers exists, all produced via relatively straightforward processes. These synthetic pathways begin with the linking together of large numbers of small-molecule *monomers* to form long polymer chains.

Three general types of monomers are utilized for the synthesis of polymers: (1) small molecules with double or triple bonds that can undergo addition–polymerization reactions; (2) molecules with two functional end groups that allow condensation reactions to take place; and (3) cyclic compounds that undergo ring-opening polymerization processes. The monomer and polymers to be mentioned in this chapter are shown in Table 6.1. In addition to these polymerization methods, some polymers are modified by subsequent reactions—the so-called macromolecular substitution reactions. The crosslinking of polymer chains can also be accomplished by secondary processes. These aspects are discussed in turn.

2. Addition Polymerization

Addition polymers represent the largest class of synthetic macromolecules in terms of total volume produced. This class includes materials such as polyethylene, polypropylene, polystyrene, poly(methyl methacrylate), polyisobutylene, styrene–butadiene copolymers (SBR rubber), poly(vinyl chloride), and poly(tetrafluoroethylene) (Teflon®) (see Table 6.1). Thus, addition polymerization is one of the main methods for the production of organic polymers.

a. The Overall Polymerization Mechanism. Addition polymerization processes can be understood in terms of the following general steps:

1. *Initiation.* An unsaturated organic monomer is treated with an initiator compound to give a reactive species that will initiate the polymerization process.

TABLE 6.1. Monomers and Polymers Mentioned in This Chapter

Monomer A	Monomer B	Product Polymer

Addition monomers

$CH_2{=}CH_2$
Ethylene

$\left[\!CH_2{-}CH_2\!\right]_n$
Polyethylene

$\underset{\displaystyle CH_2{=}CH}{\overset{\displaystyle CH_3}{|}}$
Propylene

$\left[\!CH_2{-}\underset{|}{\overset{CH_3}{CH}}\!\right]_n$
Polypropylene

$CH_2{=}CH$ (phenyl)
Styrene

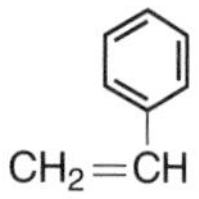

$\left[\!CH_2{-}CH\!\right]_n$ (phenyl)
Polystyrene

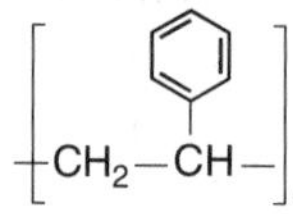

$\underset{\displaystyle CH_2{=}CH}{\overset{\displaystyle C(O)OCH_3}{|}}$
Methyl methacrylate

$\left[\!CH_2{-}\underset{|}{\overset{C(O)OCH_3}{C}}{-}\atop CH_3\!\right]_n$
Poly(methyl methacrylate) (PMMA)

$\underset{\displaystyle CH_2{=}CH}{\overset{\displaystyle OC(O)CH_3}{|}}$
Vinyl acetetate

$\left[\!CH_2{-}\underset{}{\overset{OC(O)CH_3}{CH}}{-}\!\right]_n$
Poly(vinyl acetate)

$\underset{\displaystyle \underset{CH_3}{|}}{\overset{\displaystyle \overset{CH_3}{|}}{CH_2{=}C}}$
Isobutylene

$\left[\!CH_2{-}\underset{CH_3}{\overset{CH_3}{C}}{-}\!\right]_n$
Polyisobutylene

$CH_2{=}CH$ (phenyl)
Styrene

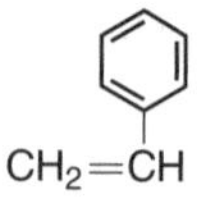

$CH_2{=}CH{-}CH{=}CH_2$

$\left[\!CH_2{-}CH{-}CH_2{-}CH{-}\right]_n$ (phenyl, $CH{=}CH_2$)
SBR rubber

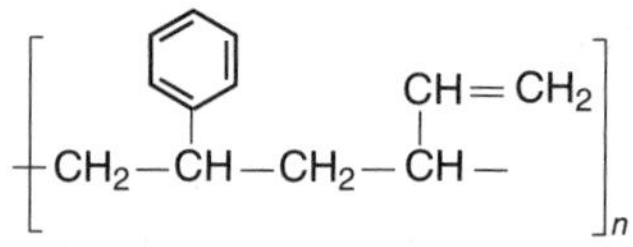

$\underset{\displaystyle CH_2{=}CH}{\overset{\displaystyle Cl}{|}}$
Vinyl chloride

$\left[\!CH_2{-}\underset{}{\overset{Cl}{CH}}{-}\!\right]_n$
Poly(vinyl chloride) (PVC)

$CF_2{=}CF_2$
Tetrafluoroethylene

$\left[\!CF_2{-}CF_2\!\right]_n$
Poly(tetrafluoroethylene) (PTFE)

$\underset{\displaystyle CH_2{=}CH}{\overset{\displaystyle C{\equiv}N}{|}}$
Acrylonitrile

$\left[\!CH_2{-}\underset{}{\overset{C{\equiv}N}{CH}}{-}\!\right]_n$
Polyacrylonitrile

$CH{\equiv}CH$
Acetylene

$\left[\!CH{=}CH\!\right]_n$
Polyacetylene

TABLE 6.1. *Continued*

Monomer A	Monomer B	Product Polymer
Polyurethanes HO—R—OH	O=C=N—R—N=C=O	O=C=N—[—R—NH—CO—R—O—CO—NH]$_n$— R—N=C=O

Condensation monomers

H$_2$NRCOOH Amino acid		+(H)NRC(O)+$_n$ Poly(amino acid), polypeptide
H$_2$N(CH$_2$)$_6$NH$_2$ Hexamethylenediamine	HOOC(CH$_2$)$_4$COOH Adipic acid	+(O)C(CH$_2$)$_6$C(O)NH(CH$_2$)$_4$HN+$_n$ Nylon 66
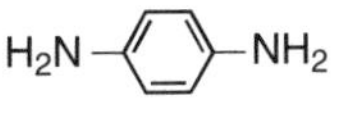 H$_2$N—⬡—NH$_2$ P-Phenylenediamine	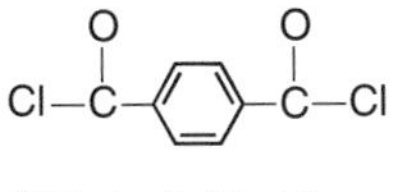 Cl—C(O)—⬡—C(O)—Cl Phthaloyl chloride	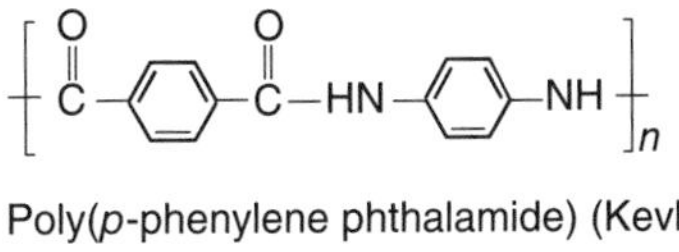 Poly(p-phenylene phthalamide) (Kevlar™)
HO(CH$_2$)$_2$OH Ethylene glycol	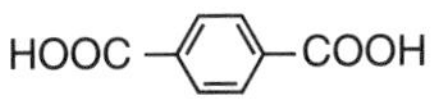 HOOC—⬡—COOH Terephthalic acid	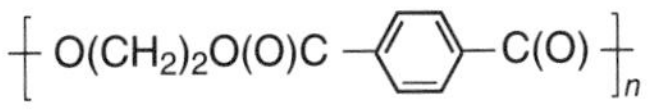 Poly(ethylene terephthalate) (Mylar, Dacron)
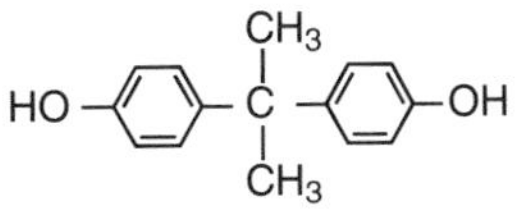 HO—⬡—C(CH$_3$)$_2$—⬡—OH Bisphenol A	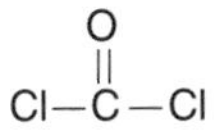 Cl—C(O)—Cl Phosgene (dangerous)	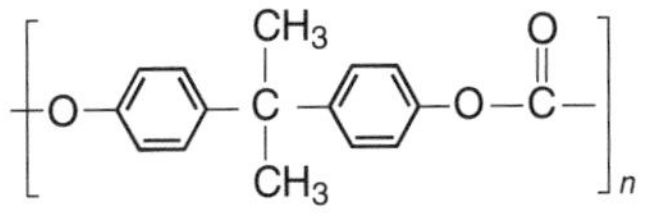 Polycarbonate
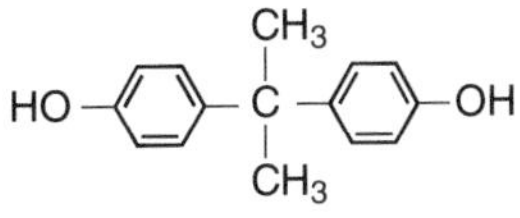 HO—⬡—C(CH$_3$)$_2$—⬡—OH Bisphenol A	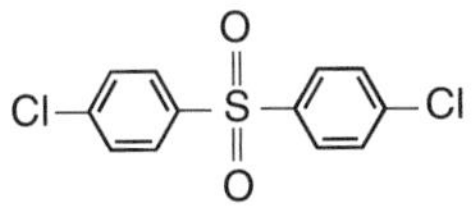 Cl—⬡—S(O)$_2$—⬡—Cl 4,4′-Dichlorodiphenyl sulfone	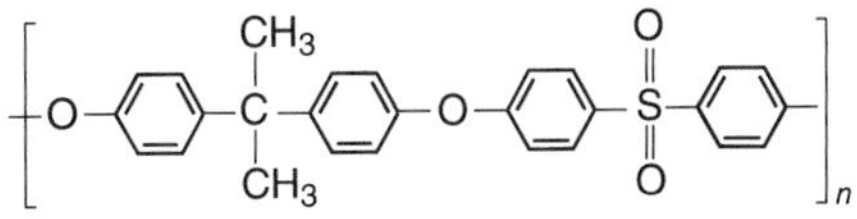 Polysulfone
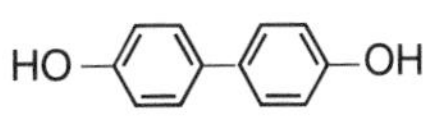 HO—⬡—⬡—OH 4,4′-Dihydroxybiphenyl	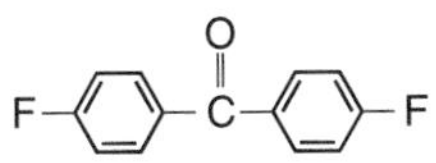 F—⬡—C(O)—⬡—F 4,4′-Difluorodiphenyl ketone	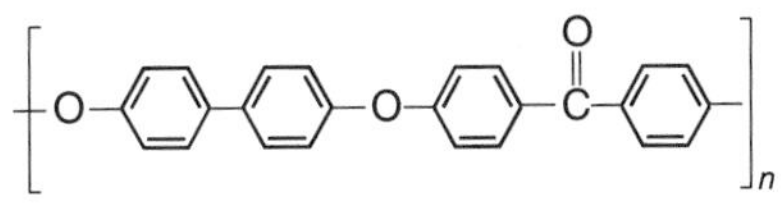 Poly(ether ether ketone) (PEEK)
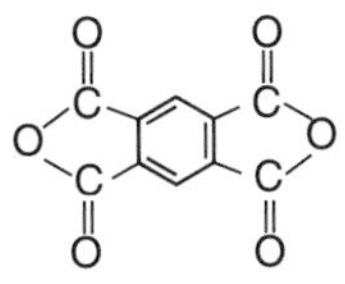 (pyromellitic dianhydride)	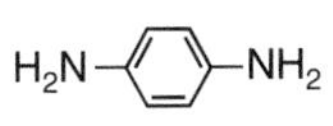 H$_2$N—⬡—NH$_2$ P-Phenylenediamine	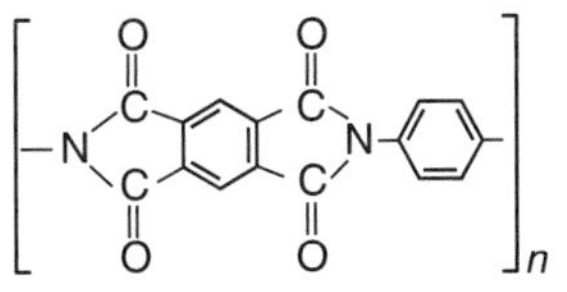 Polyimide
HO—⬡—C(CH$_3$)$_2$—⬡—OH Bisphenol A	CH$_2$—CHCH$_2$Cl (epoxide) 	Epoxy resin

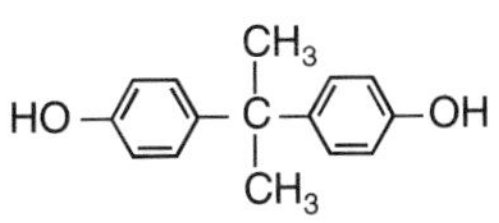

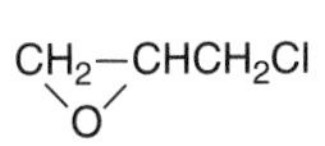

TABLE 6.1. Continued

Monomer A	Monomer B	Product Polymer

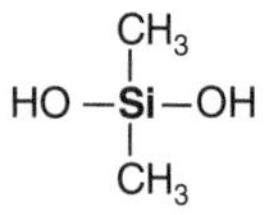

CH₃
HO—Si—OH
CH₃

Dimethylsilanediol

$$\left[O-\underset{\underset{CH_3}{|}}{\overset{\overset{CH_3}{|}}{Si}} \right]_n$$

Poly(dimethylsiloxane) (silicone)

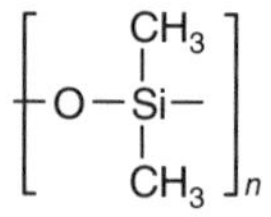

Ring-opening monomers

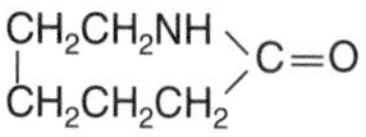

CH₂CH₂NH
              C=O
CH₂CH₂CH₂

Caprolactam

$$+(CH_2)_5\,NHC(O)\,]_n$$

Nylon 6

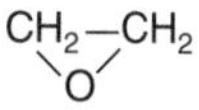

CH₂—CH₂
      O

Ethylene oxide

$$+CH_2CH_2O]_n$$

Poly(ethylene oxide)

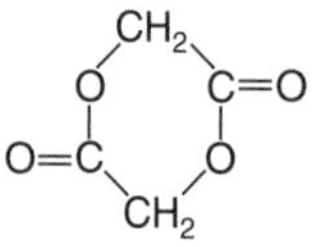

Glycolide dimer

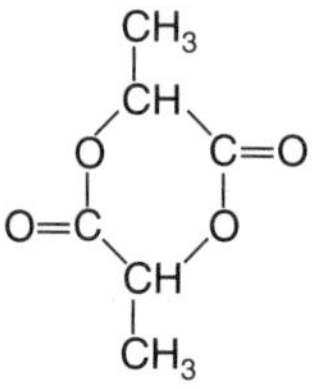

Lactide dimer

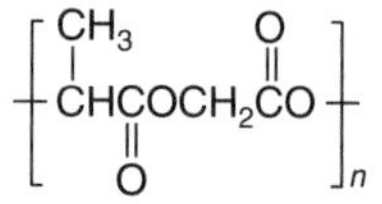

$$\left[\begin{array}{cc} CH_3 & O \\ | & \| \\ CHCOCH_2CO \\ \| \\ O \end{array} \right]_n$$

Poly(lactic-glycolic acid) (PLGA)

Norbornene

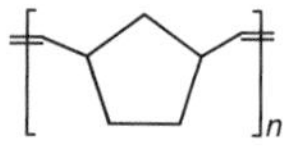

Polynorbornene

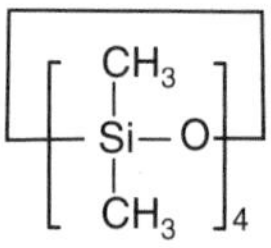

$$\left[\begin{array}{c} CH_3 \\ | \\ Si-O \\ | \\ CH_3 \end{array} \right]_4$$

Octamethylcyclotetrasiloxane

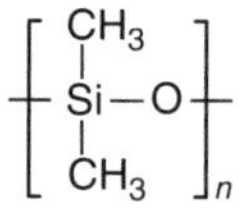

$$\left[\begin{array}{c} CH_3 \\ | \\ Si-O \\ | \\ CH_3 \end{array} \right]_n$$

Poly(dimethylsiloxane) (silicone)

R₂
    Si
R₂Si   SiR₂
R₂Si   SiR₂
    Si
    R₂

Cyclic silane

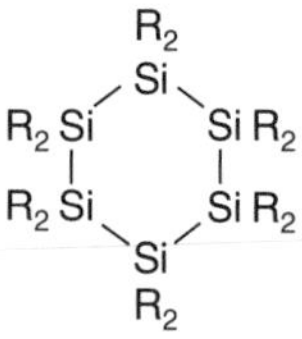

$$\left[\begin{array}{cc} R & R \\ | & | \\ Si-Si \\ | & | \\ R & R \end{array} \right]_n$$

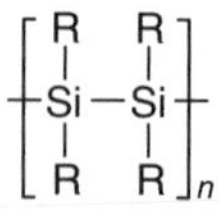

Polysilane

TABLE 6.1. *Continued*

Monomer A	Monomer B	Product Polymer

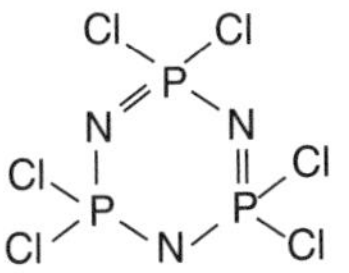

Chlorophosphazene cyclic trimer

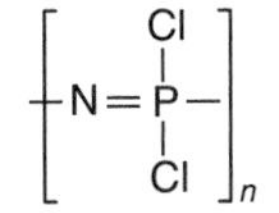

$$\left[-N=P- \right]_n$$
with Cl, Cl substituents

Poly(dichlorophosphazene)

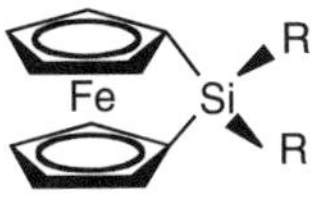

Ferrocenophane

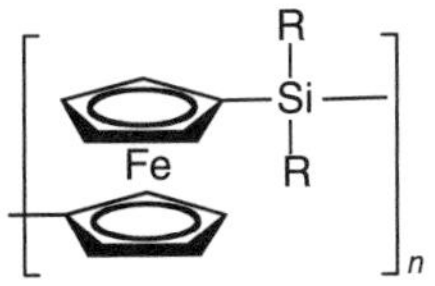

Polyferrocenophane

$(S=N)_4$
Tetrasulfur tetranitrude

$(S=N)_n$
Polythiazyl Poly(sulfur nitride)

The action of heat or radiation on the monomer may also yield an activated species.

2. *Chain Propagation.* The initiated monomer adds to an unactivated monomer, and the resultant active dimer adds to another monomer to start an addition chain reaction. The polymer chains continue to grow until side reactions terminate the chain growth process or the monomer is depleted.

3. *Chain Termination.* This is a process in which the active chain ends are deactivated to bring chain growth to a halt.

4. *Chain Transfer.* The active end of a growing chain attacks the middle units of another chain or a residual monomer molecule to produce a new activated site and simultaneously terminates the original chain. If the attack takes place at a middle unit of another polymer, it may generate a branched structure.

These steps are summarized in reactions 1–4, where the asterisk (*) indicates an activated unit such as a free radical, anion, or cation, and T is a terminator molecule. The overall course of a polymerization depends on the chemical structure of the monomer, the type of initiator used, and the reaction conditions (solvent, dilution, temperature, etc.). The following discussion revolves around the different types of initiators.

$$CH_2=\overset{\overset{\displaystyle R}{|}}{CH} \xrightarrow{\text{initiator}} CH_2-\overset{\overset{\displaystyle R}{|}}{\underset{\underset{\displaystyle H}{|}}{C}} * \qquad (1)$$

$$
\begin{array}{c}
\text{CH}_2-\underset{\overset{|}{\text{H}}}{\overset{\text{R}}{\text{C}}}* + \text{CH}_2{=}\overset{\text{R}}{\text{CH}} \longrightarrow \text{CH}_2-\underset{\overset{|}{\text{H}}}{\overset{\text{R}}{\text{C}}}-\text{CH}_2-\underset{\overset{|}{\text{H}}}{\overset{\text{R}}{\text{C}}}*
\end{array}
$$

$$
\text{CH}_2-\underset{\overset{|}{\text{H}}}{\overset{\text{R}}{\text{C}}}-\text{CH}_2-\underset{\overset{|}{\text{H}}}{\overset{\text{R}}{\text{C}}}* + n\,\text{CH}_2{=}\overset{\text{R}}{\text{CH}} \longrightarrow
$$

$$
\longrightarrow \left[\text{CH}_2-\underset{\overset{|}{\text{H}}}{\overset{\text{R}}{\text{C}}}\right]_n \text{CH}_2-\underset{\overset{|}{\text{H}}}{\overset{\text{R}}{\text{C}}}* \tag{2}
$$

$$
\left[\text{CH}_2-\underset{\overset{|}{\text{H}}}{\overset{\text{R}}{\text{C}}}\right]_n \text{CH}_2-\underset{\overset{|}{\text{H}}}{\overset{\text{R}}{\text{C}}}* \xrightarrow{\;\text{T}\;} \left[\text{CH}_2-\underset{\overset{|}{\text{H}}}{\overset{\text{R}}{\text{C}}}\right]_n \text{CH}_2-\underset{\overset{|}{\text{H}}}{\overset{\text{R}}{\text{C}}}-\text{T} \tag{3}
$$

$$
\left[\text{CH}_2-\underset{\overset{|}{\text{H}}}{\overset{\text{R}}{\text{C}}}\right]_n \text{CH}_2-\underset{\overset{|}{\text{H}}}{\overset{\text{R}}{\text{C}}}* + \left[\text{CH}_2-\underset{\overset{|}{\text{H}}}{\overset{\text{R}}{\text{C}}}\right]_n \longrightarrow
$$

$$
\left[\text{CH}_2-\underset{\overset{|}{\text{H}}}{\overset{\text{R}}{\text{C}}}\right]_n \text{CH}_2-\underset{\overset{|}{\text{H}}}{\overset{\text{R}}{\text{C}}}-\text{H} + \left[\text{CH}_2-\underset{\overset{|}{\text{H}}}{\overset{\text{R}}{\text{C}}}\right]_n \overset{*}{\underset{\overset{|}{\text{H}}}{\overset{\text{R}}{\text{CH}}}}-\underset{\overset{|}{\text{H}}}{\overset{\text{R}}{\text{C}}}\left[\text{CH}_2-\underset{\overset{|}{\text{H}}}{\overset{\text{R}}{\text{C}}}\right]_n \tag{4}
$$

b. Free-Radical Initiators. This is one of the easiest ways to carry out an addition polymerization. Monomers such as styrene, methyl methacrylate, or ethylene are polymerized on a large scale by free-radical processes. The monomer is used undiluted, or it may be dissolved in an organic solvent or be part of an emulsion of water and an organic medium. Introduction of a small amount of an initiator, often followed by heating, then starts the polymerization process.

The initiator is a compound such as a peroxide or azo derivative that can decompose under mild or moderate conditions to generate free radicals as shown in reaction 5.

$$
\text{R}'{-}\text{O}{-}\text{O}{-}\text{R}' \;\text{ or }\; \text{R}'{-}\text{N}{=}\text{N}{-}\text{R}' \longrightarrow \text{R}'{-}\text{O}\cdot \;\text{ or }\; \text{R}'\cdot + \text{N}_2 \tag{5}
$$

Typical free-radical initiators are aliphatic azo compounds, such as azo-*bis*-isobutyronitrile (AIBN) or peroxides such as benzoyl peroxide. Different initiators are chosen for use at different temperatures. Alternatively, ultraviolet light or γ rays may be employed to initiate polymerization because these sources of energy can generate free radicals from the monomer, the solvent, or added photosensitizers.

Radiation polymerization has the advantage that it minimizes the number of impurities that are formed in the polymer—for example, from initiator residues.

The radicals formed will attack a few of the monomer molecules to convert them to free radicals, and these, in turn, will add to other monomer molecules to initiate a chain reaction as in reactions 6 and 7. Termination can occur via several different mechanisms, but radical recombination (reaction 8) is an ever-present possibility. Chain transfer, which can lead to branching, may occur by hydrogen radical abstraction from a nearby polymer molecule (reaction 9).

$$\underset{\underset{H}{|}}{\overset{\overset{R}{|}}{CH_2=CH}} \xrightarrow{R'\text{-}O^{\cdot}\ or\ R'^{\cdot}} R'CH_2-\underset{\underset{H}{|}}{\overset{\overset{R}{|}}{C}}\!\cdot \tag{6}$$

$$R'CH_2-\overset{\overset{R}{|}}{\underset{\underset{H}{|}}{C}}\!\cdot\ +\ \overset{\overset{R}{|}}{CH_2=CH}\ \longrightarrow\ CH_2-\overset{\overset{R}{|}}{\underset{\underset{H}{|}}{C}}-CH_2-\overset{\overset{R}{|}}{\underset{\underset{H}{|}}{C}}\!\cdot$$

$$CH_2-\overset{\overset{R}{|}}{\underset{\underset{H}{|}}{C}}-CH_2-\overset{\overset{R}{|}}{\underset{\underset{H}{|}}{C}}\!\cdot\ +\ n\,\overset{\overset{R}{|}}{CH_2=CH}\ \longrightarrow$$

$$\longrightarrow\ \left[\!CH_2-\overset{\overset{R}{|}}{\underset{\underset{H}{|}}{C}}\!\right]_n\!CH_2-\overset{\overset{R}{|}}{\underset{\underset{H}{|}}{C}}\!\cdot \tag{7}$$

$$\left[\!CH_2-\overset{\overset{R}{|}}{\underset{\underset{H}{|}}{C}}\!\right]_n\!CH_2-\overset{\overset{R}{|}}{\underset{\underset{H}{|}}{C}}\!\cdot\ +\ \cdot\overset{\overset{R}{|}}{\underset{\underset{H}{|}}{C}}-CH_2\!\left[\!\overset{\overset{R}{|}}{\underset{\underset{H}{|}}{C}}-CH_2\!\right]_n\!\longrightarrow$$

$$\left[\!CH_2-\overset{\overset{R}{|}}{\underset{\underset{H}{|}}{C}}\!\right]_n\!CH_2-\overset{\overset{R}{|}}{\underset{\underset{H}{|}}{C}}-\overset{\overset{R}{|}}{\underset{\underset{H}{|}}{C}}-CH_2\!\left[\!\overset{\overset{R}{|}}{\underset{\underset{H}{|}}{C}}-CH_2\!\right]_n \tag{8}$$

$$\left[\!CH_2-\overset{\overset{R}{|}}{\underset{\underset{H}{|}}{C}}\!\right]_n\!CH_2-\overset{\overset{R}{|}}{\underset{\underset{H}{|}}{C}}\!\cdot\ +\ \left[\!CH_2-\overset{\overset{R}{|}}{\underset{\underset{H}{|}}{C}}\!\right]_n\!\longrightarrow$$

$$\left[\!CH_2-\overset{\overset{R}{|}}{\underset{\underset{H}{|}}{C}}\!\right]_n\!CH_2-\overset{\overset{R}{|}}{\underset{\underset{H}{|}}{C}}-H\ +\ \left[\!CH_2-\overset{\overset{R}{|}}{\underset{\underset{H}{|}}{C}}\!\right]_n\!CH-\overset{\overset{R}{|}}{\underset{\underset{H}{|}}{C}}\!\left[\!CH_2-\overset{\overset{R}{|}}{\underset{\underset{H}{|}}{C}}\!\right]_n \tag{9}$$

Emulsion polymerizations are free-radical reactions that take place in an emulsion of water and monomer, or water and an immiscible organic solvent. The initiator (typically a metal ion redox system) is soluble in the aqueous component. The monomer in the organic phase is initiated at the organic/water interface. Emulsion-polymerized polymers often consist of spherical particles that grow within the aqueous medium.

Free-radical polymerizations have certain characteristics: (1) they are generally accelerated by increases in temperature; (2) unless special initiators are used, they give very broad molecular weight distributions, because not all the initiator molecules decompose to free radicals at the same time, so some chains are started sooner than others; and (3) free radicals are so reactive that they undergo side reactions such as abstracting hydrogen radicals from monomer, polymer, or solvent molecules. This will terminate a growing chain and generally start a new one. This same process leads to chain branching and sometimes to crosslinking. Hence, most free-radical processes do not give clean products but generate a broad mixture of linear and branched species. Nevertheless, this is a process widely used in industry to produce polymers such as polyethylene, polystyrene, poly(methyl methacrylate) (Plexiglas®), poly(vinyl chloride) (PVC), and many others. A more recently developed method to overcome the problems described above uses initiators that induce "atom transfer radical" processes. This process can be traced through references given at the end of this chapter.

b. Anionic Initiation. Compounds such as *n*-butyllithium or sodium naphthalenide also initiate addition chain reactions, but via a different mechanism. The process is shown in reactions 10 and 11.

$$
\underset{\underset{\displaystyle CH_2=CH}{|}}{R} \quad \xrightarrow{R'Li} \quad \underset{\underset{\displaystyle H}{|}}{\overset{\overset{\displaystyle R}{|}}{R'CH_2-C^-\ Li^+}} \tag{10}
$$

$$
\underset{\underset{\displaystyle H}{|}}{\overset{\overset{\displaystyle R}{|}}{R'CH_2-C^-\ Li^+}} + \underset{\underset{\displaystyle CH_2=CH}{|}}{R} \longrightarrow \underset{\underset{\displaystyle H}{|}}{\overset{\overset{\displaystyle R}{|}}{R'-CH_2-C-CH_2-}}\underset{\underset{\displaystyle H}{|}}{\overset{\overset{\displaystyle R}{|}}{C^-\ Li^+}} \tag{11}
$$

etc.

The organometallic reagent adds across the double bond of a vinyl compound to generate an anion that attacks another monomer molecule, and then another, and so on. Anionic polymerizations, particularly when carried out in solution at temperatures below 25 °C, are much cleaner than free-radical reactions. It is essential that the solvent, glassware, and the atmosphere above the reaction mixture be free from impurities such as water or carbon dioxide, which would react with the active site at the growing chain end.

These reactions are described as "living" because, in the absence of an added terminator, they proceed without side reactions to give linear polymer chains that are still active even when all the monomer has been consumed. The consequences of this are quite profound:

1. The initiation process is very rapid, and all the chains begin to grow at the same time. Thus, by the time the monomer has been consumed, all the chains have approximately the same length. In other words, the molecular weight distribution, given by M_w/M_n, is close to 1. Imagine this being like a footrace where all the runners start simultaneously when the gun is fired, run at exactly the same speed along a straight track, and all burst through the finish-line tape at exactly the same instant.

2. Because each initiator molecule starts only one chain, it is possible to control the chain length of the polymer with some precision. Thus, the length of each polymer chain is given by the number of monomer molecules in the reaction vessel divided by the number of initiator molecules. The fewer initiator molecules present, the longer will be the chains.

3. Because the chains are still active after the monomer has been consumed, it is possible to add more of the same monomer to continue the process and grow longer chains. Alternatively, a different monomer may be added to generate a diblock copolymer (reaction 12).

$$\text{R}'-\text{CH}_2-\underset{\underset{\text{H}}{|}}{\overset{\overset{\text{R}}{|}}{\text{C}}}\left[\text{CH}_2-\underset{\underset{\text{H}}{|}}{\overset{\overset{\text{R}}{|}}{\text{C}}}\right]_n\text{CH}_2-\underset{\underset{\text{H}}{|}}{\overset{\overset{\text{R}}{|}}{\text{C}^-}}\ \text{Li}^+ + \text{CH}_2=\underset{\underset{\text{H}}{|}}{\overset{\overset{\text{R}''}{|}}{\text{CH}}} \longrightarrow$$

$$\longrightarrow \text{R}'-\text{CH}_2-\underset{\underset{\text{H}}{|}}{\overset{\overset{\text{R}}{|}}{\text{C}}}\left[\text{CH}_2-\underset{\underset{\text{H}}{|}}{\overset{\overset{\text{R}}{|}}{\text{C}}}\right]_n\left[\text{CH}_2-\underset{\underset{\text{H}}{|}}{\overset{\overset{\text{R}''}{|}}{\text{C}}}\right]_n\text{CH}_2-\underset{\underset{\text{H}}{|}}{\overset{\overset{\text{R}''}{|}}{\text{C}^-}}\ \text{Li}^+$$

$$(12)$$

4. A living chain end can be deliberately terminated by the addition of a reagent that puts a functional site at the chain terminus. For example, treatment with carbon dioxide places a carboxylate group at the chain terminus. This may then be employed to couple that chain to another polymer molecule by, for example, a condensation reaction.

A variant of the classical anionic polymerization method is provided by the use of sodium naphthalenide as an initiator. This initiator is a radical anion (reactions 13 and 14).

$$(13)$$

$$\text{Na}^+\ ^-\text{CH}_2-\underset{\overset{|}{\text{R}}}{\text{CH}}\cdot + \cdot\text{CH}_2-\underset{\overset{|}{\text{R}}}{\text{CH}}^-\ \text{Na}^+ \longrightarrow \text{Na}^+\ ^-\text{CH}_2-\underset{\overset{|}{\text{R}}}{\text{CH}}-\text{CH}_2-\underset{\overset{|}{\text{R}}}{\text{CH}}^-\ \text{Na}^+$$

$$(14)$$

In the presence of an unsaturated monomer, the initiator transfers an electron to the monomer to convert it to a radical anion. The radical end of the initiated monomer then combines with its counterpart on another initiated species to form a dimer, both ends of which bear an anion. Monomer molecules are then polymerized from both ends of the initiated species, and all of the characteristics described above for anionic polymerizations are retained. The only structural difference is that there is a discontinuity (a "tail to tail" structure) in the middle of the final chain. The existence of two active chain ends on the final polymer molecule means that the addition of a different monomer then leads to the formation of a *tri*block copolymer.

Many different monomers can be polymerized by anionic initiation, but the technique works best for unsaturated organic compounds that have electron-withdrawing side groups such as phenyl, cyano, or ester units. Thus, narrow-distribution polystyrene, polyacrylonitrile, and poly(methyl methacrylate) are synthesized by anionic initiation. The monomer should not have a side group such as chlorine, bromine, iodine, hydroxyl, or amino that reacts with organolithium compounds.

c. Coordination Initiation. Other organometallic-type initiators also play a major role in the synthesis of addition polymers. Two of the best known are Ziegler–Natta and metallocene initiators. Ziegler–Natta initiators are prepared by the reaction of a metal halide such as titanium or vanadium chloride with an aluminum alkyl such as triethylaluminum (structure **6.1** in Figure 6.1). The structure of these catalysts and their reaction mechanisms are a subject of ongoing debate. One proposed mechanism is shown in Figure 6.1 in which an olefin coordinates to a vacant site on the transition metal and then inserts itself into the adjacent metal-carbon bond (structures **6.2** and **6.3** in Figure 6.1). Some Ziegler-type catalysts are soluble in organic solvents, but others are microparticulate materials that function as heterogeneous initiators. For these last species, the surface structure of the particles is crucial to their function. Ziegler–Natta initiators have one marked advantage over classical anionic initiators, namely, that they generate stereospecific polymers. For example, in a Ziegler–Natta reaction, propylene can enter a growing polymer chain with the pendent methyl group on only one side of the polymer skeleton (*isotactic*). The importance of this is that, whereas the analogous polymer formed by free-radical methods, with a random disposition of side-group configurations (*atactic*), is a gum, the isotactic form of polypropylene is the tough solid material that is widely used in technology. Different initiators and different monomers may generate polymers in which a side group alternates regularly from one side to the other along each chain. This is called a *syndiotactic* configuration. The importance of these stereoregular polymer structures is that they allow the chains to become oriented and crystallize. A disadvantage of Ziegler–Natta initiators compared to the classical anionic reactions is that the polymer molecular weights are usually not as high, and the molecular weight distributions may be broad. High-density (unbranched) polyethylene is produced by the Ziegler–Natta technique. Acetylene is also polymerized by this technique to give the metal-like polyacetylene. This polymer is discussed in Chapter 10.

Metallocene initiators are related to Ziegler–Natta catalysts, but they contain metals such as zirconium, titanium, or hafnium sandwiched between two 5-

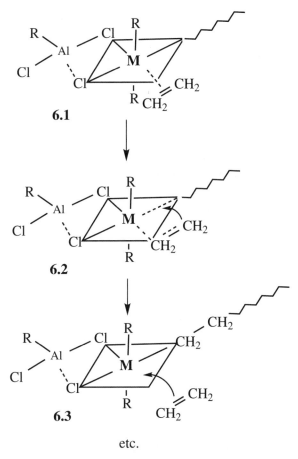

Figure 6.1. *A possible mechanism for chain growth in Ziegler–Natta polymerization.*

membered aromatic rings. Example catalysts are shown as **6.4–6.6**. Cocatalysts such as organoaluminum compounds are also employed. The use of metallocene catalysts has had a major impact on the production of polyethylene and polypropylene. The advantage of metallocene catalysts over Ziegler–Natta initiators is their ability to give pure isotactic or syndiotactic polypropylene and to allow better control over the chain length and molecular weight distribution.

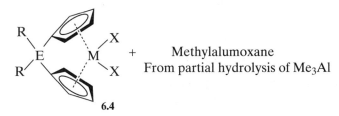

M = Zr, Ti, Hf
X = halogen or alkyl   1 : 500–1 : 15,000 ratios
E = silicon or carbon

6.5

6.6

Gives issotactic polypropylene   Gives syndiotactic polypropylene

d. Cationic Initiation. Unsaturated organic compounds with electron-supplying side groups close to the double bond, such as 1-butene, isobutylene, isoprene, 1,3-butadiene, vinyl ethers, α-methylstyrene, or aldehydes undergo addition polymerization when treated with Lewis acid or protonic acid initiators. Some examples are shown in Table 6.1. The active species at the chain ends are carbonium ions. The overall reaction sequence is shown in reactions 15 and 16.

$$
\underset{\substack{| \\ \text{CH}_2=\text{CH}}}{\overset{R}{|}} \xrightarrow{\text{H}^+\text{X}^-} \underset{\substack{| \\ \text{H}}}{\overset{R}{\underset{|}{\text{HCH}_2-\text{C}^+}}}\ \text{X}^- \tag{15}
$$

$$
\underset{\substack{| \\ \text{H}}}{\overset{R}{\underset{|}{\text{HCH}_2-\text{C}^+}}}\ \text{X}^- + \underset{\text{CH}_2=\text{CH}}{\overset{R}{|}} \longrightarrow \underset{\substack{| \\ \text{H}}}{\overset{R}{\underset{|}{\text{HCH}_2-\text{C}}}}-\underset{\substack{| \\ \text{H}}}{\overset{R}{\underset{|}{\text{CH}_2-\text{C}^+}}}\ \text{X}^- \tag{16}
$$

etc.

Examples of such initiators include BF_3OH^- H^+, H_2SO_4, HCl, $AlCl_3$, or $SnCl_4$. The characteristics of many classical cationic polymerizations are as follows: (1) exposure of the monomer and solvent to water must be controlled carefully because minute traces of water serve as cocatalysts and change both the rate of polymerization and the average molecular weights; (2) temperatures below 25 °C are preferred for these reactions to minimize chain transfer and premature termination of active chains and also, paradoxically, to speed up the rate of polymerization; (3) although the copolymerization of two or more different monomers is possible, the monomers must be of similar reactivity; (4) although "living" cationic polymerizations are possible, they are not as common as in anionic polymerizations; and (5) cationic polymerization is used on a large scale to produce a variety of synthetic rubber materials.

3. Condensation Polymerization

a. General Features. Close similarities exist between organic polymer synthesis by condensation reactions and the synthesis of oxide ceramics. Condensation reactions take place when difunctional monomers react and split out a side product, which is usually water. Organic monomers such as diols, diamines, dicarboxylic acids, amino alcohols, and amino acids undergo these reactions. Several diverse examples are shown in Table 6.1. The earliest synthetic polymers such as Nylon 66 and poly(ethylene terephthalate) (Dacron® or Mylar®) were made by condensation reactions, and many of the most advanced modern polymers such as Kevlar®, Nomex®, and polyimides, are also produced by this technique. Condensation reactions involve initiation, propagation, chain transfer, and termination steps, but are also capable of undergoing the reverse process—hydrolytic depolymerization. Removal of the water formed by the condensation minimizes this reverse process. Catalysts such as acids or bases are often used to accelerate the reactions. The overall steps are shown in reactions 17–19.

$$HO-R-OH \xrightarrow{\quad -H_2O \quad} \left[R-O \right]_n \qquad (17)$$

$$H_2N-R-COOH \xrightarrow{\quad -H_2O \quad} \left[R - \overset{\overset{\displaystyle O}{\|}}{C} - \overset{\overset{\displaystyle H}{|}}{N} \right]_n \qquad (18)$$

$$HO-R-COOH \xrightarrow{\quad -H_2O \quad} \left[R - \overset{\overset{\displaystyle O}{\|}}{C} - O \right]_n \qquad (19)$$

To summarize, the key characteristics of condensation reactions are as follows:

1. The driving force for condensation is the removal of water (or another small molecule such as hydrogen chloride or a salt) from the polymerization reaction mixture. Unless this is accomplished, an equilibrium system will be established between condensation and chain cleavage (depolymerization), severely limiting the chain lengths that can be achieved.
2. The molecular weight distributions are generally broad because of the multiplicity of reactions that can take place. For example, not only do the end units of monomers react with other monomers; monomers also react with polymers, polymers react with other polymers, rings react with monomers and polymers, and so on.
3. The maximum molecular weights attainable are usually well below those obtained in addition or ring-opening polymerizations. Moreover, deliberate limitation of the chain lengths is possible by the addition of monofunctional reagents, such as alcohols or amines that cap the chain ends.
4. The highest molecular weights can be achieved only by ensuring an exact 1:1 ratio of the functional groups (i.e., equimolar amounts of diacid and diamine or diacid and diol), or by fixing this ratio within the monomer as in the case of an amino acid or a hydroxy acid. If one functional group predominates in the reaction mixture, the growing chains will be terminated at an early stage in the reaction.
5. Crystalline polymers are accessible, especially when the polymers contain aryl groups in the backbone.
6. Chain branching and crosslinking are accomplished by the presence of trifunctional reagents such as triols or triamines.

Condensation polymers are synthesized via the reactions of a wide variety of monomers (Table 6.1). Examples include the reactions of dicarboxylic acids with diamines (polyamides), dicarboxylic acids with diols (polyesters), elimination of hydrogen chloride from diols and phosgene (polycarbonates), elimination of sodium chloride from aromatic dichloro compounds and sodium salts of diols (polysulfones), self-condensation of amino acids (polyamides), self-condensation of silanediols (silicones), or elimination of trimethylchlorosilane from phosphoranimines (polyphosphazenes). Examples of specific condensation polymers are given later in this chapter.

4. Ring-Opening Polymerization

The third general polymerization technique involves the opening of organic, organometallic, or inorganic rings under conditions that lead to polymerization. Examples are given in Table 6.1. In many cases, ring-opening polymerizations of cyclic esters, ethers, or amides give polymers that might otherwise be produced by condensation reactions. The advantages and disadvantages of ring-opening polymerizations are as follows:

1. Unlike condensation polymerizations, this process involves no loss of a side product as the reaction proceeds. Thus, the need to remove water or another side products throughout the polymerization to drive the reaction to completion does not apply. This is a significant advantage.
2. Especially when catalysts or initiators are present, the polymer molecular weights can be significantly higher than from condensation reactions.
3. It is a characteristic of many ring-opening polymerizations that polymerization and depolymerization occur at the same time. This means that if the reaction is allowed to proceed long enough to allow an equilibrium to be established, the molecular weight distribution may be very broad. This problem arises because, unless there is appreciable ring strain in the cyclic monomer, no strong driving force exists that favors polymer chains over small rings, and so the forward and back reactions may be almost equally preferred. However, the initial kinetics of the process frequently favor polymerization, so that in the early stages of the reaction the product mixture contains mainly unreacted cyclic monomer and a high polymer. Only during the later stages, when thermodynamic equilibration takes over, does the depolymerization process become significant. Thus, a knowledge of when to terminate the polymerization is the key to generating the highest molecular weights.

A wide variety of polymers are produced by ring-opening reactions, including poly(ε-caprolactam) (Nylon 6), poly(ethylene oxide) (PEO), polyesters from lactones, polynorbornenes, poly(organosiloxanes) (silicones), polysilanes, polyphosphazenes, and poly(ferrocenophanes) (Table 6.1). Examples of polymers produced by ring-opening polymerization are given later in this chapter.

An increasingly important form of ring-opening polymerization is ring-opening metathesis polymerization (ROMP) shown in reaction 20. This is a process in which an unsaturated organic cyclic compound such as norbornene is treated with an organometallic catalyst such as those shown in structures **6.7–6.9** to give a polymer that retains unsaturation in the skeleton.

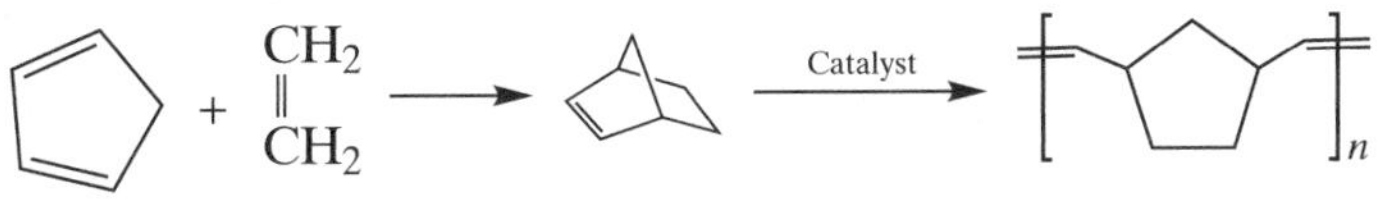

$$\text{(20)}$$

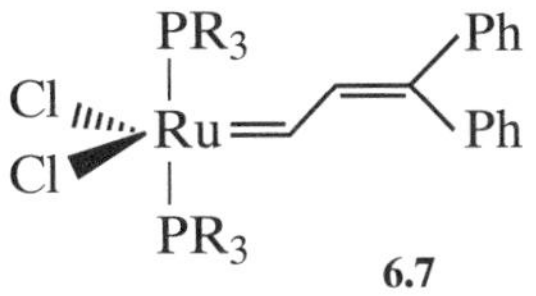

6.7

Grubbs Ru catalysts

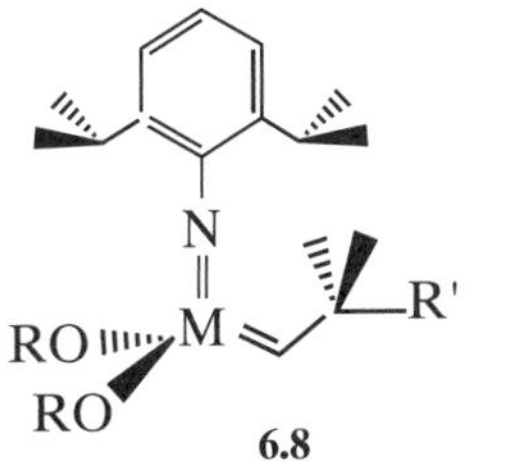

6.8 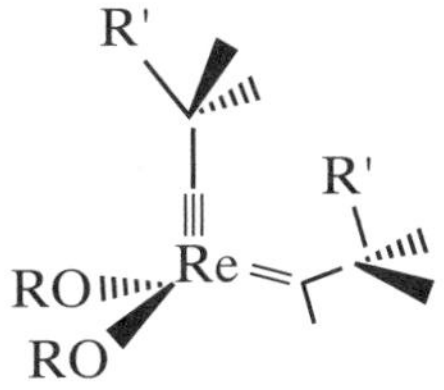   **6.9**

Schrock W, Mo, and Re catalysts

5. Electrochemical Polymerization

Yet another method for the synthesis of polymers involves electrochemical oxidation or reduction. Electronically conductive polymers such as polythiophene, polypyrrole, and polyaniline are produced by this technique. Typically this is a method for the linkage of organic rings to give linear, cyclolinear or cyclomatrix polymers. As an example, the formation of polypyrrole is described in Chapter 10.

6. Secondary Reactions

a. Modification of Polymer Structure. In addition to the direct methods of polymer synthesis just described, the possibility also exists that preformed polymers can be altered by chemical reactions carried out on their side groups (reactions 21–25). This process at is not widely applicable to most organic polymers because of the unreactivity of their side units. However, polystyrene can be nitrated or sulfonated, poly(vinyl acetate) can be hydrolyzed to poly(vinyl alcohol), and cellulose may be acetylated or nitrated. Reactions carried out on the surfaces of solid organic polymers are more common. However, in one polymer system, the polyphosphazenes, macromolecular substitution constitutes the main method by which a wide variety of different polymers are produced (see discussion later).

Macromolecular Substitution in Solution

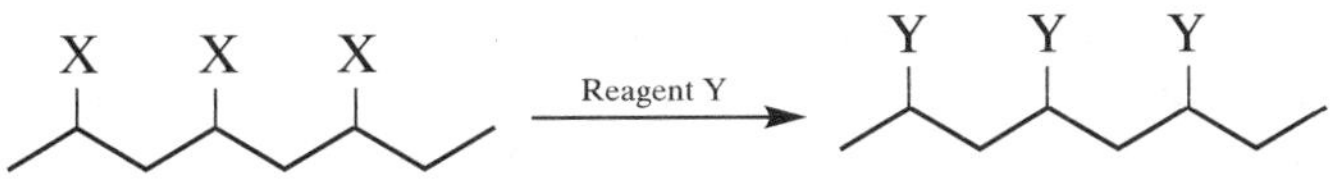

$$\text{(21)}$$

Crosslinking

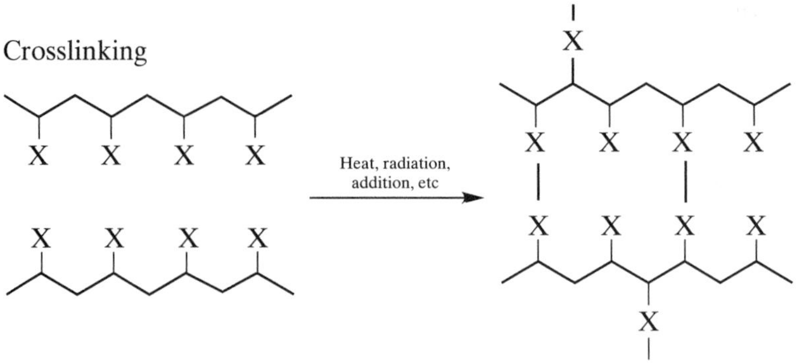

(22)

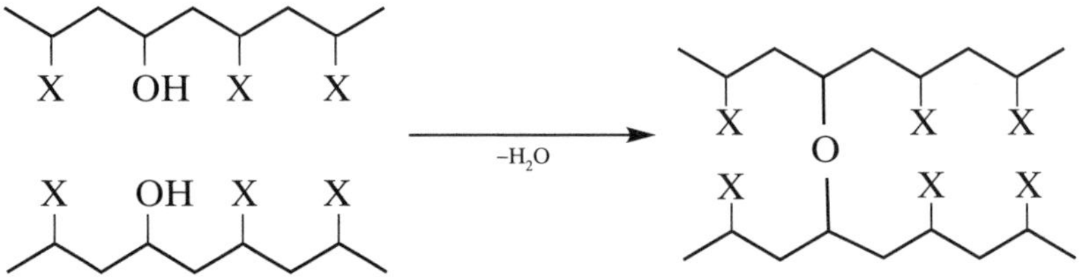

(23)

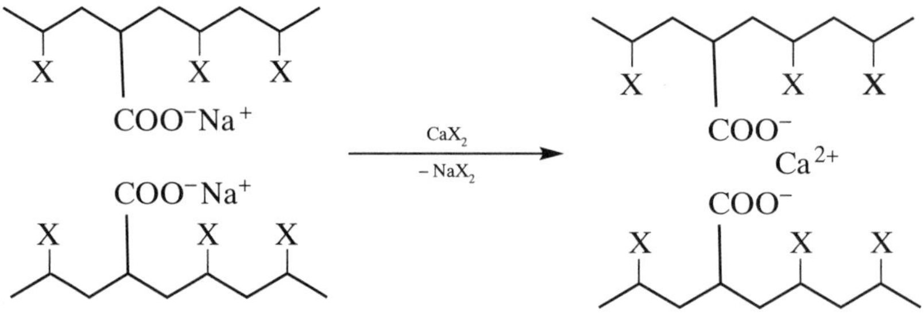

(24)

Surface Reactions

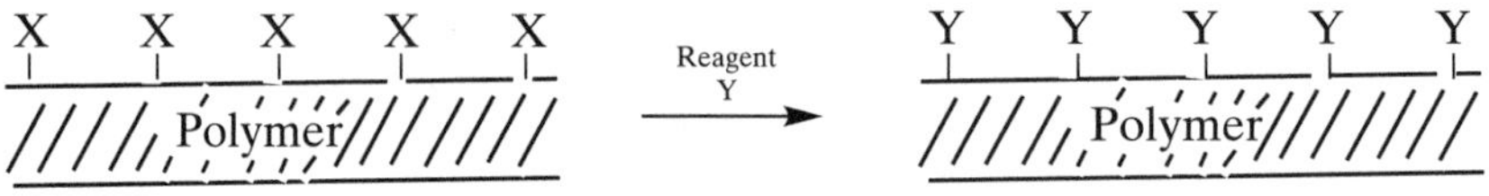

(25)

b. Crosslinking. The utilization of polymers as useful materials frequently requires that the macromolecules be *crosslinked* to reduce their solubility, improve their strength, or enhance their thermal stability. This is usually accomplished by secondary reactions (reaction 22). For example, polymers that bear unsaturated aliphatic side groups, such as vinyl or allyl units, become crosslinked when exposed to free radicals. Similarly, polymers with saturated aliphatic side groups undergo hydrogen radical abstraction in the presence of decomposing peroxides or azo compounds, and the resultant radical-bearing side groups react with their counterparts on other chains to form covalent crosslinks. Exposure of polymers to ultraviolet, X-ray, or gamma radiation has the same effect.

Alternatively, a polymer may be synthesized with a side group that reacts with water to form crosslinks (reaction 23). Poly(dimethylsiloxanes) with acetyl side units hydrolyze in the atmosphere to give acetic acid and a polymer with a few hydroxyl side groups. These cross-condense with similar units on other chains to form crosslinks. Polymers with side groups that bear the sodium salts of acidic side groups (SO_3Na, COONa, etc.) become crosslinked when exposed to di- or trivalent cations such as Ca^{2+} or Al^{3+} (reaction 24).

c. Surface Reactions. The surface character of a solid polymer has a profound influence on its uses. For example, the biomedical compatibility of an implantable material frequently depends on whether the surface is hydrophobic or hydrophilic. Adhesion is highly dependent on surface character. So, too, are abrasion resistance, frictional properties, membrane behavior, and the compatibility of one polymer with another in polymer laminates. It often happens that a specific application requires one set of properties for the bulk material and another set for the surface. This can be accomplished by selection of a polymer for its bulk properties and then modifying the surface by chemical reactions—usually reactions on the polymer side groups (reaction 25). As an example, fluorination of polymer surfaces is an important process for converting hydrocarbon-swellable polymers such as polyethylene to materials that are unchanged in the presence of gasoline or oil. However, such modification reactions should not penetrate deep into the material in order to preserve the bulk properties.

C. STRUCTURE–PROPERTY RELATIONSHIPS AND POLYMER DESIGN

1. Influence of Molecular Architecture

The properties of a polymer depend not only on the skeleton and side groups but also on the architecture of the polymer chains (Figures 6.2 and 6.3).

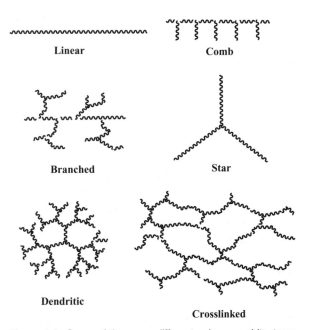

Figure 6.2. *Some of the many different polymer architectures.*

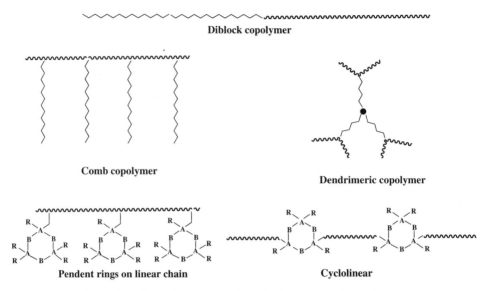

Figure 6.3. *Examples of different architectures for block copolymers.*

a. Linear Polymers. Linear, extended chain polymers represent the classical architecture normally associated with synthetic macromolecules. Several properties are a direct result of the long-chain structure. In general, unlike crosslinked polymers, linear macromolecules are soluble in appropriate solvents because the solvent molecules are able to penetrate into the bulk polymer and ease the chains apart. The solutions so formed usually have a high viscosity because the polymer chains become entangled and retard the flow of the liquid. This is why linear polymers are added to motor oils to maintain viscosity at the operating temperature of the engine. Chain entanglement in a solid polymer converts what might otherwise have been a gum into a solid material (see later). A disadvantage of many linear polymers is that they decompose at high temperatures more easily than materials that are heavily crosslinked.

Linear polymers in the solid state can be aligned when the material is stretched. This in turn facilitates ordering of the chains and the formation of crystallites. Linear polymers may also give rise to structural "tacticity," meaning that different monomer residues along the chain have different *configurations*. Consider a polymer of formula $-(CH_2CHR)_n$. The group R can, in principle, lie on either side of the backbone. For many free-radical polymerizations, the group R is oriented randomly on either side of the main chain. However, some organometallic initiators have the ability to favor the formation of polymers with all the R groups on the same side of the chain (isotactic), or alternating from side to side (syndiotactic). Isotactic and syndiotactic polymers have a higher tendency to crystallize than do their random counterparts because the order along the polymer chains assists shape-fitting.

b. Branched Structures, Stars, and Dendrimers. The existence of branched, star, or dendritic structures in a high polymer (Figure 6.2) has two effects. First, branching will enhance the opportunities for chain entanglement and hence increase the viscosity in both the solid and solution states. However, if the branching is so extensive that it leads to dendrimer formation, the structure may have an overall spherical architecture, in which the peripheral branch points are sterically prevented from forming entanglements with their counterparts on other molecules. In such a case the viscosity will be lower than that of a linear or lightly branched polymer with the same molecular weight. Branching lowers the tendency for crystallization because it makes it more difficult for neighboring chains to pack together efficiently.

c. Combs and Grafts. Comb or graft architectures (Figures 6.2 and 6.3) also discourage crystallization, but can markedly increase solid or solution viscosity compared to a linear counterpart because of the enhanced opportunities for chain entanglements. The most important comb or graft structures are copolymers in which the main backbone is constructed from one monomer type and the teeth or grafts are made up of another.

d. Combinations of Rings and Chains. Materials in which rings are linked together by chains or chains with pendent rings are quite common, as are polymers in which the entire backbone consists of linked rings. Often the rings will increase the material's rigidity and may enhance thermal stability. These are architectures that are often found in preceramic polymers (see Chapter 7) or in polymers for special applications such as ion conductive membranes or photonic materials.

e. Copolymers. Polymer architecture can also be diversified through the synthesis of copolymers, which are produced from two or more different monomers or by the linkage of two or more different macromolecules, some variations of which are shown in Figure 6.3. Copolymers and hybrid systems allow the development of combinations of properties derived from the two or more different components— for example, a copolymer of a hard polymer with an elastomeric counterpart can yield a material that is both strong and impact resistant.

2. Molecular Weights and Distributions

Short-chain molecules form mobile liquids or brittle solids because of the limited number of molecular entanglements that are possible. Materials properties improve steadily as the chain length is increased until they reach a plateau at about 1000 repeating units. The longest chains are needed for maximum solid-state dimensional stability, good elastomeric behavior, and the formation of strong fibers. On the other hand, medium-length chains are often preferred if the polymer is to be fabricated by melt extrusion or injection molding, where high melt viscosities would be detrimental. Broad molecular weight distributions (with M_w/M_n values higher than ~1.5) tend to lower the softening point of a polymeric material and accelerate dissolution in solvents. Narrow molecular weight distributions ($M_w/M_n \sim 1$) facilitate crystallization.

3. Chain Flexibility

The fundamental properties of a polymer depend on the nature of the backbone, the size and polarity of the side groups, and the free volume between the side groups (Figure 6.4).

Many materials properties are related to molecular chain flexibility. Polymer chains change their shape mainly through torsional (twisting) motions of the skeletal bonds. In general, the more flexible the polymer chains, the more flexible and extensible is the bulk material. Some backbone bonds have more torsional mobility than do others. For example, the carbon–carbon single bond has more torsional mobility than does the carbon–carbon double bond because the latter must surmount a torsional barrier as the π orbitals move into and out of the maximum overlap position. Aryl rings in a polymer backbone inhibit skeletal torsional motions. The silicon–oxygen bond in silicones has one of the lowest torsional barriers known, and so, too, have the phosphorus-nitrogen bonds in the polyphosphazene skeleton (Table 6.2).

However, the side groups sometimes exert an equal or more profound influence on backbone flexibility. Polymers with large side groups or two side groups on every skeletal atom incur high torsional barriers because the side groups experience

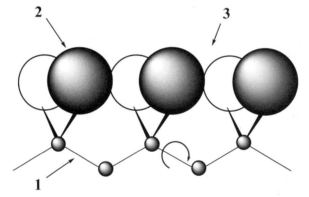

Figure 6.4. *A model for understanding polymer structure–property relationships. Properties can be analyzed in terms of the skeletal bonds (1), the side group structure (2), and the free volume (3).*

TABLE 6.2. Influence of Different Elements and Bonds in the Polymer Main Chain[a]

—C—C—	High torsional flexibility (low T_g); sensitive to thermooxidative cleavage
—C=C—	Restricted rotation; high chemical sensistivity; electrical conductivity
⬡	Restricted rotation (high T_g); chemical resistance
⬡—C=C—	Restricted rotation; colored; electrical conductivity
—C—N—	Restricted rotation (moderate T_g); chemical sensitivity
—C—O—	High chain flexibility; stable as ether linkage
—C—O—C—	High chain flexibility; soluble in water
—S—	Chain flexibility; sensitive to oxidation
—Si—O—	Very high chain flexibility (very low T_g); stable to heat and oxidation
—P=N—	Very high chain flexibility; wide window of optical transparency

[a]Note that the main influence of the backbone is on chain flexibility and side-grouip geometry.

mutual repulsions as the chains twist. Energy is needed for the bonds to surmount such barriers, and this stiffens the chains. If the skeletal bonds are long or if the skeletal bond angles are wide, the steric influence by large side groups may be reduced. Chain stiffening often results in higher glass transition temperatures.

4. Influence of Different Skeletal Elements and Backbone Bonding

The skeletal elements control the underlying properties of a macromolecule. They influence backbone torsional mobility as just described through their control of skeletal bond lengths and bond angles. They also affect properties such as flammability, photolytic stability, and radiation resistance. For example, a polymer with a backbone consisting mainly of aliphatic carbon atoms will be prone to combustion. Phosphorus atoms in the backbone inhibit combustion. Thermal stability is a property closely associated with the types of skeletal elements present; inorganic elements such as silicon, phosphorus, or metals in metallocene units are generally associated with high-temperature polymers. Aliphatic carbon atoms are more prone to photooxidative chain cleavage than are aryl units or inorganic species. Conjugated carbon–carbon double bonds in the skeleton may give rise to electronic conductivity, color, and electroluminescence. On the other hand, phosphorus–nitrogen double bonds yield electrical insulators and colorless polymers. Etheric oxygen atoms favor torsional mobility and chain flexibility, as do oxygen atoms flanked by silicon. Nitrogen atoms double-bonded to carbon confer rigidity. Some of these influences are summarized in Table 6.2.

The skeletal atoms also define the *orientation* of the polymer side groups through their bond angles and number of bonds to the side groups. For example, whether there are zero, one, or two side groups linked to a particular skeletal site will affect molecular crowding and molecular flexibility.

5. Specific Influence of Different Side Groups

Side groups have a strong influence on the solubility, flammability, hydrophilicity or hydrophobicity, and reactivity of a polymer. They also control crystallinity and, through their steric and polar properties, the flexibility of the polymer chain. Some examples of side-group influences are summarized in Table 6.3.

For example, organic side groups that contain phenyl rings, fluorine, or trimethylsilyl groups confer hydrophobic properties on the solid material. Side groups with hydroxyl, sodium carboxylate, or amino units confer hydrophilicity and even solubility in water. Aromatic side groups favor solubility in aromatic solvents. Aliphatic side groups are hydrophobic and often confer solubility in organic solvents, but they are sensitive to oxidative free-radical attack, especially at high temperatures. Chlorine, bromine, or iodine in organic side groups are associated with thermo- and photodegradation, but may provide fire resistance. Side groups that bear vinyl or allyl unsaturated groups participate in crosslinking. The size and flexibility of the side groups is also an important factor. Rigid, bulky side groups like naphthyl or biphenyl units raise glass transition temperatures and introduce materials stiffness. Colored side groups carry their optical properties into the polymeric material. So, too, do liquid crystalline or nonlinear optical side groups and units with high refractive indices.

TABLE 6.3. Properties Asscociated with Different Polymer Side Groups

Side Group	Properties
CH_3, C_2H_5, C_3H_7, etc.	Hydrophobic; soluble in hydrocarbons
CF_3, CH_2CF_3, $CH_2CF_2CF_3$, etc.	Extremely hydrophobic; soluble only is special solvents
$Si(CH_3)_3$	Strongly hydrophobic
Phenyl rings	Hydrophobic; soluble in aromatic solvents
Naphthyl or biphenyl groups	Hydrophobic; high T_g
Phosphate groups	Fire resistance
Pyridine groups	Coordination to metals
NH_2	Hydrophilic; hydrogen-bonding, therefore materials rigidity
COOH	Hydrophilic
COONa	Hydrophilic and water solubilizing
OH	Hydrophilic and water solubilizing
$OCH_2CH_2OCH_2CH_2OCH_3$	Hydrophilic and water-solubilizing
⬡—SO_3H	Hydrophilic or water-solubilizing

6. Effects of Crosslinking

The formation of crosslinks between polymer chains has a striking effect on materials properties. First, it converts a soluble polymer to one that does not dissolve. If the crosslink density is low, the polymer may absorb solvent and swell to form a gel, but if heavy crosslinking is present, the material will be unaffected by solvents. Crosslinking also enhances thermal stability because more bonds must be broken to bring about decomposition and because the crosslinks may hold the ends of a broken skeletal bond close enough to recombine or "heal."

A major influence of crosslinking is to convert a gum into an elastomer. In a gum the individual chains are free to move within the material (reptate) and to slip past each other when stress is applied. But when the polymer is crosslinked, the chains are no longer free to slide past each other when the material is stretched. Any distortion produced by a stress is reversed when the stress is removed, and the polymer retracts to its original shape and dimensions. In its natural state rubber is a permanently extensible gum, but the introduction of crosslinks turns it into the material that is familiar as a rubber band or a bicycle or automobile tire.

D. POLYMERS IN THE SOLID STATE

1. Chain Entanglement

Unless the polymer chains in a solid occupy a regular extended conformation and become organized into crystallites, they will assume random conformations that inevitably result in chain entanglement. Chain entanglement underlies many of the valuable properties of solid polymers because it reduces the tendency for chains to be drawn past each other when stress is applied and provides a "shock absorber" network when the material is subjected to an impact. Entanglement in solution generates high viscosities. In general, those polymers with the lowest barriers to backbone torsion are the ones with the highest tendency to form random conformations and are most likely to become entangled with their neighbors. Entanglement can become so pervasive that the chains become tied in knots and are difficult to separate at a later time. Knots may retard the dissolution of a polymer in what otherwise would be a good solvent. Highly entangled, lightly crosslinked, or knotted networks are the basis of a number of important rubbery elastomers.

2. Crystallinity

An uncrosslinked amorphous linear or branched polymer at temperatures above its T_g is a viscoelastic solid or a gum, which means that, when stress is applied, the polymer chains will unwind and slide past each other, and the material will eventually flow. For many engineering applications, viscoelastic behavior is a bad thing, and this can be avoided if the polymer contains crystallites. As discussed above, the presence of microcrystallites or crosslinks will stiffen the material, lower the tendency to flow under stress, and perhaps generate elasticity.

Linear polymers that have a symmetric arrangement of side groups on each side of the chain and/or a rigid backbone structure may have a tendency to assume extended chain conformations in the solid state, and these favor the formation of microcrystallites (Figure 6.5).

Microcrystallites are held together by weak van der Waals attractions or sometimes by ionic or hydrogen bonding forces. Microcrystallites serve as temporary crosslinks. In other words, they stiffen the material, reduce its flexibility, and prevent the chains from being pulled past each other when the material is subjected to tensile stress. High-density polyethylene is microcrystalline, as are isotactic polypropylene, Nylon 66, poly(ethylene oxide), several polyphosphazenes, and numerous other polymers. Note that only rarely does the crystallinity extend throughout the material. It is much more common for the microcrystallites to be separated from each other by regions of random coil amorphous polymer, which provide impact resistance. A major advantage for the use of microcrystallites rather than covalent linkages as crosslink sites is the fact that crystallites can be melted by raising the temperature. Thus, when heated above the melting temperature, the polymer can be fabricated by melt extrusion or injection molding. When subsequently allowed to cool, the polymer will crystallize to a dimensionally stable material. Side groups that favor crystallinity include small organic units such as methyl groups or rigid moieties such as phenyl or phenoxy groups. Highly flexible side chains, such as oligoethyleneoxy or branched alkyl groups, tend to prevent crystallization. Random copolymers rarely crystallize.

3. Liquid Crystallinity

Some polymers also have *liquid crystalline* character within a limited temperature range. In the liquid crystalline or mesogenic state, which occurs above the glass transition temperature but below the final melting point, there are regions of partial order that may allow pressure fabrication but that increase the viscosity compared to the truly molten state.

Two types of liquid crystalline polymers are known—those in which the backbone structure is responsible for the liquid crystalline properties, and those where mesogenic side groups (see Chapter 5) give rise to the liquid crystalline properties. A representation of a main-chain liquid crystalline polymer is given in Figure 6.6a.

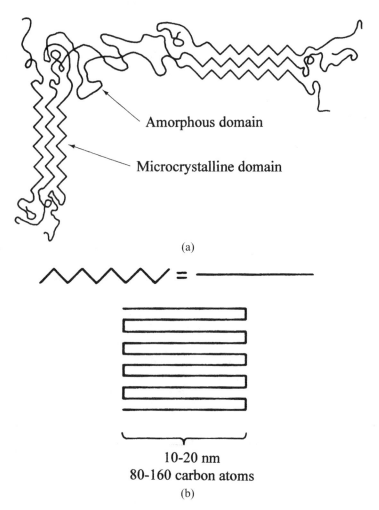

(a)

(b)

10-20 nm
80-160 carbon atoms

Figure 6.5. *Microcrystallites are formed in a polymer when sections of the chains become coaligned, as shown in (a) or bend back and forth to generate larger structures as shown in (b). In both cases the crystalline regions serve as stiffening domains. Crystallite formation can be encouraged by stretching or annealing (raising and lowering the temperature).*

Here the rigidity of segments in the chain causes adjacent chains to become coaligned within a specific temperature range. The effect on bulk physical properties is often to provide increased viscosity in the quasiliquid state, and to confer increased strength due to macromolecular alignment when the polymer is cooled below the liquid crystalline transition. Specifically, such polymers can often be fabricated under viscous flow conditions near the T_{lc} to give high-strength fibers that contain highly oriented polymer molecules.

Side-chain liquid crystalline polymers (Figure 6.6b) have the mesogens in the side groups, connected to a flexible main chain via flexible spacer groups. The flexibility of both the main-chain and the spacer groups provides the freedom for the mesogenic side units to undergo self-assembly. Rigid main chains or the absence of spacer groups allows the main chain to dominate the solid-state behavior to the detriment

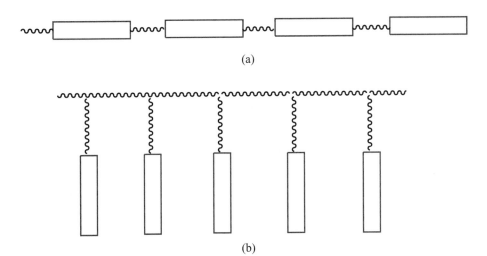

Figure 6.6. *Crystallinity in polymers can arise from (a) rigid segments in the main chain, or (b) rigid units in the ride groups.*

of liquid crystallinity. As mentioned in Chapter 14, some mesogenic small molecules that are not liquid crystalline become liquid crystalline when linked by a flexible spacer to a flexible polymer. The polymer chain and spacer groups provide just enough restriction of movement in the liquid state to permit the mesogenic side groups to become stacked or coaligned.

E. FABRICATION OF POLYMERS

The conversion of polymers into shaped articles and devices is a major component of industrial materials technology. Many different methods have been developed and the following techniques illustrate some of the options.

1. Solution Casting of Films

This is an excellent method for the laboratory preparation of films. The polymer is first dissolved in a suitable solvent, and the solution is spread out on a flat horizontal surface such as a piece of plate glass or a sheet of poly(tetrafluoroethylene) (Figure 6.7). The solvent is then allowed to evaporate slowly to leave a coherent film, which can be peeled from the substrate. The final film thickness is controlled in one of three ways. (1) the depth of the solution on the flat substrate may be controlled by use of a "knife" or a glass rod with cylindrical spacers on each end that is rolled over the solution "Dikes" surrounding the solution may also allow some control of its depth; (2) the more concentrated the solution, the thicker will be the final film; or (3) a "spin-caster," which consists of a rapidly rotating disk onto which is dripped a solution of the polymer, may be employed. Centrifugal forces spread the solution from near the center of the disk to the perimeter as the solvent evaporates. The process can be repeated until the required film thickness is obtained. This method is employed in semiconductor integrated fabrication procedures.

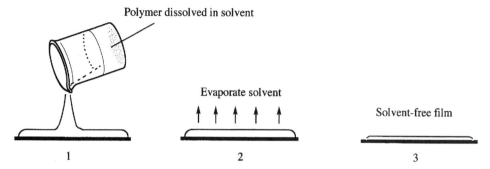

Polymer dissolved in solvent

Evaporate solvent

Solvent-free film

1   2   3

Figure 6.7. *Solution casting of polymer films. The thickness of the final film depends on the concentration of polymer in the solution and the depth of solution on the casting surface. Complete removal of the solvent often requires heating or exposure of the film to a vacuum.*

Note that, for most of these techniques, rapid solvent evaporation should be avoided since this could lead to *precipitation* of the polymer as a particulate material due to cooling or water condensation, rather than formation of a coherent film.

2. Melt-Fabrication of Films

This is a high-volume technique for film manufacture. The molten polymer is extruded through a series of polished metal rollers, sometimes oriented by stretching, and then cooled before being wound onto a roller.

3. Fabrication of Fibers

The conversion of polymers into fibers is a major component of both research and polymer technology. It is accomplished in the following ways. First, the polymer may be melted and extruded through a perforated disk (a spinerette). The extruded fibers are cooled and stretched on a roller system to orient them and increase their strength. Alternatively, a concentrated solution of the polymer may be extruded through a spinerette and the solvent removed by a stream of hot air, or the solution is extruded into a nonsolvent where the fibers coagulate as solid fibers. These, too, are subsequently stretched to orient the polymer chains. A less common but increasingly important laboratory technique is the process of *electrostatic spinning*, in which a globule of polymer solution is extruded from the end of a hypodermic needle and is induced by means of a multikilovolt electric field to form a fiber that whips out and deposites a nanofiber mat on a grounded target (Figure 6.8). The electrostatic potential causes a rapid stretching of the fiber while solvent is evaporating and this is responsible for the resultant reduction in fiber diameter often to the nanometer level.

4. Injection Molding

Injection molding is a process in which a molten polymer is forced into a mold, cooled below its liquefaction temperature, and the formed solid is then released by opening of the mold into two halves. This is a rapid, mass-production method of fabrication that is widely used for thermoplastics. It requires a low melt viscosity,

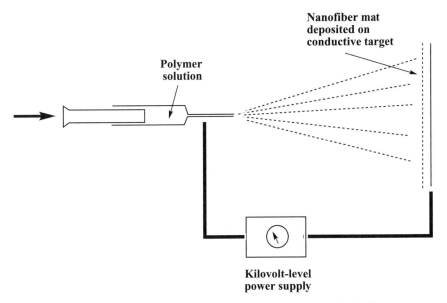

Figure 6.8. *Fabrication of polymer nanofibers by electrostatic spinning.*

polymer, and this means that the molecular weight of the polymer must not be excessively high. Obviously, it cannot be used for polymers that decompose near their melting points, although it can be employed for polymers that crosslink within the mold.

5. Thermoforming

Thermoforming is a low-volume process used to fabricate large objects like boats or housings. A sheet of polymer is softened by heating and pressed into a mold where it cools and retains the shape of the mold.

6. Blow Molding

This is a method employed for the fabrication of bottles. A tube of semimolten polymer is forced under internal air pressure to expand and cool inside a mold.

7. Sintering

Sintering is a process in which powder particles are compressed in a mold and heated below the melting point. This process fuses the surfaces of the particles into a solid mass. It is used mainly for the fabrication of high-melting-melting polymers such as Teflon. Sintering is a major process for the fabrication of ceramics (see Chapter 7).

8. Polymerization Combined with Fabrication

Polymer windows are produced by polymerization of the liquid monomer between two parallel glass plates. Poly(methyl methacrylate) (Plexiglas) windows are produced in this way.

F. EXAMPLE POLYMERIC MATERIALS

Monomers and polymers described here are listed in Table 6.1.

1. Polymers Produced by Addition Reactions

a. Polyethylene. Polyethylene is one of the most widely used polymers in technology. It exists in two forms: (1) branched, low-density polyethylene (T_g −125 °C) and the linear, high density, high crystallinity form (T_g −80 °C, T_m 137 °C). There is also a 100% crystalline form produced by special precipitation/orientation techniques, but this is not widely available. Low-density polyethylene is produced via high-pressure free-radical methods, whereas the high-density modification is obtained by Ziegler–Natta or metallocene polymerizations.

The low-density form was first manufactured during World War II in the United Kingdom as an insulator for radar applications. It is currently produced on a very large scale for conversion into films, containers, toys, and a wide variety of household and technological articles. Its attributes include low cost, toughness, ease of thermal fabrication, excellent electrical insulation properties, resistance to microorganisms, insolubility in many solvents, moderately good transparency, and resistance to attack by water, acids, and alkalis. Its main disadvantages are a relatively low softening point, the swelling or solubility in some hot organic solvents, and a tendency to undergo photochemical decomposition when exposed for long periods of time to sunlight.

The high-density form is tougher, is more rigid and opaque, and has a higher softening point than does the amorphous version, but it is sometimes slightly colored by the Ziegler or metallocene catalyst residues that are difficult to remove. It has the properties needed for the production of monofilament fibers, pipes, and bottles. Polyethylene fibers are an alternative to Kevlar™ or ® for the production of bullet-proof vests. The molecular weights of the high density form can range from 200,000 to 3–6 million, with the latter being most useful for fibers. The solid-state structure of high-density polyethylene consists of closely packed planar zigzag chains that bend back and reenter each crystallite as shown in Figure 6.5. The 100% crystalline form has planar zigzag chains that extend along the length of each fiber.

b. Poly(tetrafluoroethylene) (PTFE or Teflon™ or ®). This polymer is produced commercially by the free-radical emulsion polymerization of tetrafluoroethylene. It was discovered by Plunkett in 1962 when he cut open a seemingly empty cylinder of tetrafluoroethylene to find out why it had failed to deliver the full amount of gas. The cylinder was partly filled with powdered PTFE. The polymer is highly crystalline, with a melting point at 327 °C. The T_g has been variously reported as 130 °C or −113 °C. Thermal decomposition begins at 250–350 °C, which is why melt fabrication is difficult. However, investigators have reported that short nanofibers can be produced by extrusion at very high pressures, especially when the polymer is "plasticized" by nitrogen or carbon dioxide.

The presence of fluorine atoms has a dramatic effect on the properties compared to polyethylene. Solid PTFE has a relatively high density near 2.3 g/cm³. It has excellent lubricity and is, therefore, a valuable material for bearings and nonadhesive surfaces. It is also highly resistant to photochemical decomposition. PTFE is one of the most hydrophobic materials known, and this property underlies its use as a biomedical material. Because of its insolubility and high melting point, PTFE is difficult to fabricate. Films or blocks of this polymer may be obtained by sintering techniques in which the powdered material produced by the emulsion synthesis is heated in a mold under pressure. The polymer is insoluble in all solvents at room temperature, but it is softened by and will even dissolve in hot perfluorokerosene at temperatures near 300 °C. Such swollen or dissolved materials are fabricated into films and fibers. When the solvent is removed under appropriate conditions, the polymer is left as a porous material known by the trade name, Gore-Tex®.

c. Polystyrene. Polystyrene is the archetypical multipurpose glassy polymer. It is usually produced by free-radical techniques, and can be melt-fabricated into transparent but brittle, inexpensive plastic lenses for automobile lights, toys, picnic cups and drinking glasses, computer housings, food containers, and housings for videotapes, compact disks, and electrical devices. This material is amorphous and has a T_g near 100 °C. Hence, it can be melt-processed above this temperature. A second modification is expanded polystyrene foam, which is the material used in thermally insulated coffee cups, lightweight "peanuts" used in packaging, and molded packaging such as the impact-absorbing material used for protecting computers during shipping. Expanded polystyrene is made by allowing pentane or carbon dioxide to dissolve in the polymer and then, by heat and reduced pressure, permitting the solvent or "blowing agent" to expand to form bubbles. After removal of the solvent (and recycling of the pentane) 95% of the material is air—hence the lightweight character.

A crystalline form of polystyrene (syndiotactic) is produced by metallocene catalysis, but this material is much more expensive than the normal atactic form. However, it has the advantage of a higher T_g (270 °C) and therefore a higher use temperature. Note also that polystyrene is a component of polystyrene–polybutadiene (SBR) graft copolymers—high-impact materials that are phase-separated into rubbery and rigid domains.

d. Poly(methyl methacrylate) (PMMA). This transparent, water-white glassy polymer has been produced commercially since the 1930s. It can be fabricated by the free-radical, bulk polymerization of methyl methacrylate monomer within a mold. Thus, manufactured objects, especially sheets of polymer for window glass, can be shaped *during* the polymerization reaction. Alternatively, the polymer is fabricated by injection molding or thermoforming. Its T_g is near 100 °C, and it is amorphous. Under the name of "acrylic," this material is widely employed as a lightweight shatter-resistant alternative to silicate glass in eyeglasses and camera lenses, and "light pipes" for illuminated signs. It is also a preferred material for applications as diverse as dentures, shower stalls and bathtubs, and molded components for automobiles.

2. Polyurethanes

Polyurethanes are widely used for applications that range from foam elastomers to bowling balls. The basic structure is shown in Table 6.1. They are formed by the addition reactions that take place when hydroxyl groups at the ends of an oligomeric or polymeric diol add across the terminal N=C double bonds of an oligomeric or monomeric diisocyanate. Many different types of polyurethanes are known, depending on the organic units that lie between the functional groups. These units can be aliphatic (for flexibility) or aromatic (for stiffness). The assembly reactions often use organotin compounds as catalysts and tertiary amines for crosslinking.

3. Polymers Produced by Condensation Reactions

a. Poly(ethylene terepthhalate) (PET, Dacron®, Terylene®, Mylar®). This polyester is a thermoplastic that is stable to aqueous acids but unstable to strong base. The polymer is synthesized by the condensation polymerization of ethylene glycol and terephthalic acid. The product is either an amorphous or a semicrystalline material with a T_g of 76 °C and a T_m at 250 °C. It is widely used as a melt-spun textile fiber and as a material for food and beverage containers. Melt-fabricated films of the same polymer are employed as a base for photographic film, audio/videotapes, and in packaging. The films and fibers have a relatively high refractive index of 1.58–1.64 because of the presence of the aryl rings.

b. Poly(hexamethylene adipamide) (Nylon 66). This was one of the first synthetic polymers to be commercialized. Indeed, it was the demonstration of the feasibility of this and similar reactions by Carothers and coworkers in the 1940s that provided the first proof that synthetic polymer molecules could be produced. The overall process is summarized in Table 6.1 for the classical reaction of hexamethylenediamine with adipic acid in the molten state. Water must be removed from the reaction mixture to ensure high molecular weights. The molten polymer is drawn into fibers and stretched to give a highly crystalline material with high tensile strength, which is a requirement for strong textile fibers. Many variations on this synthesis are known with different diamines and different dicarboxylic acids to give polymers with a wide range of properties. Polyamides are also used to fabricate bulk objects by injection molding. Note that nylons can also be prepared via the ring-opening polymerization of cyclic amides (see discussion later).

c. Poly(p-phenylene terephthalamide) (Kevlar®). This polymer is produced by the condensation polymerization of phthalic acid with *p*-phenylenediamine (Table 6.1). The melting temperature of the polymer is around 500 °C, which precludes the melt spinning of fibers. It is insoluble in nearly all solvents and is spun into textile fibers from a solution in concentrated sulfuric acid! The high strength of these fibers is partly a consequence of both the high level of crystallinity and the aromatic structure of the backbone. Kevlar fibers are used in bulletproof vests; in composite materials for tennis racquets, golf clubs, kayaks, helmets, and skis; and as high-strength ropes. A related fiber, Nomex®, is made from *m*-phenylenediamine instead of the para-isomer. This introduces a bent structure into the backbone, which reduces crystallinity and makes the material easier to process.

d. Synthetic Polypeptides. Many amino acids have the 1:1 ratio of functional groups that is needed for the formation of high polymers. The synthesis of polyamides via the enzyme-induced condensation of amino acids is one of the fundamental processes that take place in living cells. However, synthetic polymers can also be produced from amino acids by condensation reactions in the laboratory, but the existence of side reactions, and the thermal instability of many natural amino acids, creates problems in large-scale syntheses. Most polypeptide copolymers of biological interest are produced by complex stepwise protection–deprotection–condensation reactions, with one end of the polymer anchored to a solid substrate, a process known as *Merrifield synthesis*. Synthetic polypeptides derived from only one monomer tend to be highly crystalline and only poorly soluble in many solvents.

e. Polyimides. These are aromatic ring-based polymers specially designed for challenging high-temperature applications. A typical synthesis involves the reaction of phthalic dianhydride with *p*-phenylenediamine at elevated temperatures (Table 6.1). Water is removed as the condensation takes place. Typically the condensation takes place in two stages, the first to form amide bonds, and the second (above 300 °C) to form the cyclic imide structure. The intermediate products formed in the first step are still liquid at high temperatures and can be fabricated into fibers, films, or laminates by melt processing. After the second step the polymers become insoluble and infusible; hence fabrication is no longer possible at the final stage.

f. Polycarbonates. Polycarbonates can be viewed as esters produced from diols and carbonic acid [HOC(O)OH], although they are not made directly from this acid. Instead they can be prepared either via the reactions of aromatic diols with phosgene (ClC(O)Cl) or of a diol with diphenyl carbonate (Table 6.1). Lexan® plastics, Nalgene® bottles, and other food containers are manufactured from poly-carbonates made from Bisphenol-A. Polycarbonates are used as "engineering poly-mers", ie. for melt-fabrication into clear, transparent, high refractive index (n = 1.58), impact-resistant solid objects. Laminated sheets are used as bullet-resistant windows.

g. Polysulfones. Another high temperature engineering polymer system based on Bisphenol-A results from the reactions of this compound with 4,4′-dichlorophenyl sulfone (Table 6.1). Polysulfones are transparent solids, have high strength, possess good electrical insulation properties, have high softening temperatures, and are somewhat resistant to combustion. They are fabricated by injection molding or extrusion, and can be reinforced with glass fiber.

h. Polyether Ketones (PEK) and Polyether Ether Ketones (PEEK). This is another class of engineering condensation polymers produced, in this case, by the elimination of metal fluoride from a diol and a difluoro monomer (Table 6.1). These polymers are prized for their strength and thermal stability.

i. Epoxy Polymers. These are colorless, transparent, crosslinked materials produced by the dehydrohalogenation reactions of Bisphenol A (see earlier) with epichlorohydrin (Table 6.1). The terminal epoxy groups may then serve as ring-opening crosslinking units through reactions with pendent or terminal hydroxyl units in the presence of "curing agents". Epoxy resins are well-known adhesives, components of glass or carbon-fiber-reinforced laminates, or as resins for the fabrication of bulk objects.

j. Condensation Resins. Many crosslinked polymer resins, such as ureaformaldehyde or melamine–formaldehyde resins, are produced by condensation processes. For example, melamine–formaldehyde resins (used in heat-resistant countertops, plastic cups, and dinnerware) are made by allowing melamine to react with formaldehyde to form a methylol derivative. Subsequent heating drives off water and yields a rigid three-dimensional (3D) construct that resembles a ceramic rather than a conventional polymer.

4. Polymers Produced by Ring-Opening Polymerizations

a. Poly(lactic-glycolic acid) (PLGA). Glycolides and lactides are cyclic dimers formed by condensation of glycolic and lactic acids. Polymers can be produced from the monomeric hydroxy acids by condensation methods, but ring-opening polymerization of the cyclic dimers gives cleaner reactions and higher molecular weights. The starting materials and products are shown in Table 6.1. Poly(glycolic acid) and poly-D,L-(lactic acid) homopolymers are produced by the ring-opening polymerization of the appropriate glycolide or lactide. The copolymers are prepared by allowing both cyclic monomers to polymerize together in the same reaction vessel in the presence of tin(II) 2-ethylhexanoate catalyst. This allows the ratios of the two components in the polymer to be controlled by the ratio of the two cyclic dimers in the mixture.

This is a polymer system that would normally be rejected for any practical application because the macromolecules are hydrolytically unstable. Hydrolytic breakdown results in the conversion of the polyester to the two hydroxy acid monomers, and these lack the strength and cohesion that is typical of a polymer. However, two applications exist for which hydrolytic instability is a requirement. These are in the fields of bioabsorbable polymers in medicine, and for uses in packaging that will degrade in the environment. Poly(glycolic acid) hydrolyzes at a faster rate than does poly(lactic acid). The hydrophobic α-methyl group in poly(lactic acid) retards attack by water on the backbone ester linkage. Hence the rate of degradation of the copolymer is controlled by the composition ratio. A copolymer with 90:10 PGA:PLA degrades in aqueous media in 9–20 days, depending on pH and temperature. Crystalline poly-L-lactic acid (mp 170 °C) hydrolyzes at a slower rate, with lifetimes up to 1.5 years reported. These polymers are fabricated into fibers by melt-extrusion and drawing techniques, and into shaped devices such as surgical clips and clamps by injection molding.

For biomedical applications [surgical sutures, clips, staples, or tissue regeneration mats (see Chapter 16)] the hydrolysis products (glycolic and lactic acids) are chemically compatible with mammalian tissues and are soluble in water. However, they are also acidic, and this is detrimental to efficient wound healing. For environmentally acceptable packaging, such as garbage bags, this is not a problem, and strong commercial efforts are being made to develop technology based on poly(lactic acid).

b. Polycaprolactam (Nylon 6). Caprolactam is a cyclic amide (Table 6.1), which, when initiated with a strong base or water, undergoes ring-opening polymerization to Nylon 6. The mechanism of this polymerization depends on the initiator. Initiation by water may involve cleavage of a ring followed by attack by the resultant amino acid on another ring molecule, and so on.

c. Polytetrahydrofuran. Another example of a ring-opening reaction is the polymerization of the common solvent, tetrahydrofuran (THF) to poly(tetramethylene oxide), by cationic initiators such as phosphorus trichloride or pentachloride to give poly(tetramethylene oxide). Apparently there is enough ring strain in THF to favor ring-opening polymerization, whereas cyclic ethers with larger rings may be less reactive.

d. Poly(ethylene oxide) (PEO). Poly(ethylene oxide) is a water-soluble polymer produced by the ring-opening polymerization of ethylene oxide (Table 6.1). The polymer has widespread uses as a viscosity enhancer for aqueous media. It has FDA approval for some biomedical uses, and has been investigated in detail for surface grafting to other polymers and for conversion to hydrogels. It has a T_g at $-67\,°C$, and is microcrystalline, with a T_m near $80\,°C$. The solid polymer has been investigated intensively as a solid solvent for lithium salts for uses in solid, rechargeable lithium batteries (see Chapter 11).

e. Polynorbornenes. Other ring-opening reactions include the polymerization of unsaturated cyclic hydrocarbons such as norbornene or cyclopentene to give polymer with double bonds in the backbone. These reactions occur under the influence of organometallic catalysts such as structures **6.7–6.9**, and are known as *ring-opening metathesis polymerizations* (ROMPs) (reaction 20).

f. Poly(dimethylsiloxane) (PDMS, Silicone Rubber). This was the first hybrid inorganic–organic polymer to be synthesized and developed commercially. It is produced by the catalyzed, thermal ring-opening polymerization of a cyclic tetramer, octamethylcyclotetrasiloxane (Table 6.1), which itself is one of the products formed by the hydrolysis of dimethyldichlorosilane. The molecular weights can be extremely high ($>1 \times 10^6$) but with very broad distributions. The average chain lengths are maximized by termination of the polymerization before thermodynamic equilibrium is reached (see earlier), and control of the molecular weight is also achieved by the addition of chain transfer agents such as $Me_3SiOSiMe_3$. Crosslinking, which is essential for elastomeric properties, is achieved via free-radical methods. Alternatively, crosslinking is accomplished by the addition of siloxane oligomers that bear silicon–hydrogen bonds to a PDMS that has a few vinyl side groups per chain. The presence of a platinum catalyst causes the Si—H bonds to add across the $CH{=}CH_2$ bonds to form crosslinks. Side groups other than methyl units are introduced via copolymerization with cyclosiloxanes that bear other side groups such as fluoroalkyl, phenyl, vinyl, or allyl. Hydrolytically sensitive side groups provide crosslinking when the polymer is exposed to a moist atmosphere.

Dimethylsilicone elastomers have very low glass transition temperatures ($-130\,°C$) and low crystalline melting points ($\sim{-}30\,°C$). They are also stable in the atmosphere up to temperatures near $200\,°C$. However, silicones are flammable. PDMS elastomers are used in biomedical materials such as contact lenses or catheters, and for nonmedical applications as seals and O-rings, hydrophobic surfaces, and caulking materials. Because silicone polymers are more expensive than most other elastomers, they are often employed when no other materials can provide the combination of water repellency, low glass transition temperature, and heat stability required for high-performance applications. Low-molecular-weight polymers or oligomers are "silicone oils."

g. Polyphosphazenes (Ring-Opening Polymerization Followed by Macromolecular Substitution).

Polyphosphazenes provide the main example of an entirely different method of polymer synthesis. Whereas the side groups of most polymers are introduced at the monomer stage and are then carried through the polymerization process, those of polyphosphazenes are introduced after polymerization has taken place. This allows a much wider range of side groups to be utilized. The synthesis process is summarized in reactions 26–28. Thus, several hundred different macromolecules exist, with different side groups and different combinations of two or more different side groups. Different side groups are employed to control high or low glass transition temperatures, solubility, optical properties, and numerous other characteristics. In addition to linear polymers, other molecular architectures are known, including branched, dendritic, cyclolinear, and comb structures as well as block copolymers with organic polymers and polysiloxanes. Just three representative linear polyphosphazenes are described below to illustrate the range of properties that are accessible:

$$\tag{26}$$

$$\tag{27}$$

$$\tag{28}$$

1. If all the side groups are trifluoroethoxy units (reaction 26), the polymer is a white or opalescent crystalline thermoplastic that readily forms nonflammable films, coatings, and fibers by solution casting or solution spinning techniques. The T_g ranges from −66 to −73 °C, and the T_m is 234 °C. This material is radiation-resistant and highly hydrophobic (contact angle > 100°). It has similar properties to Teflon, but is far easier to fabricate. However, if the trifluoroethoxy groups are progressively replaced by octafluoropentoxy units to give $[NP(OCH_2CF_3)x(OCH_2CF_2CF_2CF_2 CF_2H)y]n$, the polymers become amorphous elastomers once 13–75% of the larger side groups are present. This occurs because the larger side groups disrupt the solid-state order and prevent crystallization. These elastomers are resistant to hot oils,

hydraulic fluids, and fuels; remain flexible and elastomeric at temperatures down to -60 or $-70\,°C$; are thermally stable up to $250\,°C$; and have unusual vibration damping characteristics. They have been used for the manufacture of O-rings and seals for aerospace and automotive applications, as shock absorbers, as radiation-resistant coatings, and as nonthrombogenic biomedical materials.

2. By contrast, if the side groups are alkyl ether units such as $-OCH_2CH_2OCH_2CH_2OCH_3$ (reaction 27), the polymers become water-soluble. When radiation-crosslinked, they give rise to hydrogels (see Chapter 16). If $-OC_6H_4COOH$ groups are also present as cosubstituents, the gels expand and contract with pH changes to serve as responsive membranes. In the absence of water the alklyl ether–substituted polymers are solid ionic conductors for use in rechargeable lithium batteries (see Chapter 12).

3. The linkage of amino groups to the skeleton (reaction 28) gives another range of polymers, some of which are bioerodible and useful in medical applications (Chapter 16).

h. Polysilanes (Condensation and Ring-Opening Polymerizations).

Short-chain and cyclic oligomeric silanes (Table 6.1) are accessible through the condensation reactions of R_2SiCl_2 with molten sodium or potassium in a suitable solvent. However, polymers are obtained mainly by the thermal ring-opening polymerization of cyclosilanes. These polymers, with an all-silicon backbone, have several unique characteristics. They are hole transport semiconductors (see Chapter 10), and they undergo backbone cleavage when irradiated with ultraviolet light. The different types of organic side groups along the silicon chain disrupt crystallinity and favor solubility in organic solvents. As discussed in Chapter 7, poly(dimethylsilane) is converted to silicon carbide when heated at high temperatures.

i. Polyferrocenophanes.

Polyferrocenophanes (Table 6.1) illustrate one of the possibilities for the incorporation of transition metals into a polymer backbone. Most of the known polymers that bear the highly stable ferrocene unit in the backbone are produced by thermal or anionic ring-opening polymerization. The T_g values for these polymers are surprisingly low, with values that range from $33\,°C$ when both side groups are methyl units to $-51\,°C$ when two hexoxy side groups are present per repeat unit. This flexibility is attributed to the ease of torsion of the cyclopentadienyl rings around the iron atoms. These are known as "swivel groups." Doping with iodine converts them from insulators into semiconductors. Pyrolysis of these polymers generates ceramics that contain magnetic iron particles. Block copolymers with poly(dimethylsiloxane) form cylindrical micelles. Although no commercial products have yet been reported, this system provides a further illustration of the materials-related possibilities for polymers with inorganic elements in the main chain.

j. Poly(sulfur nitride) (Polythiazyl).

Finally, a very specialized example of ring-opening polymerization is provided by the unusual behavior of the all-inorganic poly(sulfur nitride system). Here, the unstable cyclic tetramer $(S=N)_4$ is heated in a vacuum to form the heavily strained cyclic dimer $(S=N)_2$, which condenses on a cooled surface and then polymerizes in the solid state to give the gold-colored, metal-like polymer $(S=N)_n$. All three products are susceptible to spontaneous detonation, but the polymer occupies the unique position of being the first polymeric electronic conductor and (when cooled to near absolute zero) the first polymeric superconductor (Chapters 10 and 11).

G. FUTURE CHALLENGES IN POLYMERIC MATERIALS SCIENCE

Hundreds of different polymers are now known, with a wide range of properties and uses. Yet, there are many potential applications for which no polymeric materials are available with the required combination of properties. High-temperature stability is a continuing problem for polymeric materials either because of low melting points or low glass transition temperatures or because of thermooxidative degradation or depolymerization. This is a serious challenge for aerospace and automotive applications, and it is a reason for the expansion of interest in hybrid materials that incorporate inorganic elements either into the molecular structure of polymers or in composite materials.

A need also exists for polymers that have better electronic and ionic electrical conductivity than do existing materials for uses in light-emitting devices, batteries, fuel cells, and solar cells (see Chapters 10 and 12). At the opposite end of the scale, new polymers are required with low dielectric characteristics for uses in advanced integrated circuits. Very few polymers that are specially designed for use as biomedical materials are currently available, and this will be a productive area for future research (Chapter 16).

In addition to these challenges, a need exists for better polymerization initiators to allow superior control over polymer architecture—especially for stereospecific addition and ring-opening polymerizations and for the polymerization of functional (i.e., reactive) monomers. At the polymer morphology level, the increasing trend toward miniaturization of devices provides opportunities for new methods for polymer fabrication at the nanoscale.

Another area of growing importance is the development of "shape memory" polymers that are fabricated in one shape but change to another shape after mild heating. Uses for these materials in sensors, packaging, and in biomedicine are increasingly important. All the signs point to a continuing expansion of polymer science to meet challenges in nearly all areas of materials technology.

H. SUGGESTIONS FOR FURTHER READING

1. Allcock, H. R.; Lampe, F. W.; Mark, J. E., *Contemporary Polymer Chemistry*, 3rd ed., Pearson/Prentice-Hall, Upper Saddle River NJ, 2003.

2. Mark, J. E.; Allcock, H. R.; West, R., *Inorganic Polymers*, Prentice-Hall, 1992, Englewood Cliffs, NJ, 1992; 2nd ed., Oxford University Press, 2005.

3. Allcock, H. R., *Chemistry and Applications of Polyphosphazenes*, Wiley-Interscience, Hoboken, NJ, 2003.

4. Flory, P. J., *Principles of Polymer Chemistry*, Cornell University Press, 1969 (the classical original text on polymer science).

5. Odian, G., *Principles of Polymerization*, 2nd ed. (and later editions), Wiley-Interscience, New York, 1981.

6. Billmeyer, F. W., *Textbook of Polymer Science*, 3rd ed., Wiley-Interscience, New York, 1984.

7. Mark, J. E.; Ngai, K., Graessley, W, Mandelkern, L., Samulski, E., Koenig, J., Wignall, G., *Physical Properties of Polymers*, 3rd ed., Cambridge University Press, Cambridge, UK, 2004.

8. Mark, H. F.; Bikales, N. M.; Overberger, G. C.; Menges, G., *Encyclopedia of Polymer Science and Engineering* (16 vols.), Wiley-Interscience, New York, 1985–1989.

9. Matyjaszewski, K., *Controlled Radical Polymerization*, American Chemical Society, Washington, DC, 1998.

10. Braun, D.; Cherdron, H.; Rehahn, M.; Ritter, H.; Voit, B., *Polymer Synthesis: Theory and Practice*, 4th ed., Springer, 2005.

11. Naka, K.; Nakahashi, A.; Chujo, Y. "Periodic terpolymerization of cyclooligoarsine, cyclo-oligostibine, and acetylenic compound," *Macromolecules* **40**:1372–1376 (2007).

I. STUDY QUESTIONS (for class discussions or essays)

1. Refer to Table 6.1 and write down the overall reactions and reaction conditions that lead from monomer to polymer for each entry, including catalysts that may be needed.

2. Compare and contrast the advantages and challenges of the different methods for the conversion of small molecules to polymers, paying special attention to the different types of initiators that are used, and the reaction temperature ranges that are most effective.

3. How can polymers be recycled after they are used and discarded? Give examples chosen from addition, condensation, and ring-opening polymerization materials. Are there any polymer types that cannot be recycled?

4. A modern ocean-going racing yacht uses polymers for many key components. Assume that you have been retained by a yacht designer to suggest new materials that will bring about major improvements in performance. Write down your recommendations stating why each change would be a significant improvement. Assume that cost is not a serious limitation.

5. It is sometimes erroneously assumed that the higher the molecular weight of a polymer, the better is its materials performance. Discuss applications where this may not be valid.

6. Polymers that contain silicon or phosphorus in the backbone account for the largest number of hybrid inorganic–organic polymers currently known. Speculate about future developments in this area, and mention which other elements might offer significant property advantages.

7. The two main methods available to bring about molecular structural changes in polymers are (a) using a different monomer or (b) carrying out substitution reactions on a preformed polymer. Discuss the advantages and disadvantages of each method. Give examples.

8. The same addition polymer produced by free-radical, anionic, or Ziegler initiation may have entirely different properties. Explain this.

9. You are a member of a team that is designing a new supersonic airliner. Your task is to choose or find a way to synthesize the material to be used in the cockpit windows. Write a brief proposal describing the design parameters, specific materials properties needed, and what your recommendations are and why.

10. Why are some polymers more stable to heat, light, or high-energy radiation than others? Use specific examples to discuss some of the factors that underlie the differences.

7

Glasses and Ceramics

A. OVERVIEW

The technology of glass and ceramics is one of the oldest human activities. The fabrication of food and liquid storage containers from clay and glass, and the production of arrowheads and other weapons from flint and other minerals, marks the beginning of technological skill in the human race. Glass and ceramics remain an important component of technology today, and large numbers of the applications on which we depend for modern living utilize ceramics in one form or another. For the most part, ceramics are totally inorganic materials, typically high-melting, often heavily crosslinked, and relatively intractable once they are fabricated.

In contrast to classical chemistry, mineralogical and ceramics science uses a special way to describe molecular and materials structures. It is based on the concept of connected tetrahedra, octahedra, and other structures through edges or corners rather than by drawing specific bonds between the participating elements. It allows a visual simplification of structures without loss of detailed information. The method works as follows. Elements such as silicon, which use four tetrahedrally oriented bonds to connect to other atoms, are represented by the center of a tetrahedron, with the four atoms to which the silicon is connected represented by the corners of the tetrahedron (Figure 7.1). Thus, rings and chains of alternating silicon and oxygen atoms can be depicted as a series of tetrahedra connected through their corners, with the remaining two corners at each silicon being the location of side groups attached to silicon (see the examples in Figures 7.2 and 7.3). Ladder silicate structures are then represented by two chains of tetrahedra connected through corners (Figure 7.4). Sheet structures or three-dimensional arrangements are shown as tetrahedra connected through edges (i.e., utilizing two oxygens linked to each silicon).

Introduction to Materials Chemistry, by Harry R. Allcock
Copyright © 2008 by John Wiley & Sons, Inc.

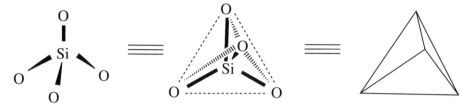

Figure 7.1. The tetrahedral orientation of bonds to silicon, and the use of a tetrahedron as a building block in the depiction of silicate structures.

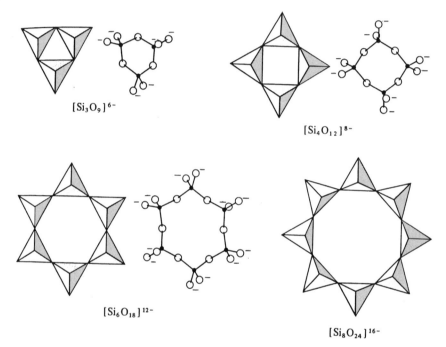

$[Si_3O_9]^{6-}$

$[Si_4O_{12}]^{8-}$

$[Si_6O_{18}]^{12-}$

$[Si_8O_{24}]^{16-}$

Figure 7.2. Illustration of the use of tetrahedra to describe cyclic metasilicate molecules. (From Greenwood, N. N.; Earnshaw, A., Chemistry of the Elements, *Pergamon Press, 1984. Reproduced with permission.*)

Elements such as aluminum, which can occupy sites in a solid structure connected to six oxygen atoms, are depicted at the center of an octahedron, with connections to adjacent octahedra or tetrahedra being through corners or edges.

Ceramics and glasses can be classified into three types: (1) oxide-type materials that are produced from mineralogical materials by means of relatively simple chemical processes or no chemical modification at all, (2) oxide ceramics that resemble their mineralogical counterparts but are synthesized and fabricated from small-molecule precursors via low- or moderate-temperature processes, and (3) nonoxide ceramics—materials that only rarely exist in nature but are produced by high-temperature chemical processes either from the elements or from polymeric precursors. These three types of materials are discussed in turn.

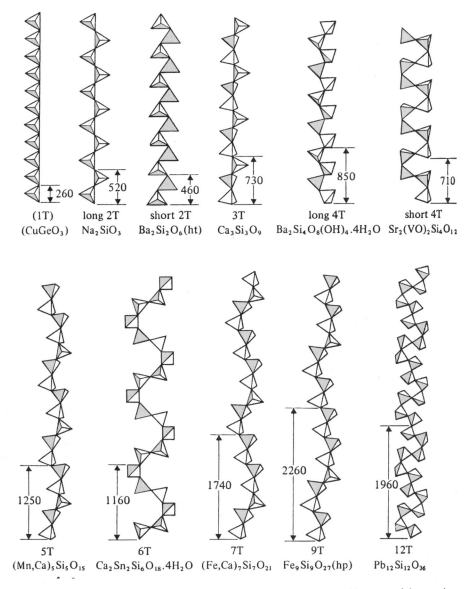

Figure 7.3. The structures of various metasilicate polymer chains depicted in terms of the number of tetrahedral silicate units per repeat distance along each chain. The repeat distances are given in picometers.

B. OXIDE CERAMICS AND GLASSES OBTAINED OR PRODUCED DIRECTLY FROM MINERALOGICAL MATERIALS

1. General Observations

Oxide ceramics and glasses are the classical materials that are the basis of a large part of traditional technology. They include various forms of glass, objects derived from clays such as brick and fine china, mica furnace windows, alumina and titania

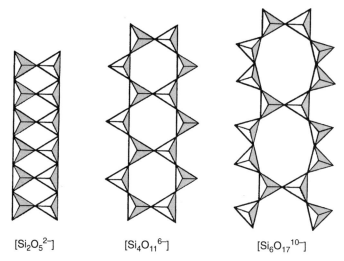

$[Si_2O_5{}^{2-}]$ $[Si_4O_{11}{}^{6-}]$ $[Si_6O_{17}{}^{10-}]$

Figure 7.4. *Ladder silicates depicted in terms of silicate tetrahedra. (From Wells, A. F.,* Structural Inorganic Chemistry, *4th ed. Oxford University Press, 1975, p. 816. Reproduced with permission.)*

insulators, and the minerals used in buildings such as limestone, granite, sandstone, and concrete blocks,

2. Silica, Silicates, and Aluminosilicates—General Characteristics

Oxygen and silicon are the two most abundant elements in the earth's crust. Consequently, an enormous number of different forms of silicon–oxygen-based minerals exist, estimated to number more than 25% of all known minerals, and these represent more than 90% of the earth's crust. Yet, despite this diversity, there is only one building block for all the multiplicity of known silicates—the tetrahedral arrangement of four oxygens linked to a central silicon. In nearly all of these tetrahedra the Si—O distance is close to 1.6 Å and the O $\cdots$ O distances at the boundaries of each tetrahedron are near 2.7 Å. Depending on the ratio of Si—O—Si linkages to the number of charged Si—O-units, cations are also present to balance the charges. Note that aluminum (Al^{3+}) often becomes incorporated into a silicate lattice to give aluminosilicates. When this happens, changes in the cation numbers are required to balance the charges.

Silicates are classified into six related structural groups that are easily understood in terms of the tetrahedral silicate building block:

1. Those species with a single-silicate tetrahedron, SiO_4^{4-}, associated with cations such as Na^+, K^+, and Mg^{2+}. An example structure is tetrasodium silicate, depicted schematically in Figure 7.5.

2. Short chains or small rings of tetrahedra linked through corners, with cations to balance the charges of those oxygen atoms not involved in the chain or ring structure. Examples are shown in Figure 7.2.

3. Long, single-stranded or double stranded (ladder) polymer chains that consist of tetrahedra, again linked through corners (Figure 7.3). Single-stranded polymers are known as *pyroxenes*. Ladder or ribbon polymers (Figure 7.4) are

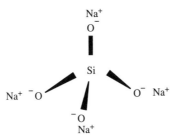

Figure 7.5. A monosilicate with all four oxygen atoms bearing anionic charges. Such species, with no Si—OH units, are unable to undergo condensation to rings, chains, or three-dimensional ultrastructures.

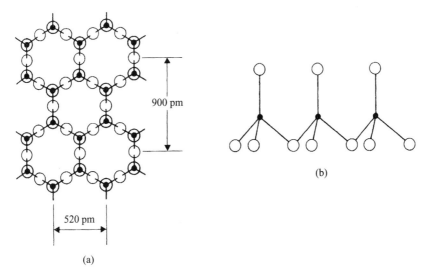

Figure 7.6. Sheet structures formed by extended two-dimensional condensation of silicate hydroxyl groups. Sheet structures are favored if one of the four oxygen atoms per silicon atom is in the form of a salt [i.e., (HO)$_3$SiO$^-$M$^+$]. (a) View at right angles to the sheet structure; (b) view along a sheet.

called *amphiboles*, an example of which is chrysotile (white) asbestos: $Mg_2Si_2O_5(OH)_4$.

4. Sheet structures are formed by connections between three of the four corners of each tetrahedron; the remaining corner oxygen atoms is associated with charge-balancing cations (Figure 7.6). Many examples exist of sheet structures that form multilayer sandwiches separated by galleries that contain the cations and sometimes water molecules. This type of structure is found in many clays, mica, and vermiculite. If those cations are divalent, such as calcium or magnesium, the layers may be connected by ionic crosslinks. If the cations are monovalent, the layers will not be connected and can often be separated under suitable conditions. Some minerals have pendent hydroxyl groups that do not condense, either because the solid-state structure provides too great a separation or because the interlayer regions contain trapped water. These layers can often be separated easily.

5. Cuboid-type nanostructures in which eight silicon atoms exist at the corners of a cube, connected through Si—O—Si bonds. These are called "sesquisiloxanes." Most of these species are produced in the laboratory, but they can be visualized as intermediates in the mineralogical formation of more extensive structures.

6. Ceramics in which all four corners of the silicate tetrahedra are connected to other tetrahedra through Si—O—Si linkages to give a three-dimensional, charge-neutral framework. Silica (SiO_2), often found in the form of sand, falls into this category. Crystalline silica exists in three different modifications— quartz, tridymite, and crystobalite. All three exist with two different modifications (α and β), and have different melting points and slightly different arrangements of the tetrahedra in the lattice. However, quartz is by the far the most common variety. The structure of christobalite is shown in Figure 7.7.

Another way to visualize the structure of silicates is as species formed when some of the Si—O—Si connections in silica are broken by an aqueous base such as sodium hydroxide. As these connections are progressively broken, the system becomes decreasingly crosslinked and changes from a three-dimensional network into a system of chains and rings (Figure 7.8).

Still another way to understand silicate structures is as polymers formed by the condensation of Si—OH units derived from silicic acid, $Si(OH)_4$. From the discussion in Chapter 6, it will be clear that the products formed by condensation will depend on the number of functional OH groups attached to silicon. Hydroxyl groups that have been converted to Si—O^- M^+ units are essentially unreactive and will remain as side groups to the chains, rings, sheets, or ultrastuctures formed during condensation. For example, condensation of the difunctional $(NaO)_2Si(OH)_2$ will give a linear silicate polymer of the type shown in Figure 7.3. Alternatively, condensation of $NaOSi(OH)_3$ can yield ribbon or ladder polymers or sheets of rings, as shown in Figure 7.4, with the Si—O^- Na^+ groups oriented above or below the plane of the fused rings It follows that, if the sheet structure bears pendent OH rather than

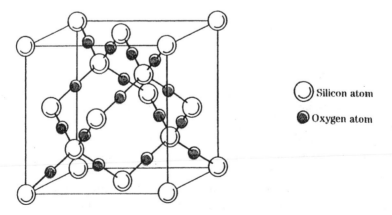

Figure 7.7. The covalently linked, three-dimensional atomic arrangement in cristobalite (SiO_2). (From Kingery, W. D., Bowen, H. K., and Uhlmann, D. R., Introduction to Ceramics, 2nd ed., Wiley, 1976.)

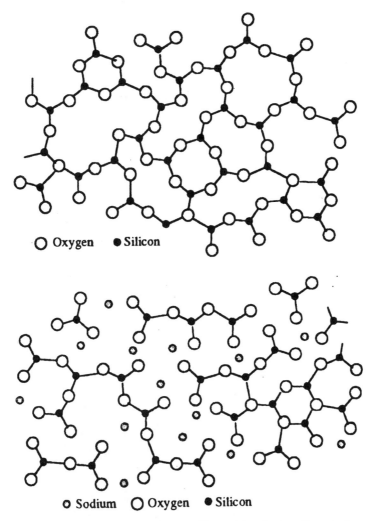

○ Oxygen   ● Silicon

○ Sodium   ○ Oxygen   ● Silicon

Figure 7.8. *Conversion of silica glass to a lower-melting-point soda glass through cleavage of Si—O bonds by sodium hydroxide or sodium carbonate. (From Thrower, R. A., Materials in Today's World 2nd ed. McGraw-Hill, 1996.)*

O—Na$^+$ groups, the sheets themselves can be joined by condensation to form a three-dimensional structure like silica. If a template unit such as water or large ions are present, these may cause the framework to grow around them and give rise to tunnels or cavities.

An additional level of complexity is reached when aluminum atoms are incorporated into the structure in place of some of the silicon atoms to give aluminosilicates. Because aluminum hydroxide contains only three hydroxyl groups rather than the four of silicic acid, the normal condensed silicate tetrahedral arrangement becomes interrupted. This has far-reaching consequences, including the formation of discontinuities in both structure and charge neutralization. Moreover, aluminum can form six coordinate bonds to oxygen, and often does so in three-dimensional lattices. In this way the presence of aluminum in a silicate lattice leads to the formation of layered structures in which the galleries between the layers are occupied by water; monovalent cations, to maintain charge neutrality; or divalent cations, which ionically bind the layers together but that can be disrupted by ionic reagents. Zeolites are aluminosilicate variants of these structures in the sense that, instead of layers, they contain cavities or tunnels in which ions, water, or other small molecules can be accommodated. They are discussed later in this chapter.

3. Aluminosilicate Clays and Related Minerals—Properties and Structure

Clay minerals are finely divided particulate materials found in almost every region of the globe. They are an essential component of soil. The most familiar clays are layered aluminosilicate materials such as china clay (kaolinite) and fuller's earth (montmorillonite or bentonite). Vermiculite and mica are layered aluminosilicate minerals that do not form clays but are found in igneous rocks. Talc is a closely related layered material formed from a silicate rather than an aluminosilicate.

Kaolinite is one of the main materials used for the manufacture of china clay ceramics. It is also a key component in the manufacture of high-quality paper since it provides gloss and body to what would otherwise be a highly porous product. Prodigious amounts of these minerals are mined and distributed each year; the main sources are in the United States, United Kingdom, Russia, and China.

Clays have an aluminosilicate sandwich structure based on laterally connected silicate tetrahedra connected (crosslinked) to a layer of laterally connected alumina octahedra. There may be just two layers per sandwich unit—one silicate and one aluminate, or three layers—one aluminate flanked by two silicate or vice versa. The silica and alumina layers are connected covalently through oxygen atoms, but the two- or three-layer plates are stacked together, separated by "galleries." Each aluminum atom is connected to six neighboring oxygen atoms (rather than the expected three) because the coordination sphere of aluminum can readily accommodate six oxygen donor units. Charge balancing usually requires that cations also be present in the lattice, usually in the galleries that separate the covalently bonded layers. These cations can be hydrogen ions, alkali metal cations, or multivalent cations such as Ca^{2+} or Mg^{2+}. Mutivalent cations in the galleries may hold the layers together firmly by ionic crosslinking. The size of the cations determines the thickness of the galleries. One type of cation can be replaced by another.

These minerals fall into two different structural categories—those with double-layer plates separated by galleries (bilayer systems), and those with triple-layer plates also separated by galleries. Water molecules or cations occupy the galleries. A schematic representation of a bilayer arrangement is shown in Figure 7.9, and an example of this structure in the form of the mineral $Ca_2Al_2SiO_8$ is shown in Figure 7.10. Schematic representations of some typical trilayer systems are shown in Figure 7.11, and the building block atomic arrangements for representative systems are shown in Figure 7.12.

Those systems with weak forces between the plates, specifically, with monovalent cations or water in the galleries, often allow the layer plates to separate in liquids to form colloids. For example, the presence of monovalent sodium or hydrogen ions in the galleries can lead to inclusion of water and partial or complete separation of the layers. This may be one reason why the character of the clay in a garden soil or in an artist's studio undergoes a dramatic change in properties if the pH is altered.

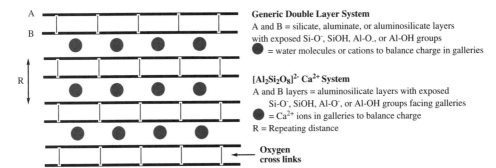

Generic Double Layer System

A and B = silicate, aluminate, or aluminosilicate layers
with exposed Si-O⁻, SiOH, Al-O⁻, or Al-OH groups

● = water molecules or cations to balance charge in galleries

[Al₂Si₂O₈]²⁻ Ca²⁺ System

A and B layers = aluminosilicate layers with exposed
Si-O⁻, SiOH, Al-O⁻, or Al-OH groups facing galleries

● = Ca²⁺ ions in galleries to balance charge

R = Repeating distance

**Oxygen
cross links**

Figure 7.9. *Schematic representation of a two-layer galleried silicate or aluminosilicate. The galleries contain cations or water. If the cations are di- or trivalent, the layer-to-cation attractions will stabilize the three-dimensional structure. If the cations are monovalent, such as Na⁺, K⁺, or H⁺, the double-layer sheets may be induced to separate them from their neighbors as water or other small molecules penetrate the galleries and ease the sheets apart. Also shown is a representation of the structure of $Ca_2Al_2Si_2O_8$.*

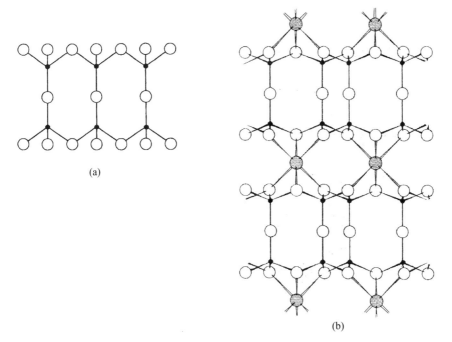

(a)

(b)

Figure 7.10. *(a) View along the double-layer aluminosilicate sheets in $Al_2Si_2O_8^{2-}$ held together by oxygen crosslinks (open circles). The sites marked with closed circles are occupied by equal numbers of aluminum and silicon atoms. (b) Structure formed by the insertion of 6-coordinated Ca^{2+} ions into the galleries between the double layers. (Modified from Greenwood, N. N.; Earnshaw, A.,* Chemistry of the Elements, *Pergamon Press, 1984.)*

Separation of the plates is also facilitated when the original inorganic cations are displaced by large organic cations such as tetrabutylammonium ions. The ceramic may then disintegrate into individual plates that form as colloids in liquid media. A change of pH or the presence of another cation can then induce restacking of the

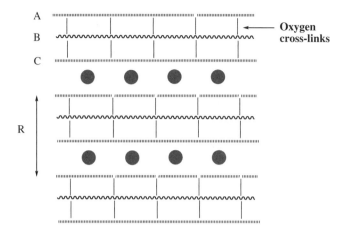

Kaolinite
A and C = aluminate layer with exposed Al-OH groups
B = silicate layer
● = water molecules associated with Al-OH groups that line the galleries
Repeating distance (R) = 7.15 A

Montmorillonite
A and C = silicate layer with no free Si-OH groups
B = Magnesium aluminosilicate layer with OH groups and negative charges
● = water molecules or labile cations in the galleries
Repeating distance (R) = 15 A

Mica
A and C = aluminosilicate layer $[(Si_{1.5}Al_{0.5})O_5]_n$
B = aluminate layer with hydroxyl groups
● = potassium ions in the galleries
Repeating distance (R) = 9.9 A

Vermiculite
A and C = aluminosilicate layer $[(Si_{1.5}Al_{0.5})O_5]_n$
B = $[Mg_3(OH)_2]_n$ layer
● = water molecules and Mg^{2+} ions in the galleries
Repeating distance (R) = 14 A

Talc
A and C = silicate sheets
B = magnesium oxide layer,
● = potassium ions in the galleries

Figure 7.11. *Schematic view along the sheets of three-layer galleried silicates and aluminates. The variations in structure and properties of different minerals arise from the different elemental compositions of each of the three layers and the nature of the molecules or ions in the galleries. In all cases the three layers in each assembly are held together by oxygen crosslinks.*

plates, but not always in the original pattern. For example, although it is possible for the plates to restack in a face-to-face structure, it is also possible for them to form edge-to edge aggregates or "house of cards" arrangements. Widening of the galleries by tetraalkylammonium ions also facilitates entry of water-soluble organic or inorganic polymers into the gallery space, a change that can strengthen the materials' properties.

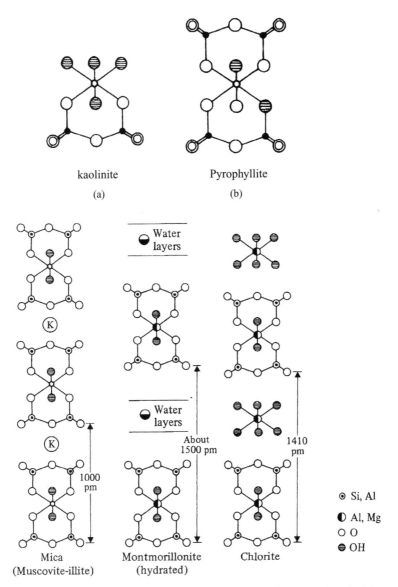

Figure 7.12. *(a) Aluminosilicate building blocks found in common clay-type minerals; (b) organization of these units in muscovite mica [K₂Al₄(Si₆Al₂)O₂₀(OH)₄], hydrated montmorillonite [Al₄Si₈O₂₀(OH)₄ xH₂O], and chlorite [Mg₁₀Al₂(Si₆Al₂)O₂₀(OH)₁₆] galleried structures. (From Greenwood, N. N.; Earnshaw, A., Chemistry of the Elements, Pergamon Press, 1984. Reproduced with permission.)*

"Pillared clays" are galleried clays with more permanent structures that separate the layers. An example is the structure formed by replacement of the original cations in the galleries by large aluminate cations. Steps are first taken to ensure that the aluminate pillars are uniformly distributed throughout the galleries. Then the face-to-face galleried ceramic is heated to "fix" each pillar in place as a permanent feature. The final structure can be viewed as resembling a multilevel parking garage with pillars in fixed positions on each level, or as a coal mine in which the "pit props" support the roof of each gallery. The size of the pillars defines the height of the galleries. Once formed, a galleried clay can absorb other molecules into the galleries, and the height of the gallery and the number of pillars per unit area control the size of the molecules that can enter or leave. Thus, galleried clays may be employed to separate small molecules in much the same way as in zeolites. Moreover, polymer molecules can be intercalated into the galleries and used to strengthen the ceramic or to provide some other property such as ionic conductivity. β'' Alumina is a pillared ceramic that has no silicate component.

Kaolinite clay is the basis of a very large industry. Its three-layer structure is shown schematically in Figure 7.11, and the basic building block is depicted in Figure 7.12. The structure can be visualized as a stack of three-layer plates with each plate consisting of one layer of sheet silicate sandwiched between and covalently bonded on both sides through oxygen bridges to a layer of sheet aluminate. The exposed aluminate layers bear Al—OH units that point into the galleries between the plates. The repeating distance normal to the sheets is 7.15 Å. The trilayer plates in kaolinite are apparently held together by weak van der Waals and hydrogen bonding forces, and they can be separated fairly easily by the absorption of water. The rheology of kaolinite clay depends on pH, presumably because of the amphoteric (acidic or basic) nature of Al—OH groups and its influence on water absorption and layer displacement.

Montmorillonite clays have been studied in great detail. They have the formula $[M_2Si_4O_{10}(OH)_2]^{0.33-}$ 0.33 Na^+ xH_2O, where $M = Al_{1.67}$ and $Mg_{0.33}$. This class includes several minerals that share a general structure but differ in minor ways. The basic structure is shown schematically in Figures 7.11 and 7.12. Montmorillonite itself has a three-layer lamellar structure separated by galleries. Each plate within the structure consists of two layers of silicate sheets bonded to each other via oxygen links to a sandwiched layer of aluminum and/or magnesium atoms. The trilayer plates are separated from their neighbors by galleries that contain protons or monovalent cations, together with water molecules. The repeat distance from stack to stack is about 15 Å. Because any "free" hydroxyl units are in the middle of each sandwich rather than on the surface, there are no $Al—O^-M^+$ or $Si—O^-M^+$ groups facing the galleries to cause ionic crosslinking. This is a difference from the structure of kaolinite. Thus, one type of cation in the galleries is readily replaced by another, and trapped water molecules may be removed by heating and replaced by hydration. Moreover, the water is readily replaced by hydrophilic organic solvents and even by hydrophilic polymers.

Fuller's earth is a close relative of montmorillonite in which $Al—O^-H^+$ units within the lamellar structure are charge-balanced by Ca^{2+} ions in the galleries. The ability of divalent cations to form ionic crosslinks reduces the tendency of this material to undergo lamellar separation. Fuller's earth is used for the removal of impurities from liquids during filtration, presumably by ion exchange within the galleries.

Bentonite is the species in which Al—OH protons in the central layers of montmorillonite have been replaced by sodium ions, which occupy the galleries. Bentonite is used in large quantities as a drilling mud. Because the bentonite lamellae bear monovalent cations on their surface, this mineral absorbs substantial quantities of water into the galleries, which causes separation of the plates to form a colloidal dispersion. This process yields very small colloidal particles with negative charges on the flat surfaces and positive charges on the edges. The attractive forces between these oppositely charged components generates a thixotropic gel that liquefies when the colloid is sheared and gels again when the shear forces cease. This is a valuable property in oil drilling.

Pyrophillite [$Al_2Si_4O_{10}(OH)_2$] can be viewed as a parent structure from which the three-layer aluminosilicate sandwich minerals are derived (Figure 7.12). Its structure consists of a central aluminate layer covalently bonded above and below to silicate layers. The Al—OH groups are buried within the aluminate layer and are therefore less accessible for replacement of protons by metallic cations.

Vermiculite is a magnesium aluminosilicate material that is widely used as a nonflammable packing substance, especially valuable for its ability to absorb liquids from broken bottles and other sources. Vermiculites are produced by the thermal dehydration of various montmorillonite structures that contain magnesium. The dehydration process causes the mineral to form strange, porous, lightweight, worm-like shapes, hence the name *vermiculite*. An idealized sandwich structure is summarized in Figure 7.11, in which each plate consists of a condensed magnesium hydroxide layer flanked above and below by aluminosilicate layers. The galleries contain water and Mg^{2+} crosslinking cations. The absorptive powers of this material are due to both the porosity and the presence of galleries.

Micas (also known as *muscovite* or *illite*) are transparent aluminosilicate minerals that can be readily separated into the thin sheets and used in the windows of high-temperature equipment (furnaces, home heaters, etc.). The solid-state structure (Figure 7.11) consists of a lamellar arrangement with potassium, calcium, or H_3O^+ ions occupying the galleries. Micas can be visualized as being formed by replacement of one-quarter of the silicon atoms in pyrophyllite by aluminum. The requirements for charge balancing then lead to the need for different cations in the galleries. The presence of aluminum or magnesium hydroxide cations in the galleries yields another mineral known as *chlorite* (Figure 7.12).

Talc is a three-layer potassium magnesium *silicate* rather than an aluminosilicate. Its structure is shown schematically in Figure 7.11. The galleries are occupied by potassium ions, which cannot crosslink the layers. Thus, the magnesium silicate layers can slide past each other. Hence the use of this material in talcum powder.

4. Chrysotile and Other Forms of Asbestos

Asbestos is a general name for several fibrous minerals that were once widely used as nonflammable thermal insulation materials in buildings, ships, brake linings, floor and roof tiles, and electrical wiring. However, the health hazards associated with the inhalation of asbestos fibers have caused severe restrictions on the use of these materials and have prompted widespread efforts to remove asbestos from existing buildings. The health hazards are believed to be a consequence of the narrow diameter (~1 μm) and twisted morphology of the fibers, which favors their penetration into and retention in lung tissue.

Three different forms of asbestos have been used: white asbestos (chrysotile), brown asbestos (amosite), and blue asbestos (crocidolite). Of these, blue asbestos is believed to be the most dangerous form. All three are found as fibrous deposits in metamorphic formations, and they were probably generated by crystallization of the mineral from molten rock. White asbestos is the form most widely used in the United States. This is believed to be less hazardous than the blue form but is nev-

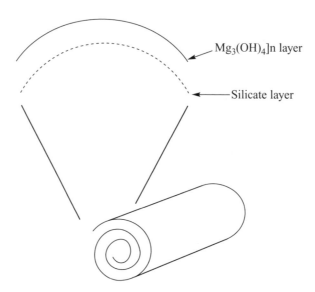

Mg$_3$(OH)$_4$]n layer

Silicate layer

Figure 7.13. *Chrysotile asbestos is a bilayer system with silicate and magnesium oxide/hydroxide layers. The greater surface area required by the magnesium oxide/hydroxide layer causes each plate to curve into a spiral straw arrangement.*

ertheless a suspected carcinogen. Asbestos fibers can be readily distinguished from glass fibers when rotated between crossed polarizers on a microscope stage. Asbestos fibers show bright birefringent colors because they are crystalline, whereas glass, which is amorphous, does not.

Chrysotile is a member of a large class of minerals known as *serpentines*, which have the general formula (Mg or Fe)$_3$Si$_2$O$_5$(OH)$_4$. Chrysotile asbestos is a two-layer fibrous magnesium silicate with the formula Mg$_3$(Si$_2$O$_5$)(OH)$_4$, and with the structure shown schematically in Figure 7.13. Thus, it resembles the layered arrangement of the clay-type species described in the preceding sections, but with magnesium oxide units *on one side only* of each plane. This causes each layered plane to bend away from the magnesium side to form coiled ribbons—ribbons that have wound inward to form a straw-like structure as shown in Figure 7.13.

Amosite is a dark-colored iron silicate with a composition such as Fe$_7$Si$_8$O$_{22}$(OH)$_2$, which is a member of a group of minerals known as *amphiboles* (not all of which are fibrous). These species are ladder-type structures that contain two linear silicate chains connected through every fourth repeating unit by oxygen atoms that lie at the corners of the silicate tetrahedra. Thus, the structure of amosite can also be viewed as ribbons of connected 16-membered rings. Crocidolite is another colored amphibole, with the formula Na$_2$Fe$_5$Si$_8$O$_{22}$(OH)$_2$. The structure is believed to be similar to that of amosite.

5. Glasses

a. General Features. *Glasses* are amorphous solids that are usually transparent. Useful glasses have glass transition temperatures (T_g) well above room temperature.

A few types of glass are derived from organic polymers, but the traditional inorganic glasses are produced mainly from molten oxides and their salts, such as silica, silicates, borates, or phosphates, many of which have been modified to disrupt crystallinity.

b. Methods of Glass Formation. Glasses are formed in several different ways. First, a crystalline material such as quartz may be heated above its melting point (~2000 °C) and then cooled rapidly ("quenched") to form a glass. Note that volcanic eruptions often provide the conditions for glass formation on a massive scale. A glass formed by quenching can be considered as a supercooled liquid that could, in theory, crystallize over a long period of time, but is prevented from doing so by the high viscosity of the system at room temperature. This method of glass formation is based on the principle that the ordered state of the crystalline phase is lost on melting, and the randomness of the liquid state is frozen into the glassy material when the system is quenched. Many other crystalline inorganic (and organic) compounds can be induced to form glasses by the same technique. The ease with which the amorphous glass reverts to the crystalline state depends on the ease of molecular movement in the glassy state. Large, unsymmetric or hydrogen-bonded molecules are less likely to revert than small, symmetric species. In the case of quartz, it seems likely that melting disrupts the material into chains or rings but not down to the level of individual tetrahedra.

A special form of silica glass is known as a *glass ceramic* or Pyroceram (Corning Ware®), which is prized for its resistance to impact and thermal shock. This material is opaque rather than transparent because it consists of fused silica that has been induced to crystallize so that only 10% remains as amorphous glass and 90% consists of silica microcrystals. This conversion from glass to glass ceramic is accomplished by seeding the molten silica before fabrication and solidification with small crystals of titanium dioxide to nucleate the growth of silica crystals within the glass. Controlled heating of the solid glass object then induces growth of silica microcrystals from the amorphous matrix. In the final product the remaining glass serves as a strong binder between the crystals, while the crystallites interrupt crack formation and increase the density of the material. The coefficient of thermal expansion is also reduced, and this, in turn, increases the thermal shock resistance.

A second type of glass is formed by deliberately disrupting the bonds that hold a supermolecular crystalline material together. Quartz is crystalline silica with a regular, three-dimensional repeating arrangement of silicon atoms bonded tetrahedrally to four oxygens throughout the lattice. The addition of reagents that break silicon–oxygen bonds at elevated temperatures, such as metal oxides or carbonates, disrupts the three-dimensional lattice and converts quartz into a random mixture of silicate chains and rings (see Figure 7.8). This creates solid state disorder, lowers the melting point, and inhibits or prevents crystallization. Ordinary window glass, or bottle glass is produced in this way by the addition of sodium or potassium carbonate and calcium oxide.

Note that, as discussed in Chapter 6, a third type of glass is formed from linear or branched polymers. Unless the polymer molecules are highly symmetric, it is difficult for the chains to pack together in crystallites. The same situation exists with totally inorganic or hybrid polymers such as linear polysilicates, polyphosphates, or polyborates, as well as some high-T_g polyphosphazenes.

c. Silicate Glasses. Silicate glasses are by far the best known, and have the longest history in technology. Glass was being made in Egypt in 1500 BC. Excellent examples of decorative glass have been found that date back to the Roman Empire approximately 2000 years ago. Arrowheads from the Stone Age are obsidian glass found naturally. The basic chemistry of silicate glass formation is as follows. Although, as mentioned above, fused silica can be quenched to a glass, the T_g and T_m of this material are so high (1200 °C and 2000 °C, respectively) that it is difficult to fabricate. However, the melting point can be lowered considerably by cleavage of some of the silicon–oxygen bonds by heating silica (usually sand) with sodium or potassium oxide or carbonate and calcium oxides. In practice, metal carbonates may be used since they are easier to handle on a large scale and decompose to the oxides at elevated temperatures. The calcium is employed to reduce the solubility of the glass in water. This process lowers the T_g to about 550 °C if the material is cooled rapidly, or to 170 °C after slow cooling. The melt-processing temperature for good fluidity falls to about 1000 °C. The product is window glass or bottle glass, also known as "soda glass" or "soda–lime glass." This material is inexpensive and chemically and photolytically stable. "Float glass" and *fiberglas* are essentially the same material. The drawbacks of common soda–lime glass are its brittleness and its relatively low refractive index. It may also be discolored by traces of transition metal impurities (especially iron) which are present in the original sand. Hence, the "white sands" found in various locations around the globe are especially prized for glassmaking.

Note that the nature of the cation used in the formation of glass has an important influence on properties. Monovalent cations such as Na^+ or K^+ do not form ionic crosslinks between rings or chains. However, Ca^{2+} or Al^{3+} are capable of coordination to two or three other chains to form ionic crosslinks. Hence, the multivalent cations may stiffen and strengthen a glass. Transition metal cations such as iron, cobalt, chromium, nickel, or uranium are incorporated into the glass to generate specific colors. This technique has been used for centuries in stained-glass windows and decorative glassware. These glasses are often applied as "glazes" (usually in the form of an emulsion of colored glass particles in water) that are applied to clay objects such as table china or decorative vases, and then heated above the T_g of the glaze to form a coherent film on the surface. Glasses that contain the colored ions uniformly distributed throughout the bulk material are used in optical filters.

The addition of lead or barium oxide or carbonate or other heavy-metal compounds to glass lowers the T_g and increases the refractive index. These high-refractive-index glasses are used in "crystal" drinking glasses, prisms, and lenses. Special low-dispersion glasses used in camera lenses utilize cations such as lanthanide ions.

Note that, although most glasses are chemically inert, cations can be exchanged between the surface regions of the solid and aqueous media. For example, exchange of alkali metal cations may occur, especially if a high surface area of the glass is exposed to the aqueous medium. Lead can be extracted by hot coffee or tea from the glazes used on some ceramic cups and mugs, and this is a well-known health hazard. Soda glass in the form of fiberglass actually dissolves slowly in water because of the high surface area. This was the reason for the now infamous claims for "poly-water" made in the 1960s, which were eventually traced to silicates extracted by water from glass capilliary tubes.

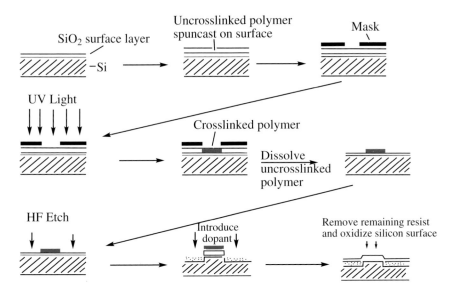

Figure 10.7. *Sequence of events in photolithography through use of a negative tone resist.*

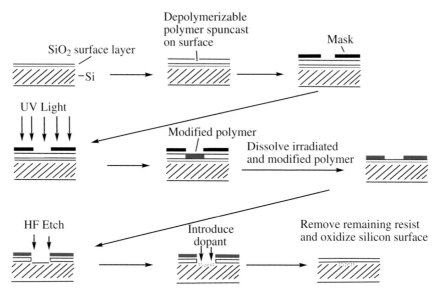

Figure 10.8. *Use of a positive tone resist.*

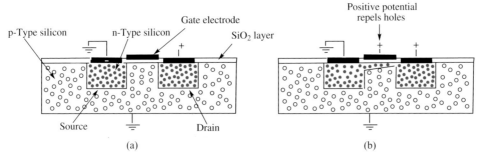

Figure 10.11. *Cross section of a metal oxide semiconductor transistor. (a) With no potential applied to the gate electrode, the source and drain n-type regions are insulated from each other by the p-type region. However, application of a positive potential to the gate electrode (b) causes the p-type region to retreat, thus allowing an "n channel" to form beneath that electrode. The current that flows through the n channel can be controlled by the positive potential applied to the gate.*

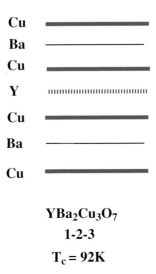

$$YBa_2Cu_3O_7$$

$$1\text{-}2\text{-}3$$

$$T_c = 92K$$

Figure 11.6. *Schematic diagram of the layer structure in $YBa_2Cu_3O_7$.*

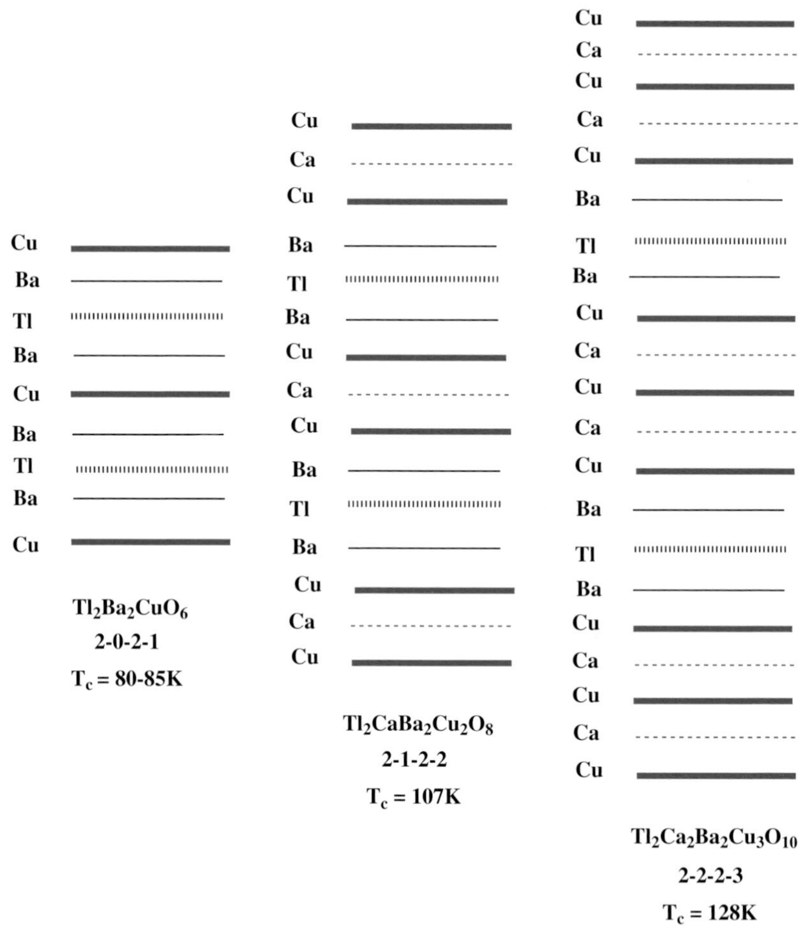

	Cu	▬▬▬▬▬
	Ca	-----------------
	Cu	▬▬▬▬▬
	Ca	-----------------
	Cu	▬▬▬▬▬
	Ba	▬▬▬▬▬
	Tl	‖‖‖‖‖‖‖‖‖‖‖‖‖

Cu ▬▬▬▬▬   Ba ▬▬▬▬▬   Ba ▬▬▬▬▬
Ba ▬▬▬▬▬   Tl ‖‖‖‖‖‖‖‖   Cu ▬▬▬▬▬
Tl ‖‖‖‖‖‖‖‖   Ba ▬▬▬▬▬   Ca -----------
Ba ▬▬▬▬▬   Cu ▬▬▬▬▬   Cu ▬▬▬▬▬
Cu ▬▬▬▬▬   Ca ----------- Ca -----------
Ba ▬▬▬▬▬   Cu ▬▬▬▬▬   Cu ▬▬▬▬▬
Tl ‖‖‖‖‖‖‖‖   Ba ▬▬▬▬▬   Ba ▬▬▬▬▬
Ba ▬▬▬▬▬   Tl ‖‖‖‖‖‖‖‖   Tl ‖‖‖‖‖‖‖‖
Cu ▬▬▬▬▬   Ba ▬▬▬▬▬   Ba ▬▬▬▬▬
                Cu ▬▬▬▬▬   Cu ▬▬▬▬▬
$Tl_2Ba_2CuO_6$ Ca ----------- Ca -----------
2-0-2-1   Cu ▬▬▬▬▬   Cu ▬▬▬▬▬
$T_c = 80-85K$ Ca -----------
                $Tl_2CaBa_2Cu_2O_8$ Cu ▬▬▬▬▬
                2-1-2-2   Ca -----------
                $T_c = 107K$ Cu ▬▬▬▬▬

$Tl_2Ca_2Ba_2Cu_3O_{10}$
2-2-2-3
$T_c = 128K$

Figure 11.7. Schematic diagram showing the layers of thallium, calcium, barium, and copper oxides in the repeating sequences of three cuprate superconductors. The superconducting transition temperatures (T_c) rise with increasing numbers of copper oxide layers.

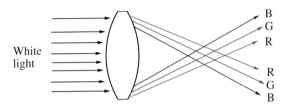

Spreading of the image due
to chromatic dispersion to give
color fringing

(a)

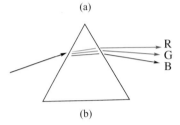

(b)

Figure 14.3. *The focus point of a lens (a) differs for the different wavelengths that make up white light due to the variation of refractive index with wavelength. This leads to color fringing of an image, which degrades overall sharpness. For a prism (b), the splitting of white light into the colors of the rainbow is a well-known phenomenon.*

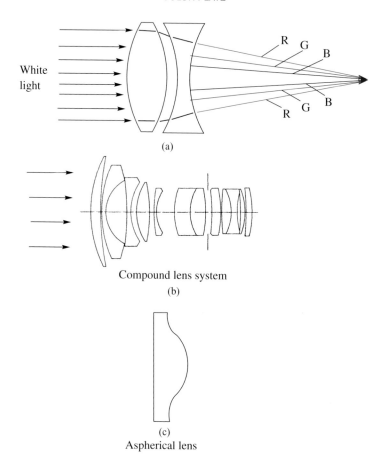

White
light

R G B
R G B

(a)

Compound lens system
(b)

(c)
Aspherical lens

Figure 14.6. *(a) Correction of chromatic dispersion can be accomplished by a combination of a high-refractive-index, high-dispersion convex lens coupled with a low-refractive-index concave lens. (b) Complex lens assemblies used to correct both chromatic and optical aberations such as coma. (c) Aspheric lenses, often made from polymers, are employed increasingly to avoid the distortion problems associated with the use of spherical profile lenses, especially in wide-angle and zoom lenses used in photography. The aspherical profile allows correction for crucial "edge effects" in which those rays passing through the outer regions of a spherical profile lens are difficult to focus when wide apertures are employed.*

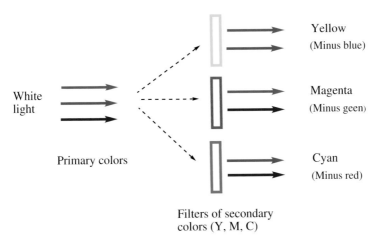

Figure 14.10. *Absorption filters* remove *certain wavelengths of light from polychromatic light. Contrast this with a nonlinear optical material that* changes *one wavelength to another. Thus, filters of the primary colors, red, green, and blue, remove all wavelegths except those colors. Filters of the secondary colors, yellow, magenta, and cyan, allow two of the primary colors to be transmitted. Filters of the primary colors that cover individual pixels are used in flat-panel liquid crystal displays (see later) and in the sensors in digital cameras.*

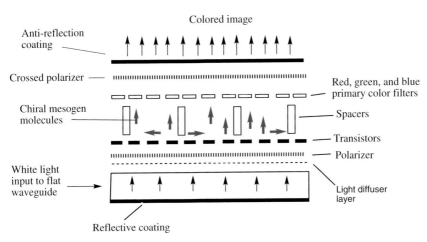

Figure 14.14. *Simplified cross section of a flat-panel liquid crystalline display showing the chiral mesogens that respond to the electric fields generated by the pixel transistors to cut off or allow passage of light through the red, green, or blue filters. The image and its colors depend on the pattern of transistors activated at any given instant.*

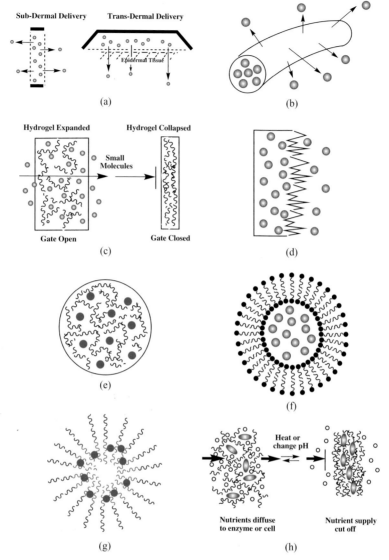

Figure 16.5. *Different designs for controlled drug delivery devices. (a) A membrane is employed to control the release of the drug. (b) A polymer tube or cylinder serves as a diffusion membrane for controlled drug release. (c) A responsive membrane that opens or closes as the pH, temperature, or ion strength is changed allows control of the drug release rate. (d) Hydrolytic erosion of a polymer provides a mechanism for drug release. (e) Microspheres are used to immobilize drugs and vaccines for protection through the stomach or for inhalation deep into the lungs. (f) Bilayer vesicles can serve as carriers for hydrophilic drugs for delivery to sites that are distant from the point of injection. (g) Micelles allow hydrophobic drugs to be solubilized and delivered to targeted sites via the bloodstream. (h) Diagram showing how a responsive hydrogel can be employed to turn on or off the action of a living cell or enzyme that either releases or decomposes a bioactive molecule.*

d. Pyrex-Type Glass. Pyrex glasses are usually based on borosilicates or alumi-nosilicates rather than on traditional silica-derived materials. They are produced by the addition of boric oxide (B_2O_3) or alumina (Al_2O_3) to a molten silicate glass at 1600 °C. This raises the glass transition temperature and provides resistance to impact and thermal shock. Hence these glasses are used in laboratory glassware and ovenware.

e. Phosphate Glasses. Phosphoric oxide (P_2O_5) forms a glass when melted and then cooled. Phosphate glasses have less utility than do silicate glasses. The ultimate crosslinked network consists of PO_4 tetrahedra, but with only three of the four oxygens serving as connection points to the surrounding lattice. The fourth oxygen is a nonbonding P=O unit. Cleavage of P—O—P bonds by water or aqueous base will reduce the crosslinking until linear polymers or small molecule fragments such as phosphoric acid [O=P(OH)$_3$] are formed. However, the preferred technique to produce phosphate glasses is to dehydrate sodium or potassium phosphate. For example, dehydration of HO—P(O)(ONa)—OH gives linear polymers that form glasses. Phosphate glasses are unstable to water, but fusion with B_2O_3 or Al_2O_3 increases the resistance to aqueous media.

f. Borate Glasses. Borate glasses are of interest because they can be designed to have low thermal expansion coefficients and have relatively low softening tempera-tures. Hence, they are used as "solder" glasses to join other glasses together in devices such as television cathode ray tubes, and to seal ceramic or metal parts to glass. Boric oxide (B_2O_3) itself forms a glass after being melted, but many other glasses are formed by the fusion of B_2O_3 with ZnO, PbO, Tl_2O, and La_2O_2, and these have specialized applications.

g. Fabrication of Glasses. Most glasses are fabricated by melt techniques. In other words, the material is heated above the melting temperature, or at least above the glass transition temperature, and is then formed into the required shape. Fiber-glas is produced by extrusion of the molten material through fine spinerettes. Float glass with high-quality surfaces is obtained by poring molten glass onto a bath of molten tin (the Pilkinton process). The sheet of glass cools as it moves from one part of the bath to the other and is drawn off in a continous ribbon. Most flat glass is now made by the Pilkinton process. An earlier process (sheet glass) involved drawing a ribbon of solidifying glass from a vat of the molten material. Plate glass is made by extrusion and roller-forming of a sheet of semimolten material, which must then be polished to remove surface imperfections. These processes all require the expenditure of large amounts of energy and massive equipment. By contrast, the sol–gel process described in the next section, which involves the formation of inorganic glass from small-molecule inorganic or organometallic precursors, takes place at lower temperatures and combines synthesis and fabrication.

C. OXIDE CERAMICS FROM SMALL-MOLECULE INORGANIC AND ORGANOMETALLIC PRECURSORS

By contrast to the materials discussed in the last section, we consider here an approach to the preparation of glasses and ceramics that does not involve the direct utilization of mineralogical materials. Instead, for various reasons that will be apparent from the following sections, the properties of the final glass or ceramic can be developed only by starting from simple chemical compounds. Of course, ultimately these are derived from the same mineralogical sources as traditional ceramics, but for reasons of purity or processing control, it is necessary to first convert minerals into pure small-molecule intermediates before synthesis of the ceramic. These additional steps inevitably increase the cost of the final products, but for high-performance applications the additional expense is acceptable.

1. Optical Waveguides (Optical Fibers)

Long-distance telecommunication is now conducted almost entirely over optical fiber networks rather than over the earlier copper wire technology (see Chapter 14). The glass needed as a starting point for the preparation of optical fibers is ultrapure silica. The presence of even minor amounts of colored impurities would reduce the optical path length along the fiber to an unacceptable degree. Hence, mineralogical silica is unsuitable for direct use, and ultra-high-purity silica, produced by chemical synthesis, is needed. This process, which involves the oxidation of highly purified silicon tetrachloride, is described in Chapter 14.

2. The Sol–Gel Process for Low-Temperature Ceramic Formation

The traditional melt fabrication of silicates into shaped glass objects requires the expenditure of large amounts of energy. Moreover, the conventional synthesis of glass uses starting materials obtained with minimal processing from geologic sources, which raises the possibility of contamination by impurities. For these and other reasons an alternative route to amorphous silicates (and other oxide ceramics) has been devised, a process that allows low-temperature processing from pure starting materials.

The usual starting material for this process is a silicon alkoxide such as tetraethylorthosilicate, $Si(OEt)_4$ (TEOS), which itself is obtained via the reaction of $SiCl_4$ with ethanol or sodium ethoxide, or by the interaction of sodium ethoxide with silica. The sol–gel process is a controlled hydrolysis of tetraethylsilicate to silica in aqueous alcohol under acidic or basic conditions at or near room temperature. The alkoxide is employed instead of the more easily hydrolyzed $SiCl_4$ because the chloride is converted to silica too rapidly and uncontrollably even in the vapor state. Other alkoxides, such as those of aluminum or titanium, may also be used, and mixtures of two or more alkoxides yield ceramics with a more uniform composition than can be obtained by sintering powders of the oxides. The incorporation of alkoxides of transition elements allows colored glasses to be formed.

A simplified outline of the basic chemistry is shown in reactions 1–4, where R is an ethyl or higher alkyl unit. Any or all of the Si—OR bonds may be hydrolyzed to Si-OH functional groups, and these condense to form Si—O—Si linkages. The reactions can be exceedingly complex and may proceed through the formation of chains or rings, all of which will crosslink at some stage in the reaction sequence. Some control can be exercised through variations in the concentration of the alkoxide, the amount of water in the solvent, the pH, and the type of alkoxy group. For example, the hydrolysis of tetrapropyl silicate is slower than that of tetraethyl silicate. Thus, the linear polymers formed in reaction 1 may undergo further alkoxide hydrolysis to give the crosslinked species also formed in reaction 2, and these crosslinked polymers can further crosslink to give an ultrastructure similar the one formed in reaction 3.

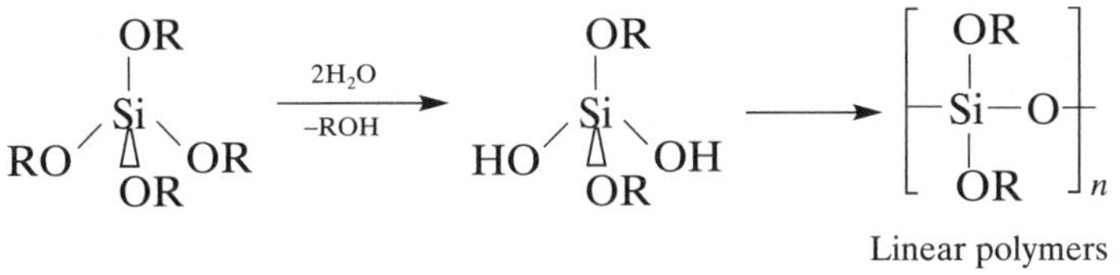
2H2O
–ROH
Linear polymers
(1)

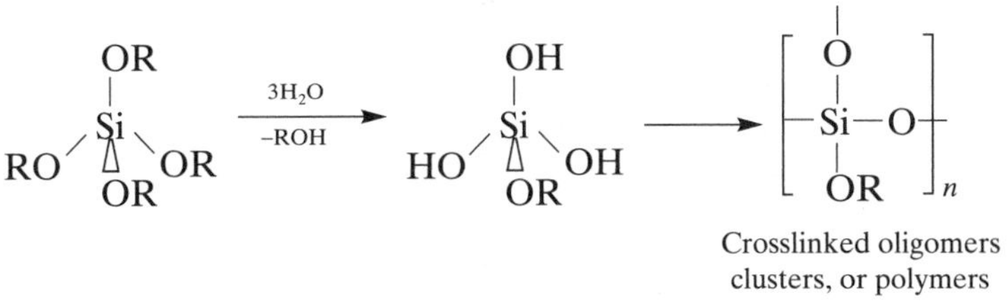
3H2O
–ROH
Crosslinked oligomers
clusters, or polymers
(2)

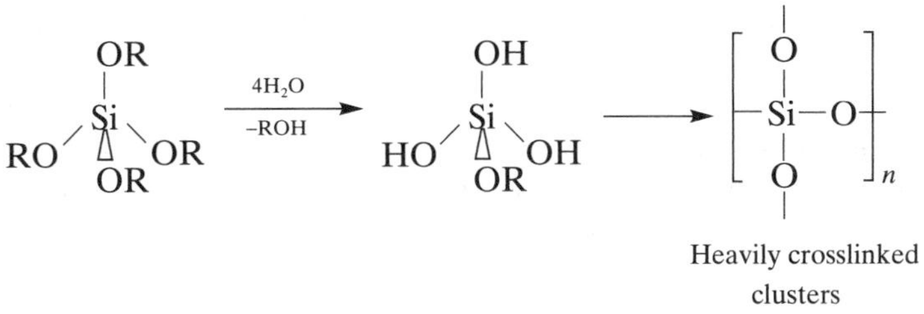
4H2O
–ROH
Heavily crosslinked
clusters
(3)

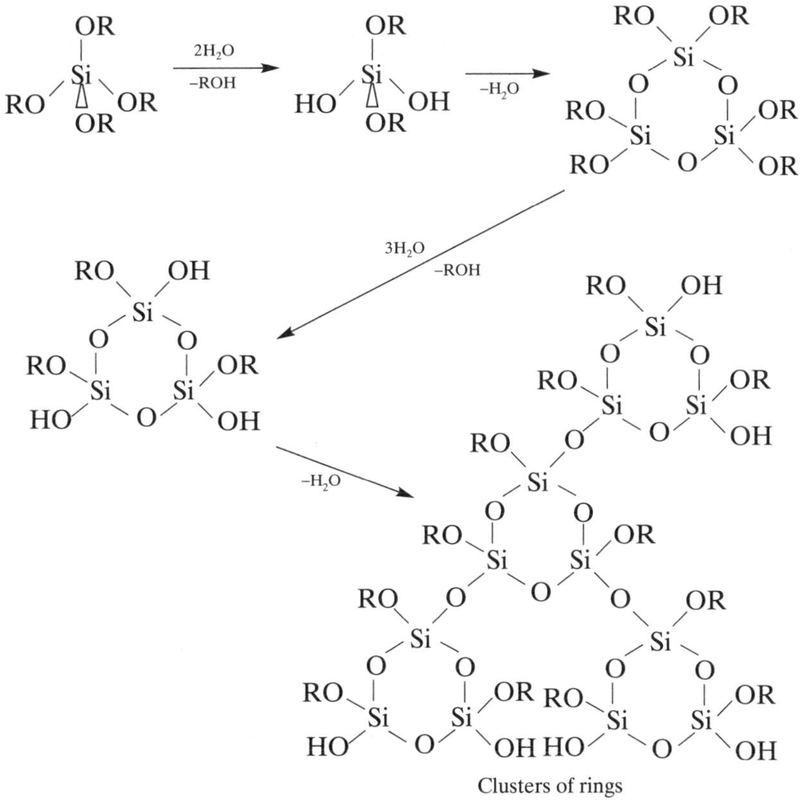
2H2O
–ROH
–H2O
3H2O
–ROH
–H2O
Clusters of rings
(4)

Step (reaction) 1 could also lead to six-, eight- or higher-membered rings instead of chains, and these rings may become joined together to yield clusters of rings (reaction sequence 4). Although cyclic trimeric siloxane rings are shown for simplicity in reaction 4, the most probable products are cyclic tetramers and higher cyclics. Moreover, reaction 4 could also generate cage structures, and these, too, could become linked to form clusters. Subsequently, the alkoxy groups on the outer boundaries of the ring or cage clusters will hydrolyze and provide covalent connections to other clusters to yield larger clusters, and so on. Ring clusters can, in principle, react with chain clusters or linear polymers to increase the structural complexity (Figure 7.14). Eventually a catastrophic gelation of the system will occur, and what was originally a solution or a colloid will become a solvent-swollen solid in the shape of the reaction vessel or mold. Subsequent heating to drive off water and alcohol will complete the condensation process and leave an amorphous form of silica. Note, however, that the removal of water and alcohol as the synthesis and fabrication proceed results in a shrinkage of a shaped object be it a fiber, a lens, or a film. The formation of transparent objects will require a heating to close to the melting point, a step that will result in further shrinkage.

Changes to the pH alter the types of products that are formed, as shown in Figure 7.15. Acidic media, high concentrations of reagents, and lower temperatures favor the formation of chains or loosely crosslinked chains. Basic media, dilute solutions, and higher temperatures favor the formation of rings, cages, and cluster networks.

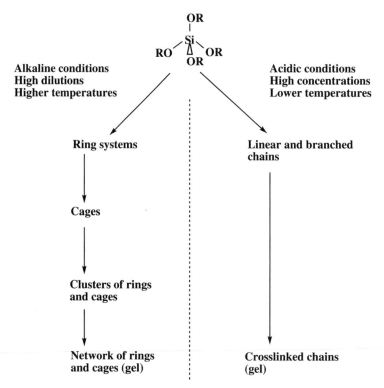

Figure 7.14. *Two different pathways and product strategies that may be followed in sol–gel reactions, depending on reaction conditions.*

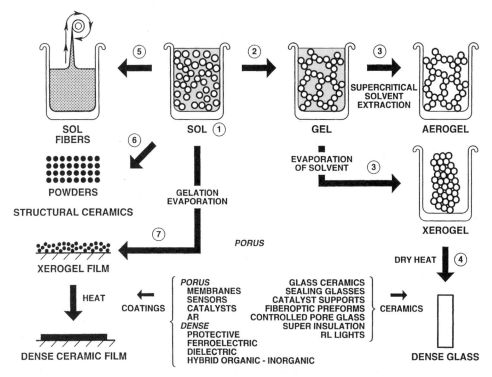

Figure 7.15. *Different products that can be produced by changes to the processing conditions in sol–gel reactions. (Reproduced by courtesy of C. J. Brinkerr.)*

Different types of materials are produced by different processing techniques (Figure 7.15). For example. fibers can be pulled from a system that contains colloidal clusters that have not yet gelled. Evaporation of a colloidal suspension or a solution can give a "xerogel" film or coating, or may yield high-purity ceramic particles with nano- or microscale diameters. For example, titanium dioxide powders for use in solar cells are produced in this way. Once a monolith has gelled, extraction of the remaining small molecules using a volatile solvent will leave an "aerogel" that has volume and shape, but is mostly unfilled space. Such materials have been used in flotation devices, lightweight structural materials, or filters. Evaporation of the solvent from a gel allows contraction of the system to a xerogel, and subsequent heating above the melting point gives a dense glass. This is a way to produce thin lens preforms. Composite materials called "ceramers" are accessible if a water- or alcohol-soluble organic polymer is included in the original reaction mixture. The organic polymer raises the impact resistance of the final ceramic.

One additional variant of the sol–gel technique is the production of discrete silicate clusters known as *silsesquioxanes or sesquisiloxanes*. As mentioned earlier in this chapter, these are silicate molecules with a cubic shape. They are produced via the hydrolysis of silanes of the type $RSi(OEt)_3$ that have three hydrolyzable groups and one nonhydrolyzable unit. Although hydrolysis might be expected to yield a three-dimensional ultrastructure product, in fact, under appropriate reaction conditions cubic molecules can be isolated (reaction 5). Cubic silsesquioxane structures can even be isolated from the hydrolysis products of tetraethoxysilane. Silsequiox-anes are important for incorporation into composite materials (see Chapter 9).

$$RSi(OEt_3) \xrightarrow[\text{-EtOH}]{\text{H}_2\text{O}}$$ (5)

To summarize, the advantages of the sol–gel route compared to traditional pro-cesses include (1) the "low temperature" synthesis, (2) the ability to incorporate different elements homogeneously at the molecular level, (3) access to lightweight ceramics, (4) the chemical purity of the final product, (5) ease of preparation of ceramic surface coatings, (6) the possibility of ceramic–polymer composites via ceramers, and (7) the ease of access to colored ceramics via transition metal alkox-ides. Disadvantages include (1) the cost of alkoxide precursors, (2) the sol–gel process is a relatively slow, multistep procedure, (3) careful processing control is needed, (4) a high temperature consolidation step may be required, and (5) although the initial shape of the object may be retained during processing, the dimensions of the final object will be much reduced by shrinkage. Overall, it is a process that favors small- to medium-scale synthesis and fabrication for objects and devices that may be unable to withstand the high temperatures needed for conventional fabrication methods.

3. Zeolites

Zeolites are aluminosilicates that, instead of having a layered structure, possess nanometer-sized tunnels or channels that penetrate through the solid. The tunnel structures arise because the ratio of silicate to aluminate residues allows formation of a three-dimensional crosslinked network in which the cations needed to balance the charge of the aluminate anion residues are confined within tunnels rather than layers. Many zeolites have interconnected tunnel structures. Often the tunnels and cavities grow around template ions or molecules that are present in the medium.

Zeolites fall into two categories—those that occur naturally as minerals, and those that are synthesized. In total, 48 different mineralogical zeolites are known, and most or all are of these are believed to have been formed when alkaline water interacted with volcanic exudates. Example structures are shown in Figure 7.16. Mineralogical zeolites are mined in large quantities in Australia, Europe, and Asia and are used for applications that require low cost, such as incorporation into con-crete, treatment of soils, components of laundry detergents and air purification equipment. These materials are produced in quantities that exceed 4 million tons per annum.

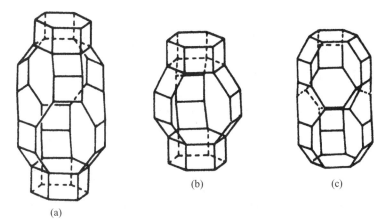

Figure 7.16. *Different-sized cavities and connecting tunnels that exist in mineralogical zeolites: (a) chabazite ($Ca_6Al_{12}Si_{24}O_{72}$ H_2O); (b) gmelinite [($Na_2Ca)_4Al_8Si_{16}O_{48}$ 24 H_2O]; (c) erionite ($Ca_{4.5}Al_9Si_{27}O_{72}$ $_{27}H_2O$). (From Cotton, F. A.; Wilkinson, G.,* Advanced Inorganic Chemistry, *5th ed., 1988. Reproduced with permission.)*

Synthetic zeolites are produced via the sol–gel process, in which silicate and aluminate residues in appropriate ratios condense around organic template molecules such as alkylammonium halides. More than 150 different synthetic zeolite structures have been made in this way. An advantage of synthetic zeolites is that the tunnel size can be controlled by the dimensions of the template molecules, which can then be burned out of the ceramic after the condensation reactions are complete. Another advantage of the synthetic products over their mineralogical counterparts is that they are not contaminated with other inorganic materials. Thus, synthetic zeolites are used for applications where purity and tunnel size are crucial, such as hydrocarbon re-forming in an oil refinery, separation or absorption of small molecules on the basis of their size and shape (size exclusion), or gas purification. Hydrocarbon cracking and reforming occur when the molecules from oil distillates enter the tunnels of a zeolite and interact with the acidic hydroxyl groups that line the tunnels. These electrophilic reactions break carbon–carbon bonds and isomerize the molecules. Shape and size exclusion are also important contributors to this process. Size exclusion also underlies the use of zeolites as "molecular sieves" to remove traces of water or other impurities from organic liquids, and to separate gases on the basis of atomic or molecular size. Water softening occurs when sodium or potassium ions, which provide charge neutralization in the tunnels, are exchanged for calcium or magnesium ions in the "hard" water.

4. Calcium Hydroxyapatite

Calcium hydroxyapatite (HAP) is the mineralogical component of bone and teeth. Calcium hydroxyapatites are various forms of calcium phosphate that are available from natural sources, but can also be synthesized. The importance of these materials is that they play a major role in dentistry (tooth restoration) and in various experimental procedures designed to repair damaged bones (see Chapter 16).

A method for the synthesis of HAP is as follows. Calcium hydroxyapatite is produced at physiological temperature from an aqueous slurry of calcium phosphate precursors or by the hydrolysis of a single phosphate salt. Specifically, hydroxyapatites are produced by the interaction of an acidic calcium phosphate, such as $CaHPO_4 \cdot 2H_2O$ with a basic counterpart such as $Ca_4(PO_4)_2O$ at 37.4 °C. The products formed in this reaction have a variable composition given by $Ca_{10-x}(HPO_4)_x(PO_4)_{6-x}(OH)_{2-x}$ with a Ca/P mol ratio varying between approximately 1.33 ($x = 2$) and 1.67 ($x = 0$). "Stoichiometric" hydroxyapatite has a Ca/P ratio of 1.67, while "calcium-deficient" hydroxyapatite has a ratio of <1.67. The 1.67 ratio in the stoichiometric form is similar to the ratio in bone and teeth. However, the biologically produced material is actually deficient in calcium because it contains some carbonate (CO_3^{2-}) in place of Ca^{2+}. Hence, interest exists in a calcium-deficient modification that has a ratio of 1.5, and in ways to incorporate carbonate into the structure. Note that biological apatites are composites with collagen. This protein provides impact resistance, while the phosphate gives strength. Aspects of this subject related to living tissue regeneration are discussed in Chapters 9 and 16.

D. NONOXIDE CERAMICS

1. General Aspects

Two approaches exist for the preparation of nonoxide ceramics. The first, and traditional, approach involves the fusing together of inorganic elements or simple compounds at very high temperatures. For example, silicon carbide is manufactured on a large scale by heating silicon and coke. The advantage of this direct approach is its simplicity. The disadvantage is the high temperature required for both the synthesis and the melt or sinter fabrication, and the resultant need for copious energy supplies. Nevertheless, ceramic abrasives and cast ceramic objects are routinely made by this method.

The second method is indirect. It requires the initial synthesis of a sacrificial polymer or oligomer that contains the required inorganic elements as an integral part of the macromolecular structure. Such polymers are easily fabricated by either solution techniques or low-temperature melt methods. Subsequent controlled pyrolysis (Figure 7.17) of the shaped polymeric material crosslinks the polymer chains to reduce their volatility; removes volatile products such as methane, ammonia, or hydrogen; and leaves the totally inorganic ceramic. The advantage of this process is that more intricately shaped objects are accessible and the energy penalty is less than in the direct processes. Fibers, "whiskers," and shaped objects of silicon carbide, silicon nitride, boron nitride, aluminum nitride, and so on are produced by this method. However, although the final *shape* of the object may be retained during heating, the *size* will be diminished due to the loss of volatile material.

The pyrolysis process requires tight control of the "temperature program"—the rate of heating, the time spent by the material at each temperature, and the maximum temperature required during the process. There are several reasons for tight temperature control. Unduly rapid heating before or during the intitial crosslinking step may cause the polymer intermediate to melt and lose its shape, or the polymer may depolymerize and the products volatilize. This is illustrated by Figures 7.18 and 7.19,

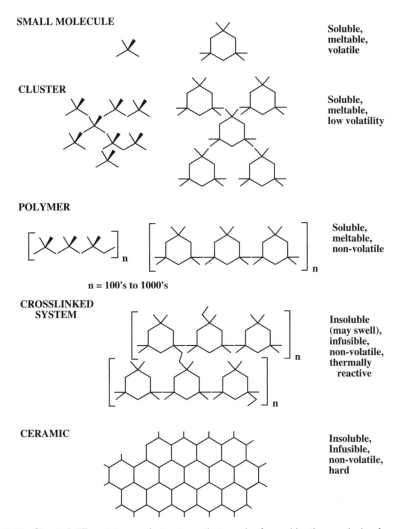

Figure 7.17. *Chart of different types of structures that can be formed by the pyrolysis of preceramic oligomers and polymers.*

in which the preliminary heating of the material is designed to introduce crosslinks, which stabilize the structure and lower the volatility compared to the starting material. Ceramics are generally accessible only through a crosslinked network. The design of polymers or oligomers that will be stable (and soluble) at room temperature, but will readily undergo crosslinking under relatively mild conditions is thus a major challenge in this field.

2. Carbon Fiber

Carbon fiber is a graphitic material with high strength, resistance to thermal decomposition, lightness of weight, and the ability to conduct an electric current. A carbon fiber cloth can be heated with a propane torch until it glows red, but it resists

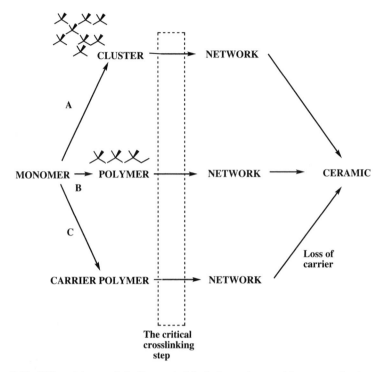

Figure 7.18. *Different types of starting materials that may be used for conversion to ceramics.*

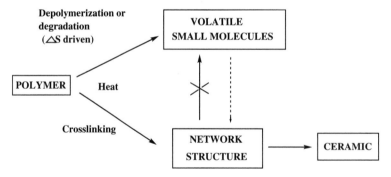

Figure 7.19. *The importance of early crosslinking in the pyrolytic conversion of clusters, and polymers to ceramics.*

decomposition for a long time. Carbon fiber is one of the main materials used to reinforce structural polymers in composites and is widely employed for aerospace and other advanced applications. As a porous form of carbon with a high surface area, it is also valuable as a material for electrodes in fuel cells (Chapter 12). Carbon fiber is produced by the pyrolysis of an organic polymer, such as poly(acrylonitrile) or regenerated cellulose (rayon), to carbon initially in the presence of air, but later under an inert atmosphere. The organic polymer is first spun into fibers and is then either woven into a cloth or handled as a skein or mesh of nonwoven fibers. Heating drives off small molecules to leave a black, graphitic material in the shape of the original fiber or cloth.

Several different organic polymers have been used as the starting point, but polyacrylonitrile has some advantages. It is easily spun into fibers, and its pyrolysis pattern favors the formation of the polyaromatic structure on which the most desirable properties depend. The pyrolysis mechanism is summarized in reaction sequence 6.

Pyrolysis follows a three-step process (reaction 6). The first step occurs when poly(acrylonitrile) fibers are heated at 200–300 °C in air. This causes minor oxidation and the formation of crosslinks. The crosslinks stabilize the structure and prepare the material for the more severe heating to follow. The second phase (known as *carbonization*) requires heating the crosslinked material in high-purity nitrogen at 1200–2000 °C. This causes cyclization of the pendent nitrile groups to form fused heterocyclic rings. The final step of "graphitization" occurs during a brief exposure to temperatures above 2500 °C in nitrogen. All or most of the nitrogen atoms in the structure are lost at this point to give fibers that contain a graphite-like sheet structure. Molecules such as NH_3, CO_2, HCN, and H_2O, are evolved throughout the pyrolysis process. An alternative starting material for carbon fiber production is pitch, which is a complex mixture of fused polyaromatic hydrocarbon species that has a sufficiently high molecular weight to allow it to be melt-spun into fibers.

(6)

The pyrolytic formation of carbon fiber provides an introduction to the conditions used to produce nonoxide ceramics from inorganic–organic macromolecules that contain silicon, boron, aluminum, and other elements.

3. Silicon Carbide

Silicon carbide (SiC) is one of the most widely used nonoxide ceramics. It is a tough, wear-resistant material with a hardness close to that of diamond. Its thermal stability is very high, and it can withstand temperatures of 2800 °C or higher. Other valuable properties include its stability to most aqueous acids, its resistance to bulk oxidation at high temperatures, and its high thermal conductivity and low coefficient of thermal expansion. In addition to these attributes, it is also an intrinsic semiconductor (Chapter 10). As a semiconductor it is incorporated into light-emitting diodes or heating elements. However, by far the largest volume of material is in the form of abrasives (carborundum) and hard coatings for machine tools and in the form of whiskers or fibers used as reinforcing agents in composite materials. Uses for silicon carbide in jet engines, high-temperature heat exchangers, and brakes are under development.

The traditional industrial method for the manufacture of α-silicon carbide is to heat silica (sand) with coke in an electric furnace at 2000–2500 °C. However, the high melting point of this product causes difficult fabrication problems. For this reason great interest has been focused on the synthesis of this material by lower temperature methods. Yajima and coworkers in 1975 were responsible for a preceramic polymer pyrolysis route that is now used commercially for the production of silicon carbide whiskers and fibers.

The process is illustrated in reactions 7 and 8. It begins with the synthesis of an organosilane via the reaction of dimethyldichlorosilane with sodium or sodium–potassium alloy in an organic solvent to yield a mixture of cyclic and linear oligomers with a composition that approximates to $(Me_2Si)_x$ (reaction 7). This mixture is heated at 350–450 °C in an autoclave to yield a "carbosilane" intermediate (reaction 7).

$$ (7) $$

$$ (8) $$

The carbosilane intermediate is a white, soluble material that can be melt-spun into fibers at 190 °C. Rather than having a linear structure as was originally thought, it has a complex configuration that contains both rings and chain segments, as shown in structure **7.1**. A brief exposure of the fibers to air 190 °C produces a surface coating of silica. This serves as an "exoskeleton" to physically stabilize each fiber during the subsequent heating to 1200–1500 °C in nitrogen. As the temperature is raised, methane and hydrogen are driven off to leave a residue of ceramic fiber (reaction 7). A drawback to this process is that the overall ratio of elements in the fiber is $SiC:C:SiO_2 = 1:0.7.8:0.22$. Thus, the ceramic has an excess of carbon, which is detrimental to the properties. The SiO_2 component is the coating that protects the bulk material against oxidation in air at temperatures up to 1000 °C.

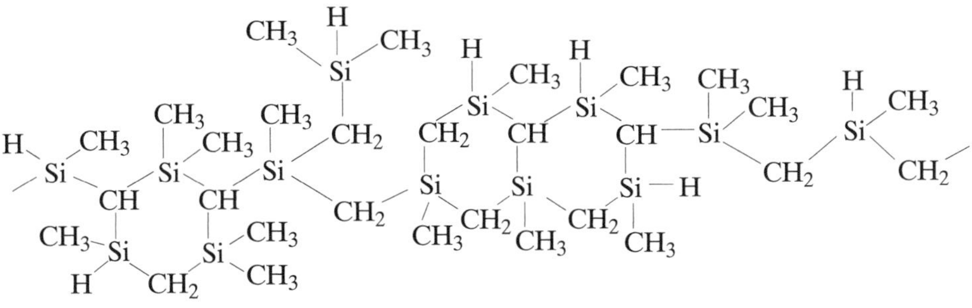

7.1

Attempts have been made to modify the chemistry in order to bring the final ceramic composition closer to $1:1$ $Si:C$ and facilitate the crosslinking step at moderate temperatures. An example of the enhanced crosslinking approach involves the pyrolysis of cyclic and linear oligosilane precursors that bear vinyl or acetylenic side groups that crosslink under relatively mild conditions. However, these ceramics also contain more carbon than silicon. Strategies explored to reduce the amount of carbon in the preceramic materials include the use of $MeSiHCl_2$ as a starting material in place of Me_2SiCl_2. However, this gives a silicon-rich ceramic. Pyrolysis of $MeSiH_3$ also gives a silicon-rich ceramic. An alternative strategy starts from the readily available $MeSiCl_3$, which is converted to $ClCH_2SiCl_3$ and dechlorinated with magnesium to give hyperbranched polysilane clusters of the types shown in structures **7.2** and **7.3**, which are soluble in hydrocarbons and are sufficiently stable to moisture that they can be handled in air.

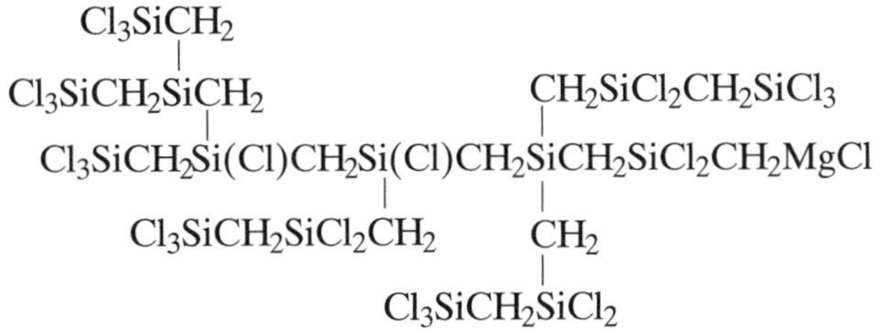

7.2

$$
\begin{array}{c}
\text{CH}_2\text{—SiH}_2\text{—CH}_2\text{—SiH}_3 \\
\hspace{2em}\text{CH}_2\text{—SiH}_2\text{—CH}_2\text{—SiH}_3 \\
\text{CH}_2\!=\!\text{CH—CH}_2\text{—SiH}_2\text{—CH}_2\text{—SiH—CH}_2\text{—Si—CH}_2\text{—SiH—CH}_2\text{—SiH—CH}_2\text{—SiH}_3 \\
\hspace{-4em}|\hspace{10em}|\hspace{10em}| \\
\hspace{-3em}\text{H}_3\text{Si—CH}_2\hspace{10em}\text{CH}_2\text{—SiH}_2\text{—CH}_2\text{—CH}\!=\!\text{CH}_2 \\
\text{H}_3\text{Si—CH}_2\text{—SiH}_2\text{—CH}_2\text{—SiH}_2\text{—CH}_2
\end{array}
$$

7.3

Species **7.3**, with allyl crosslinking groups, can be pyrolyzed to give pure silicon carbide in 89% yield. However, intermediates that contain larger amounts of allyl units give products with progressively larger amounts of carbon beyond the 1:1 Si:C ratio. The products from these pyrolyses are manufactured for use in composites for aircraft brakes and high-temperature coatings.

Finally, an alternative approach is via the ring-opening polymerization of a carbosilane cyclic dimer (**7.4**) as shown in reaction 9, or of derivatives with alkoxy groups in place of chlorine. The linear polymers ($M_n \sim 30,000$) formed in this way give 90% yields of silicon carbide after pyrolysis at 1000 °C. Crosslinking occurs by loss of hydrogen as the temperature is raised.

$$
\begin{array}{c}
\text{CH}_2\text{—SiCl}_2 \\
|\hspace{3em}| \\
\text{Cl}_2\text{Si—CH}_2
\end{array}
\longrightarrow
\begin{bmatrix}
\text{Cl} & \text{H} \\
| & | \\
\text{Si} & \text{C} \\
| & | \\
\text{Cl} & \text{H}
\end{bmatrix}_n
\xrightarrow{\text{LiAlH}_4}
\begin{bmatrix}
\text{H} & \text{H} \\
| & | \\
\text{Si} & \text{C} \\
| & | \\
\text{H} & \text{H}
\end{bmatrix}_n
\qquad [9]
$$

7.4

Silicon carbide nanotubes can be formed by the deposition of cyclic carbosilanes inside alumina nanotubes, followed by pyrolysis and removal of the alumina in hydrofluoric acid.

4. Silicon Nitride

Silicon nitride (Si_3N_4) is a hard (9 out of 10 on the Mohs scale), wear-resistant material with high mechanical strength at elevated temperatures. It has a maximum use temperature near 1800 °C in the absence of oxygen and near 1500 °C under oxidizing conditions. It melts and dissociates into the elements at 1900 °C. It has a relatively low density (3.185 g/cm³). This ceramic is an electrical insulator, and is also relatively stable to aggressive chemicals. This combination of properties has made it a useful material for components in internal combustion engines and jet engines.

Silicon nitride is manufactured by the direct reaction of silicon with nitrogen at temperatures above 1300 °C, or by heating silica with carbon (coke) in a stream of nitrogen and hydrogen at 1500 °C. However, the high temperatures needed for processing and fabrication have led to increased interest in preceramic polymer routes to this material.

Attractive starting materials for the pyrolytic synthesis of silicon nitride are inorganic–organic polymers and oligomers known as polysilazanes. An example is the product formed by the reaction of methyltrichlorosilane with methylamine, which yields the complex compound shown as structure **7.5**. These are species with a silicon–nitrogen backbone and organic side groups or hydrogen linked to each skeletal atom. They are synthesized by the reactions of organochlorosilanes with ammonia or primary amines as shown in reaction 10.

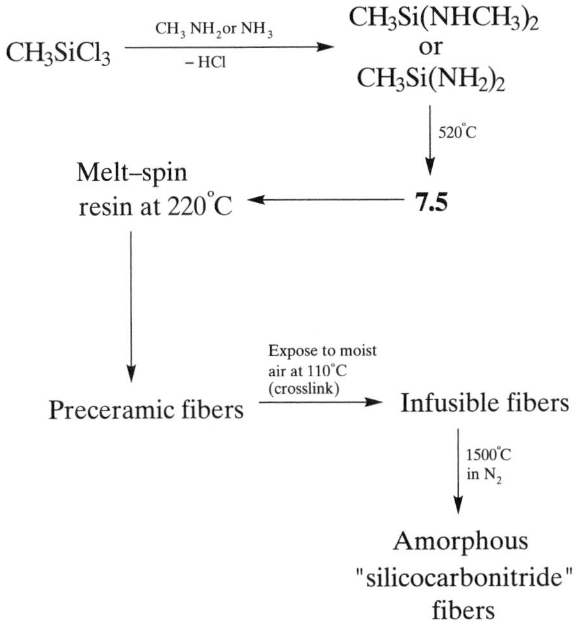

$$CH_3SiCl_3 \xrightarrow[-HCl]{CH_3 NH_2 \text{ or } NH_3} \begin{array}{c} CH_3Si(NHCH_3)_2 \\ \text{or} \\ CH_3Si(NH_2)_2 \end{array}$$

$\downarrow 520°C$

Melt–spin resin at 220°C $\longleftarrow$ **7.5**   [10]

$\downarrow$

Preceramic fibers $\xrightarrow[\text{(crosslink)}]{\substack{\text{Expose to moist} \\ \text{air at } 110°C}}$ Infusible fibers

$\downarrow \substack{1500°C \\ \text{in } N_2}$

Amorphous "silicocarbonitride" fibers

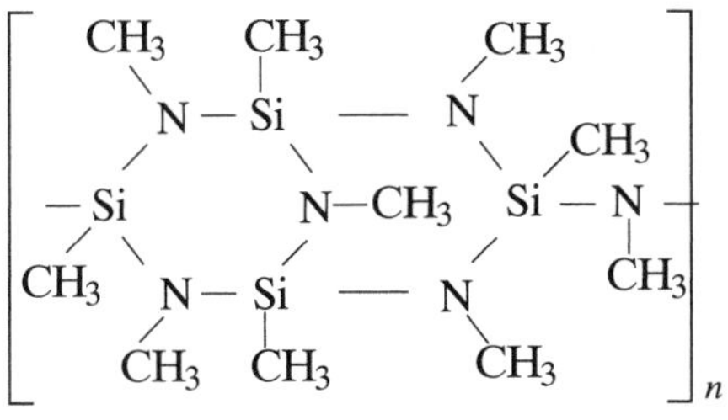

7.5

The products are then melt-spun to form fibers, which are crosslinked and pyrolyzed to give amorphous silicocarbonitride fibers. The high carbon content of the ceramic is responsible for its decomposition at temperatures above 1400 °C via a rearrangement to silicon carbide and nitrogen. Hence, a major challenge is to reduce the carbon content of the final ceramic.

This can be accomplished by the use of carbon-free starting materials such as H_2SiCl_2, but this reagent is difficult to handle. Alternatively, carbon can be tolerated during the processes that lead to the ceramic, as with the use of CH_3SiCl_3, for reactions with ammonia or amines, but the carbon must then be removed from the silico-carbo-nitride by heating with ammonia at ~1000 °C. A similar process that starts from CH_3SiHCl_2 yields spinnable preceramic polymers that contain linked eight-membered silazane rings. Other pyrolysis starting materials are cyclotri- and -tetrasilazane rings that bear vinyl side groups to facilitate crosslinking.

5. Boron Nitride and Other Boron-Containing Ceramics

Boron nitride (BN) is thermally stable at temperatures up to 2730 °C. It is also a good electrical insulator, has a high thermal conductivity coupled with excellent thermal shock resistance, and is chemically inert.

Boron nitride is normally manufactured by the fusion of urea with boric acid in an atmosphere of ammonia at 750 °C. The product is hexagonal boron nitride, which has a layer structure like that of graphite. However, unlike graphite, BN is colorless and is as electrical insulator. The hexagonal modification can be converted to the more stable cubic form by heating at 1800 °C at 85,000 atmospheres (atm) pressure.

As with the other nonoxide ceramics, considerable research has been carried out to develop alternative routes that allow processing and fabrication at moderate temperatures. There are three main approaches: (1) the pyrolysis of borazines, (2) pyrolysis of organic or inorganic polymers that bear borazine groups as side units, and (3) pyrolysis of polyhedral borane derivatives. All three involve methods to reduce the volatility of a boron-containing precursor so that the elements boron and nitrogen are retained as a pyrolysis reaction takes place.

Borazine is an inorganic analogue of benzene, with alternating boron and nitrogen atoms in a six-membered ring and a hydrogen atom linked to each ring atom (see Chapter 3 and structure **7.6**). The ring has an impressive thermal stability and, perhaps more important, has precisely the right ratio of boron and nitrogen to give boron nitride. Side groups attached to the ring atoms provide sites for reactions. Borazine is a colorless liquid obtained by the reaction of boron trichloride with ammonia, followed by reduction with lithium aluminum hydride. At elevated temperatures the hydrogen atoms are lost, with concurrent linkage of the inorganic rings, eventually to give boron nitride (structure **7.8**). The end product of this process is hexagonal boron nitride (reaction 11).

7.6   7.7   7.8

[11]

Although this is conceptually a simple way to produce the ceramic, it involves surmounting a few practical problems. These include the high volatility of borazine, which requires the use of an autoclave and high pressures, combined with the need to bleed off hydrogen gas as the reaction proceeds. An alternative is to carry out the preliminary condensation reactions to produce linked or condensed ring clusters

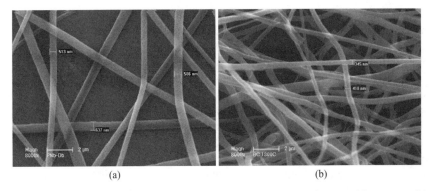

(a)   (b)

Figure 7.20. *(a) Electrospun nanofibers of a polynorbornene with carborane side groups; (b) after pyrolysis to boron carbide nanofibers. (From Sneddon, L.; Wei, X.; Bender, J.; Welna. D; Krogman, N.; and Allcock. H. R.* Mater. Res. Soc. Symp. Proc. *2005, 848, FF1.8.1–6.)*

such as structure **7.7** and then to use these less volatile products for the final pyrolysis to boron nitride at temperatures up to 1000 °C.

Similar strategies revolve around the elimination of chlorine, amines, or organosilanes linked to borazine rings as the compounds are heated to 1000 °C. The sacrificial carrier polymer route has also been employed to lower the volatility of the system. Organic polymers that bear borazine or carborane side groups have been pyrolyzed to boron nitride. If the nitrogen is not present in the polymeric precursor, it is supplied from ammonia gas used in the pyrolysis atmosphere. The carrier polymer route often causes carbon from the polymer as well as boron and nitrogen to be incorporated into the ceramic.

In the absence of ammonia the pyrolysis conditions of, for example, a polynorbornene with carborane side groups can be tuned to give *boron carbide* rather than boron nitride. The precursor polymer has been electrospun to nanofibers that are pyrolyzed with retention of the nanofiber structure (Figure 7.20). In addition, blends of the same polymer with poly(carbosilanes) have been electrospun and pyrolyzed to boron–carbon–silicon ceramics. Polyphosphazenes with borazine side groups have been pyrolyzed to boron nitride and to various boron–nitrogen–phosphorus–carbon pyrolyzates. These processes illustrate the diverse options that are available through the use of preceramic polymers.

6. Aluminum Nitride

Aluminum nitride (AIN) is one of the few materials that are both good thermal conductors and good electrical insulators. It also has a low thermal expansion coefficient, and a low dielectric constant, and does not melt or dissociate until heated above 2200 °C. However, it decomposes by oxidation in the atmosphere above 1370 °C. Other useful properties are its wear resistance, thermal shock resistance, and chemical stability to molten metals. It is also of interest as a wide-bandgap (6.3 eV) semiconductor for LED emission in the ultraviolet (see Chapter 10).

The traditional production of aluminum nitride is via the carbon-based thermal reduction of alumina in the presence of nitrogen or ammonia, or the reaction of metallic aluminum with nitrogen. The pure ceramic is a pale-blue material.

For the reasons described above, a need exists for low-temperature access routes to this ceramic. For example, $LiAlH_4$ or AlH_3 reacts with ammonia to give $Al(NH_2)_3$, which, when heated, loses ammonia and hydrogen to give AlN contaminated by carbon impurity from the processing solvent.

Ethylalazanes, formed by the reactions of triethylaluminum with ammonia, yield oligomers or polymers with the formula $[(EtAlNH)_x(Et_2AlNH_2)_y(Et_3Al)_z]_n$, which probably consist of linked aluminum–nitrogen rings and chains. These can be melt-spun and pyrolyzed in ammonia gives aluminum nitride fibers.

Aluminum alkyls, AlR_3 (R = Me, Et, or i-Bu), when treated with ammonia, initially give cyclic aluminum–nitrogen rings and chains such as $(R_2AlNH_2)_x$, together with crosslinked species RAlNH by pyrolytic loss of RH. Further pyrolysis gives a high-purity and oxygen- and carbon-free AlN in nearly quantitative yield. Alloys of SiC and AlN are accessible by the copyrolysis of precursors to both silicon carbide and aluminum nitride. Copyrolysis takes place initially at 170 °C, but later at temperatures up to 350 °C under nitrogen, and then up to 2000 °C to give a homogeneous SiC/AlN ceramic.

7. Other Ceramics Formed by the Preceramic Polymer Process

There is almost no limit to the variety of nonoxide ceramic-type materials that can be produced by the pyrolysis of inorganic or organic–inorganic polymers, provided the temperature profile of the pyrolysis can be controlled by one of the methods described earlier. For example, a ferromagnetic ceramic material has been isolated by the pyrolysis of poly(ferrocenylsilanes) (Chapter 6) in a nitrogen atmosphere. The final ceramic is a black, lustrous material that is attracted to a bar magnet. Its composition is 11% iron, 17% silicon, and 58% carbon, with the carbon concentrated at the surface. Pyrolysis of polyphosphazenes (Chapter 6) that bear amino side groups results in a loss of amine and the formation of various phosphorus nitride and phosphorus–nitrogen–carbon species.

Thus, the preceramic polymer approach provides a strong connection point between fundamental inorganic and organometallic chemistry and the more widely applied areas based on ceramics. An important component of this field is the incorporation of synthesized ceramics into composite materials in order to optimize different sets of properties. This topic is discussed in Chapter 9.

E. FABRICATION OF CERAMICS AND GLASSES

1. General Comments

As mentioned throughout this chapter, a major challenge in ceramic science is the fabrication of ceramics into shaped objects. This challenge results from the high melting or softening temperatures that are typical of ceramic materials and the severe energy penalties that are characteristic of these processes. The following summary of fabrication methods begins with the traditional processes that consume large amounts of energy, and then moves on to consider lower-temperature fabrication processes.

2. Sculpting

The early history of the human race is characterized by the development of methods for the fabrication of readily available mineralogical materials by the manual shaping of objects such as flint arrowheads and other weapons using primitive tools. Of course, these techniques, using hammer and chisel or diamond saws on granite, marble, or sandstone, are still employed by artists and artisans to produce architectural stone, countertops, sculptures, or inscriptions on gravestones, memorials, or important buildings. The process is slow, is often expensive, and requires considerable skill—aspects that are accepted because of the permanence of the final product.

3. Melting, Extrusion, and Molding

For those ceramics and glasses that have melting points below 1000 °C, a common method of fabrication is to extrude the molten material through spinerettes to give fibers, or injection of the melt into molds followed by cooling. These are heavy industrial processes that consume large amounts of energy, but that account for a substantial percentage of the fabricated ceramic objects produced commercially. The method is particularly appropriate for oxide ceramic glasses, as described earlier in this chapter. An example is plate glass cast by the "float" process on a bath of molten tin and allowed to cool below the T_g. This is a much flatter form of window glass than material extruded between rollers or, as in historical times, blown into a bubble by a glassblower and then pressed on to a flat surface. Other examples include the blow molding of glass bottles and jars, the injection molding of lens and prism preforms, and the extrusion of molten glass through spinerettes to form fiberglas used for thermal and sound insulation. The final shapes of lens and prism preforms must be refined by grinding and polishing because the molding process does not produce shapes with the needed precision. The melt extrusion of optical fibers is a specialized process that is discussed in Chapter 14. The melt coating of glasses on the surface of porous ceramics serves to prevent water absorption and provides a medium for the decoration of coffee mugs, fine china, and stained glass.

As discussed above, melt processing is also appropriate for the fabrication of preceramic polymers before they are pyrolyzed, with the understanding that the dimensions of the fiber or shaped object will be reduced during pyrolysis.

4. Powder Sintering

Aluminosilicate clays and high-melting-point oxides are fabricated by conversion to a powder, followed by sintering. The powders may be produced by grinding processes, by precipitation from solution, or the mineralogical material may already be in a microparticulate form.

Sintering is a process by which a powder is shaped in a mold or shaped by hand as a water–clay mixture, and heated to below the melting point until the outer layer of each particle softens and fuses with its neighbors (Figure 7.21). This process depends on the principle that the surface region of a solid particle has a lower melting temperature than does the interior bulk material. Thus, a coherent solid can be produced without the need to heat to the melting point. A characteristic of many sintered ceramics, such as a housebrick, concrete blocks, or unglazed china is their porosity, which lowers the weight slightly but allows water to penetrate deep into

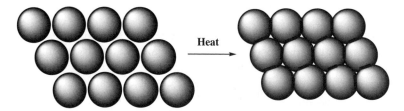

Figure 7.21. Ceramic fabrication by powder sintering. The process involves heating the powder to a temperature at which the outer layer of each particle softens or melts and fuses with its neighbors. The inner region of each particle remains unchanged.

the shaped object. The porous morphology is also a source of the brittlemess that is characteristic of such materials.

5. Sol–Gel Fabrication

The fabrication of ceramics via the sol–gel method is particularly useful for the low-temperature preparation of thin films, aerogels, or bulk objects with relatively small dimensions. In practice, the preceramic material must be shaped at the "sol" stage, while it is still able to flow. Once extensive crosslinking gets underway the shape is, for all practical purposes, fixed. Note that sol–gel fabrication involves extensive shrinkage as the reaction solvent and the eliminated alcohol molecules are lost during the heating cycle. This step must be carried out sufficiently slowly to prevent cracking of the material, which becomes increasingly likely with thicker objects. Thus, the slow-heating requirement reduces the utility of the process for the commercial production of thick objects such as lenses or prisms. However, the process is extremely useful for the fabrication of products such as thin films and fibers. It is also the basis of the production of specialized powders of high purity with micro- or nanodimensions for solar cell and other advanced applications (see Chapter 14).

F. FUTURE CHALLENGES IN CERAMICS AND GLASS SCIENCE

Ceramics and inorganic glasses have numerous advantages over other materials, including thermal stability, rigidity, and radiation resistance. However, they are heavy and often brittle. Thus, a continuing challenge is to retain their beneficial properties while attempting to minimize their weaknesses. Several approaches are proving to be especially beneficial:

1. The formation of composite materials between ceramics and organic or inorganic–organic polymers—either in the form of homogeneous or phase-separated composites or as laminates—has attracted a great deal of attention. Laminated automobile windshields are a well-known traditional example that provides a starting point for many similar applications. The coating of scratch-resistant silicate surfaces on polymeric eyeglasses is another, as is the use of glass fiber and carbon fiber–reinforced polymers. The incorporation of silsesquioxanes into polymers is receiving considerable attention.

2. Lightweight oxide ceramics such as aerogels promise to solve many of the weight-related problems of inorganic materials. They have been proposed for uses as varied as components in aircraft wings to structural materials in automobiles, boats and ships, or buildings. Their non-flammability is a powerful advantage for these applications.

3. The development of synthesized nonoxide ceramics has a strong future in the field of high-temperature materials that range from components in aircraft engines and rocket motors, to bearings, machine tools, and brakes, where high performance rather than low cost is the primary requirement. In this connection, a thorough understanding of the geologic processes that led to the formation of layered and fibrous silicates and aluminosilicates has the potential for generating new synthetic ceramics with unusual and potentially useful characteristics.

G. SUGGESTIONS FOR FURTHER READING

1. Greenwood, N. N.; Earnshaw, A., *Chemistry of the Elements*, 2nd ed., Butterworth-Heinemann, Oxford, 1997.

2. Brinker, C. J.; Scherer, G. W. *Sol-Gel Science: The Physics and Chemistry of Sol-Gel Processing*, Academic Press, New York, 1990.

3. Hench, L. L.; West, J. K., The Sol Gel Process, *Chem. Rev.* **90**:33–72 (1990).

4. Morgan, P., Carbon, Fibers, and their Composites, Taylor & Francis, ISBN 0824709837.

5. Burchell, T. D. (ed.) *Carbon Materials for Advanced Technologies*, Pergamon, 2000.

6. Sauls, F. C.; Interrante, L. V.; Shaikh, S. N.; Carpenter, L.; Gladfelter, W. L. Aluminum Nitride, Inorganic Syntheses, (Murphy, D. W.; Interrante, L. V., eds.), **20** 2007.

7. Interrante, L.V.; Moraes, K.; Liu, Q.; Lu, N.; Puerta, A.; Sneddon, L. G., Silicon-based ceramics from polymer precursors, *Pure Appl. Chem.* **74**:2111–2117 (2002).

8. Interrante, L. V.; Hampden-Smith, M. J., *Chemistry of Advanced Materials: An Overview*, Wiley, 1997.

9. Reed, C. S.; TenHuisen, K. S.; Brown, P. W.; Allcock, H. R., "Thermal stability and compressive strength of calcium-deficient hydroxyapatite composites," *Chem. Mater.* **8**:440–447 (1996).

10. Seyferth, D.; Wood, T. G.; Tracy, H. J.; Robison, J. L., Non-stoichiometric silicon carbide from an economical polysilane precursor, *J. Am. Ceram. Soc.*, **75**:1300–1302 (1992).

11. Wynne, K. J.; Rice, R. W., Ceramics via polymer pyrolysis, *Ann. Revs. Mater. Sci.*, **14**:297 (1984).

12. Wang, C.-H., Chang, Y.-H., Yen, M.-Y. Peng, E.-W.; Lee, C.-Y., Chiu, H.-T. Synthesis of silicon carbide nanostructures via a simplified process, *Advanced Materials*, **17**(4):419–422 (2005).

13. Motz, G. Synthesis of SiCN-precursors for fibers and matrices, *Adv. Sci. Technol.*, **50**:24–30 (2006).

14. Welna, D. T.; Bender, J. D.; Wei, X.; Sneddon, L. G.; Allcock, H. R., Preparation of boron carbide nanofibers from poly(norbornenyl-decaborane) single-source precursor via electrostatic spinning, *Advance Materials*, **17**:859–862 (2005).

15. Seyferth, D.; Lang, H.; Sobon, C. A.; Borm, J.; Tracy, H. J.; Bryson, N., Chemical modification of preceramic polymers, *J. Inorg. Organometall. Polymers*, **2**:59–77 (1992).

H. STUDY QUESTIONS (for class discussions or essays)

1. Cathedrals and public buildings throughout the world are disintegrating because of the effects of air pollution on the sandstone or limestone used for their construction. Assume that you have been retained as a consultant by an affected city and asked to propose a solution to the problem. What chemical reactions would you suspect are the cause of the disintegration, and what steps would you advise the city authorities to take to prevent any further damage?

2. Discuss the main challenges involved in the preceramic synthesis of nonoxide ceramics such as silicon carbide, silicon nitride, and aluminum nitride. What difficulties might be involved with scaling up these reactions to the 1000-kg/batch level?

3. Assume that you have been asked to design a method that uses the sol–gel process for the fabrication of specialized lenses. What steps would you take, and what challenges might you encounter?

4. Discuss the reasons why some silicates and aluminosilicates form solid-state structures that contain galleries, and some do not. What chemistry can you suggest to turn one type into the other?

5. Highly porous silica has been proposed as a material for the construction of aircraft wings and tail assemblies. What would be the advantages and disadvantages of such structures, and how would they be fabricated?

6. Discuss the advantages and disadvantages in the use of carbon fiber in aerospace, marine, and automotive applications, and make comparisons to glass fiber, silicon carbide, silicon nitride, and boron nitride fibers.

7. Suggest specific elements or compounds that you would add to the starting materials to prepare a colored glass optical filter, either by the direct fusion of the starting materials or via the sol–gel process.

8. What advantages exist for the powder processing of ceramics? Explain the principles involved.

9. Mineral silicates can form polymer chains, ladder structures, sheets, and extended three-dimensional solid structures. Discuss the structures of these materials and speculate how they may arise through geochemical reactions.

10. Review in detail the reactions that are employed to convert organosilicon halides to silicon carbide and silicon nitride. Stress the problem points in these syntheses, and explain why these relatively expensive processes are commercially viable.

8

Metals

A. IMPORTANT ASPECTS OF METAL SCIENCE AND TECHNOLOGY

1. Background

Most of the elements in the periodic table are metals, which is one reason why they play such an important role in materials science. Metals have been employed as useful materials since the early history of the human race. Indeed, historical eras are sometimes named in terms of the metals that had a major influence on the development of civilization—the Bronze Age, the Iron Age (Figure 8.1) and so on. What this means is that the discovery of specific metals or of methods that isolate them had a major impact on everyday life (cooking vessels, body decoration, etc.) and on warfare (swords, daggers, body armor, etc.). Our own era could well be described as the "Age of Steel and Aluminum" since these two metals dominate nearly all aspects of modern civilization, from transportation to building construction.

However, many different metals play a major role in modern materials science, including not only steel and aluminum and their alloys but also copper, silver, gold, titanium, zinc, chromium, nickel, magnesium, and tin as the most obvious examples, as well as other metals such as uranium or beryllium, which serve crucial roles in particular applications.

To a large extent, the utility of a metal depends on the availability of its ores, its ease of isolation from them, and hence its cost. Gold would be used more widely as an engineering material if it were available in quantities as large as, for example, iron or aluminum. Titanium would long ago have replaced steel or aluminum in many applications if it could be extracted more easily from its ores. The costs of

Introduction to Materials Chemistry, by Harry R. Allcock
Copyright © 2008 by John Wiley & Sons, Inc.

Figure 8.1. *Traditional use for iron and steel: vintage steam locomotive from the 1800s.*

metals vary from year to year, but the approximate values relative to steel (at 1, as baseline) are aluminum 7, copper 8, silver 600, titanium 1000, gold 90,000, and platinum 173,000. This will give some idea of the constraints that exist to expanding the uses of many metallic materials.

2. Advantages and Disadvantages of Metals as Materials

The widespread use of metals is a result of their obvious attributes such as strength, hardness, and high melting points. Some metals such as copper are flexible, malleable, and ductile. Most are good electrical conductors, and a few are magnetic. Many metals have surfaces that can be polished to give a high reflectivity and, although most metals have a "white" or gray surface color, some such as gold or copper have distinctive yellowish colors. Disadvantages also exist. Most metals are heavier than polymers or ceramics. Moreover, corrosion, surface oxidation, and instability to acids or bases are common defects. Many of the less common metals are expensive. In addition, the extraction of metals from their ores often involves penalties such as the expenditure of large amounts of energy and the production of atmospheric and groundwater pollution.

3. Scope of This Chapter

Although most of the elements in the periodic table are metals, a discussion of all of these is clearly beyond the scope of this book. Instead, the following sections deal with the 10 or so metals that are used most widely in modern technology: iron, nickel, chromium, aluminum, magnesium, titanium. tin, copper, silver, and gold. The following sections deal with the isolation of specific metals, corrosion, and solid-state structure with its relationship to properties, alloys, electrical and thermal conductivity, color, magnetism, and metal fabrication.

B. ISOLATION OF SPECIFIC METALS FROM THEIR ORES

General methods for the extraction of metallic elements are described in Chapter 3. Here we deal in more detail with the isolation of specific metals.

1. Iron and Steel

Iron is the fourth most abundant element in the earth's crust, and it has played a crucial role in the development of human skills and technology since about 1500 BC. Steel is an alloy of iron with small amounts of carbon. The properties of iron and steel that underlie their widespread use include their toughness and ductility at red heat, and the low cost of production. Until the industrial revolution in the late 1700s charcoal, obtained by the pyrolysis of wood, was employed as the reducing agent to convert iron oxides to the metal (Figure 8.2).

This placed limits on the scale of manufacturing, gave rise to serious air pollution, and caused widespread deforestation. For example, even in the 1800's, large areas of the Appalachian ridges in Pennsylvania were being clear-cut to provide charcoal for local iron furnaces. However, earlier, starting in about 1770, a new development occurred in England that revolutionized both the manufacturing process and the course of technology. This was the replacement of charcoal as a reducing agent by coke, which was readily available from plentiful supplies of coal. This resulted in a dramatic increase in the scale of production and a fivefold reduction in the price of the metal. It also initiated the industrial revolution, first in Britain and later worldwide.

The main mineralogical sources of iron are Fe_2O_3 (hematite), FeO_2H, $Fe_2(OH)_3$, and $FeCO_3$. Geologic hydrothermal reactions probably generated these ores, perhaps

Figure 8.2. *Representation of the truncated fieldstone blast furnace at Greenwood Furnace in central Pennsylvania, which operated between 1834 and 1904. Vast tracts of trees were felled in widening circles in the surrounding forest and then slow-burned to produce the charcoal used as a reducing agent. Widespread air pollution and deforestation were typical of this era. (Courtesy of the National Parks Service.)*

from FeS$_2$. Isolation of the impure metal is accomplished by reduction of the ore at high temperature using coke and carbon monoxide, itself formed by partial oxidation of the coke. The reduction process takes place in a vertical, thermally insulated cylinder known as a "blast furnace" (Figure 8.3).

The iron ore is fed into the top of the furnace together with limestone (calcium carbonate) and coke. Oxygen (air) is forced into the base of the furnace, where it oxidizes carbon to carbon monoxide. As the mineralogical charge moves down the furnace increasing temperatures cause different chemical reactions, as shown in reactions 1–6.

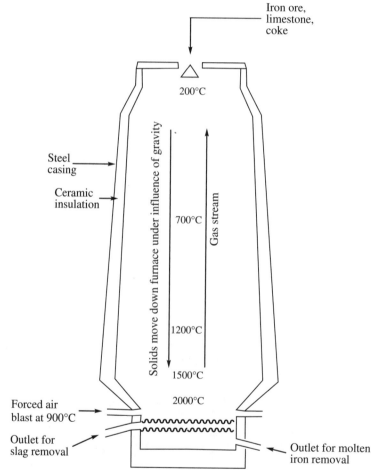

Figure 8.3. Schematic view of a modern blast furnace. The highest temperatures are at the base of the furnace, where hot, compressed air reacts with coke to generate carbon monoxide. The solid charge is fed into the top of the furnace and works it way down, increasing in temperature as various reactions (shown in reactions 1–6 in this chapter) take place. The impure iron and slag melt between 1200 °C and 1500 °C and collect at the base of the furnace. The slag, which is lighter, is tapped from the top of the molten iron. The molten metal is tapped from a lower point, and is fed into channels where it solidifies.

$$3Fe_2O_3 + CO \xrightarrow{200-700\,°C} 2Fe_3O_4 + CO_2 \tag{1}$$

$$Fe_3O_4 + CO \xrightarrow{200-700\,°C} 3Fe_xO_y + CO_2 \tag{2}$$

$$CaCO_3 \xrightarrow{200-700\,°C} CaO + CO_2 \tag{3}$$

$$C + CO_2 \xrightarrow{700-1200\,°C} 2CO \tag{4}$$

$$Fe_xO_y + CO \xrightarrow{700-1200\,°C} Fe_{solid} + CO_2 \tag{5}$$

$$2C + O_2 \xrightarrow{2000\,°C} 2CO \tag{6}$$

Finally, at ~2000 °C, molten iron is bled from the base of the furnace, followed by the lighter, floating slag. The latter consists mainly of $CaSiO_3$. The molten iron is directed into channels formed in sand, where it solidifies to an impure solid known as "pig iron." The impurities in this material, which consist of carbon, phosphates, phosphorus, sulfur, silicates, and silicon, cause it to be a brittle material, with few uses.

Refining to reduce the level of impurities, is then carried out by remelting the pig iron and blowing air or (more commonly) oxygen through it to oxidize the impurities to gaseous products or to more slag. Oxidation reduces the amount of carbon to the acceptable levels required in steel (0.5–1.5%). Further decreases in carbon content lead to the softer "wrought iron" used in decorative ironwork and similar applications. The world production of steel exceeds one billion tons per annum.

The properties of iron and steel are modified over a broad range by the addition of other metals to form alloys with Ni, Cr, Mn, or Mo (see Section D.2 later in this chapter). These alloys have improved corrosion resistance, flexibility, toughness, and impact resistance compared to pure iron or mild steel.

2. Nickel

Nickel ores include the metal sulfides and a number of mixed silicates with magnesium or iron. By far the largest known source of nickel is the sulfide deposit at Sudbury in northern Ontario, Canada, which yields more that 25% of all the nickel produced worldwide. Other sources are in Australia and Siberia. Total world production is about 1.2 million tonnes (long tons) per annum.

Nickel is isolated from its sulfide ores by the addition of sand followed by high-temperature oxidation ("roasting"), which converts iron sulfide impurities to SO_2 and iron silicates (slag). The nickel sulfide is then converted to nickel oxide, which can be used directly for mixing with iron oxides for conversion to steel alloys. Alternatively, the oxide is reduced to the metal with carbon, followed by electrolysis to give pure nickel. Another process involves reduction of the oxide by a mixture of hydrogen and carbon monoxide (a mixture formed by passing water over hot coke) and conversion of the nickel to the volatile nickel carbonyl [$Ni(CO)_4$]. Decomposition of the carbonyl at 230 °C gives 99.95% pure nickel, plus carbon monoxide, which is recycled.

Nickel-containing steel alloys are used in magnets, stainless steels, armor plating, household utensils, and hospital instruments. When alloyed with copper, nickel forms the corrosion-resistant monel metal. Alloys of nickel with chromium and iron are widely used as various stainless steels, for which the *coding number* (e.g., type *x–y*) refers to the percentage of chromium (*x*) and nickel (*y*). Nickel alloyed with aluminum is used in gas turbine blades because of its ability to withstand high temperatures (~1000 °C).

3. Chromium

The main producers of chromium ores are Australia, the countries of the former Soviet Union, and South Africa. The principal ore is chromite ($FeCr_2O_4$). This is reduced by coke in an electric furnace to yield the alloy "ferrochrome," which may be added directly to steel to produce stainless steel. Alernatively, chromium(III) oxide (Cr_2O_4), formed by alkaline air oxidation of chromite at elevated temperatures, is reduced to chromium by carbon, and is then dissolved in sulfuric acid and electrolytically deposited on steel or nickel-plated steel give corrosion-protected objects.

4. Aluminum

The metals mentioned above are typical "heavy" metals. However, lighter metals such as aluminum, magnesium, and titanium offer many advantages where weight is a crucial factor such as in aircraft, bicycles, automobile engines, electrical transmission lines, construction beams and joists, and ships and boats. Although once considered to be a semiprecious metal, aluminum is now produced in such large quantities and at such a low cost that it has replaced steel in many applications.

The principal ore from which aluminum is extracted is *bauxite*—a general name for a range of minerals that contain a mixture of iron oxides or hydroxides and aluminum oxide and hydroxides such as Al_2O_3 or $Al(O)OH$. The first step in aluminum production is extraction of the aluminum oxide or hydroxide from the ore using aqueous sodium hydroxide, followed by precipitation of $Al(OH)_3$ and dehydration at 1200 °C to alumina (Al_2O_3). The alumina is then dissolved in hot synthetic cryolite (Na_2AlF_6), itself produced from $Al(OH)_3$, HF, and NaOH. This solution is then electrolyzed at ~950 °C to yield pure aluminum.

The isolation of each ton of aluminum requires about 15,000 kilowatt-hours (kWh) of electricity, which explains why the mineral and the alumina are often produced in one location, while the electrolysis is carried out in another. Australia is the biggest supplier of ore (38%), followed by Papua New Guinea (13%) and Jamaica (11%), while regions with plentiful hydroelectricity, such as China, Russia, Canada, and the United States, are the principal producers of the metal. In principle, the largest potential source of aluminum is kaolin and related clays (see Chapter 5). In practice, the cost of extracting the metal from this source is not economically viable.

Aluminum is often alloyed with magnesium, zinc, silicon, manganese, or copper to generate the most favorable combinations of properties for different applications. The pure metal is lightweight, highly reflective, malleable, and flexible in the form of thin foils. However, it has only limited strength. It is one of the best electrical conductors. The metal has good corrosion resistance in spite of its high inherent reactivity, and this is due to the presence of a tough surface coating of the oxide.

However, the oxide coating is penetrated by salt water or strong base, and these liquids cause corrosion.

5. Magnesium

Magnesium is another metal that is prized for its lightness of weight (density $1.74 \, g/cm^3$). It is, in fact, lighter than aluminum. Hence the widespread use of magnesium and its alloys in aircraft, bicycles, rockets, and automobile components. Many mineralogical sources of magnesium exist, including dolomite (calcium magnesium carbonate), magnesite ($MgCO_3$), and brucite [$Mg(OH)_2$], plus a range of water-soluble halides, nitrates, and carbonates found in salt deposits and seawater. Isolation of the metal is either by electrolysis of molten $MgCl_2$ at $750 \, °C$ or by heating dehydrated ("calcined") dolomite above $100 \, °C$ in the presence of ferrosilicon.

6. Titanium

Titanium is often described as a "wonder metal" because it combines many of the most prized attributes of a metallic material. It is tough, has a very high melting point ($1660 \, °C$), is only half as heavy as steel, and has excellent resistance to corrosion and aggressive chemicals. Unfortunately, it is also one of the more difficult metals to isolate from its ores, which explains its high cost and its relatively low volume of production.

The main mineralogical sources are rutile (TiO_2) and ilmenite ($TiFeO_3$) found in Brazil, Switzerland, the United States, Russia, and Africa. Conventional reduction of these by heating with carbon is not feasible because titanium forms a very stable carbide rather than being reduced completely to the metal. Instead, the ore is heated with carbon and chlorine to form either $TiCl_4$ or a mixture of $TiCl_4$ and $FeCl_3$ (from ilmenite). The $TiCl_3$ is then isolated by fractional distillation, and is reduced to the metal by reaction with metallic magnesium or sodium. Finally, extraction of the product with water or hydrochloric acid removes magnesium chloride or unreacted magnesium, and the titanium is recovered as a porous "sponge." Fabrication of the metal requires the use of challenging conditions such as melting in an unreactive vessel under an atmosphere of argon or under vacuum. Powder fabrication techniques are complicated by the fact that powdered titanium, like powdered magnesium or aluminum, ignites when heated in air, which is an indication of its strong affinity for oxygen. Uses for titanium include golf clubs, armor, aerospace components, camera exteriors, and chemical reactors or autoclaves.

7. Tin

The main source of tin is SnO_2 (cassiterite), with Malaysia as the largest supplier of the ore. Metallic tin is obtained from the ore by carbon reduction at $1200–1300 \, °C$. Subsequent heating in the presence of oxygen oxidizes iron impurities to oxides, which are then removed. The principal manufacturers of the metal are China, Indonesia, and Peru.

Tin is a low-melting-point metal (mp $232 \, °C$) used for making "tin plate" (a steel sheet coated by molten tin). Tin is also a component of solder (a tin–lead alloy).

Bronze is a copper–tin alloy, and white pewter is an alloy of tin with small amounts of antimony and copper. A bath of molten tin is used as the casting surface for the fabrication of float glass, as described in Chapter 7.

8. Copper

Copper is found in geological formations in the form of sulfides such as Cu_2S or $CuFeS_2$, and as the oxide, Cu_2O and the carbonate, $Cu_2CO_3(OH)_2$. The main sources of the ores are in North and South America, Russia, and Africa. Chile, the United States, and Peru are the largest refiners of the metal. Some very pure metallic copper is found in geologic formations.

The traditional method for extraction of the copper from sulfide ores involved grinding to a powder, froth flotation. Heating in air ("roasting") to partially convert Cu_2S to Cu_2O, which then react to form impure copper. The overall process is illustrated by reactions 7 and 8.

$$2Cu_2S + 3O_2 \xrightarrow{\text{heat}} 2Cu_2O + 2SO_2 \tag{7}$$

$$2Cu_2O + Cu_2S \xrightarrow{\text{heat}} 6Cu + SO_2 \tag{8}$$

During this reaction sequence iron sulfide impurity is also oxidized and is removed as slag. The sulfur dioxide liberated by this process is an environmental problem. Some copper oxide ores can be converted directly to the impure metal by heating with coke. Final purification is accomplished by dissolution in sulfuric acid and electrolysis.

However, a more modern method that now accounts for at least 20% of all copper production employs a solvent extraction process. The crushed ore is leached with dilute sulfuric acid in the presence of ferrous salts. Sulfide ores undergo bacterial oxidation, while the acid dissolves the resultant copper oxide. The acidic solution is then extracted with kerosene that contains a copper "extractant" (presumably a chelating agent). The chelated copper is captured from the chelate using concentrated sulfuric acid, and the resultant solution is electrolyzed to give 99.99 % pure copper at the cathode. Only this purest copper is suitable for electrical applications such as household wiring, electric motor and generator windings, or long-distance power lines. However, copper has been largely replaced by aluminum for this last application. Note that large amounts of copper are alloyed with zinc to form brass, and lesser amounts with tin to form bronze.

9. Silver

Copper, silver, and gold are known as the "coinage metals" because of their use in currency. Silver is less expensive than gold, but more valuable than copper. Metallic silver is sometimes found in geologic formations in association with silver sulfide from which it was presumable formed. Deposits of silver chloride are found in Australia and Chile. However, much of the silver in circulation today is recycled metal or material recovered from photographic processing laboratories. Most new silver (10,000 metric tons per annum) is obtained as a byproduct of lead manufacture or from the insoluble material deposited below the anodes in electrolytic copper purification cells. This material (known as "anode slime"). It is dissolved in acid and electrolyzed to give pure silver. The main producers of silver are Peru, Mexico, Australia, and China.

Apart from its use in jewelry and coinage, large amounts of silver are still used in traditional photography after conversion to light-sensitive silver halides. This use is declining rapidly with the growth of digital photography, but will probably remain significant for the foreseeable future. Silver is also used in batteries, mirrors, and electrical devices.

10. Gold

The traditional sources of gold are in regions with histories of volcanic activity—Central America, California, Alaska, and New Zealand—but the major producers today are South Africa, Australia, the United States, China, and Russia. "Placer" gold is the metal in the form of nuggets or grains that, being heavier than the eroding rock, sink to the bottom of streams and are isolated by "panning." However, most gold is obtained from crushed rock by extraction with aqueous solutions of sodium cyanide coupled with air oxidation (reaction 9).

$$4Au + 8NaCN + O_2 + 2H_2O \rightarrow 4Na[Au(CN)_2] + 4NaOH \tag{9}$$

The addition of zinc dust causes precipitation of the gold, which is then refined electrolytically to give 99.95% pure metal. One of the main advantages of gold as a technological material is its resistance to surface oxidation and corrosion; hence its use in electrical switches, contacts, and other devices. Its inertness is also the main reason for its widespread use in dentistry. Among the few reagents that attack gold are cyanide ion plus a source of oxygen and aqua regia, which is a 3:1 mixture of concentrated hydrochloric and nitric acids. Mercury readily forms an amalgam with gold, which is the reason why gold jewelry must be protected against mercury in the laboratory or in dental work.

C. CORROSION

One of the main defects of metals as materials is the tendency of many of them to corrode in contact with water, brine, polluted air, or strong chemical reagents. Metals such as gold, chromium, or titanium are much more resistant than iron, magnesium, or lithium. Why are some metals prone to corrosion while a few are apparently immune, and what can be done to protect the most sensitive metals from this problem?

Two major factors determine how a metal behaves in contact with air, water, or other reactive reagents. First, it should be recognized that very few metals actually present a metallic surface to the environment. Common metals such as aluminum, iron, or magnesium undergo almost instant surface oxidation when exposed to air. Lithium, the lightest of all metals, forms a surface coating of lithium nitride in nitrogen or a mixture of oxide, hydroxide, and nitride when exposed to air. A thin layer of surface oxide can either protect against further reaction, as in the case of aluminum, or sensitize the surface to a deeper reaction, as in the case of iron. To a large extent,

TABLE 8.1. Galvanic Series of Common Metals

Metal	Approximate Voltage with Seawater as the Electrolyte
Magnesium	−1.6
Zinc	−1.0
Aluminum	−0.8
Cast iron	−0.66
Steel	−0.65
Brass	−0.35
Copper	−0.32
400-series stainless steel	−0.28
Silver	−0.12
Titanium	0.00
Hastelloy C-276[a]	+0.06

[a]An acid-resistant alloy of Mo, Cr, Fe, W, and Ni.

the outcome depends on the strength of adhesion between the surface oxide later and the metal, and on the resistance of the oxide to dissolution in weak acids, bases, or other chemicals found in the environment. For example, aluminum is quite stable in a neutral, dry atmosphere owing to the protection by the oxide layer, but that layer is removed by aqueous acid or base or by prolonged exposure to seawater.

The second factor comes into play when galvanic cells are formed at the point of contact between two different metals or a metal and a conductive oxide or hydroxide. A trace of an electrolyte at the interface, such as aqueous salts or sulfurous acid from pollution, sets in motion a corrosion reaction. The behavior of a metal in a galvanic cell can be understood in terms of a "galvanic series," a list of metals listed in an order of the voltage generated by a metal versus a saturated calomel electrode. Some values are listed in Table 8.1. Those metals listed the bottom of the table are the most resistant to galvanic corrosion. For a cell formed between two dissimilar metals, the one that is higher in the series will be the one that is corroded.

A galvanic cell is responsible for the corrosion of iron. Iron reacts with a moist, oxygen-containing atmosphere to form a coating of $Fe(OH)_3$ and/or $FeO(OH)$. In the absence of an electrolyte, the surface reaction is limited, but an electrolyte such as aqueous sodium chloride or sulfurous acid (from SO_2 and water) cause ever-deepening corrosion through the series of electrochemical reactions (shown in reactions 10–13 below). In these reactions the metallic iron functions as the anode of the electrochemical cell.

Anode reaction:

$$4Fe \rightarrow 4Fe^{2+} + 8e^- \tag{10}$$

$$4Fe^{2+} \rightarrow 4Fe^{3+} + 4e^- \tag{11}$$

Cathode reaction:

$$3O_2 + 6H_2O + 12e^- \rightarrow 12OH^- \tag{12}$$

Overall reaction:

$$4Fe + 3O_2 + 6H_2O \rightarrow 4Fe(OH)_2 \text{ or } 4FeO(OH) + 4H_2O \tag{13}$$

Corrosion can be inhibited or prevented in four different ways:

1. If a steel sheet is coated with a metal that is higher in the electrochemical series than iron, such as zinc, the zinc will form the anode of a galvanic cell and will corrode in a sacrificial manner in preference to corrosion of the steel. This is the basis for the widespread use of galvanized steel.

2. It is also possible to protect a metal against corrosion by the application of an electric current that counteracts the potential of a galvanic cell. Thus, the corrosion of iron in bridges and underground oil and gas pipelines is inhibited by the application of a negative electrical potential. The electric power for this process is often generated by solar cells (see Chapters 10 and 14).

3. However, corrosion protection is often achieved by the use of special alloys such stainless steels that contain iron, nickel, and chromium or, for example, Hastelloy C276, which is an alloy of iron, molybdenum, chromium, tungsten, and nickel. The high cost of such alloys is a reason why they are not used more widely in automobile bodies or bridges.

4. In the absence of galvanic protection or special alloys, the main defense against corrosion is to plate the sensitive metal with a corrosion-resistant metal such as chromium or to coat it with an impervious polymeric film (paint) to exclude water, often with the use of an undercoat that contains a corrosion inhibitor such as chromate. Clearly, this last option is the least expensive solution and the one that is most widely used.

D. SOLID-STATE STRUCTURE OF METALS AND ALLOYS

Many of the properties of different metals can be ascribed to differences in solid-state structure. Some of these properties and structural features are summarized in Table 8.2.

TABLE 8.2. Properties of Selected Metals

Element	Atomic Weight	Melting Temperature (°C)	Metal Atom Radius (pm)	Crystal Structure
Lithium	6.94	180.5	152	BCC
Magnesium	12	649	160	HCP
Sodium	22.99	97.8	186	BCC
Aluminum	26.98	660	143	CCP
Titanium	47.88	1667	147	HCP
Chromium	51.99	1900	128	BCC
Manganese	54.94	1244	127	BCC
Iron	55.85	1535	126	BCC/CCP
Nickel	58.69	1455	124	CCP
Copper	63.55	1083	128	CCP
Zinc	65.38	419.5	134	HCP
Tin	118.69	232	145	Tetragonal
Silver	107.87	961	144	CCP
Gold	196.97	1064	144	CCP
Mercury	200.59	−38.9	151	—

1. Packing of Spheres

The structure of a solid metal can be visualized as a series of spheres (nuclei plus inner electrons), all with the same diameter, packed together in a regular three-dimensional array, and held together by a "sea" of mobile electrons. This type of arrangement differs markedly from the structure of solids that contain molecules or of molecular solids, where the disposition of atoms in the structure depends on the orientation of covalent bonds rather than on the packing of spheres, and where electron mobility is generally restricted. However, it does bear a superficial resemblance to the structures of inorganic ionic solids, which are also based on the packing of spheres, but spheres that have different diameters as required for most anions and cations. Moreover, the electrons in ionic solids are not delocalized throughout the whole structure as they are in metals.

Thus, the solid-state structure of a metal is relatively simple compared to those of other solids. Because there are only a small number of different ways in which spheres of the same size can be packed efficiently in three dimensions, relatively few packing patterns are found in different metals. These include the three common arrangements known as hexagonal close packed (HCP), body-centered cubic (BCC), and cubic close-packed (CCP) found in the metals listed in Table 8.2. The differences between these three structures is perhaps initially best understood through a consideration of the way in which different *layers* of spheres can be organized. This is illustrated in Figure 8.4.

Starting with a single layer of close-packed spheres (the *A* layer), and placing the atoms of the second layer in the depressions between the atoms in the first layer, we generate an *AB* two-layer structure. Now it is possible to construct a third layer in two alternative ways. First, spheres in the third layer can be placed over the available depressions in the second layer such that they lie above the atoms in the first layer to give an *ABAB* arrangement. Alternatively, they can be placed in depressions in the second layer that do *not* lie above atoms in the first layer, to give an *ABCABC* sequence. Note that they cannot occupy *both* of these positions in the same layer because insufficient space exists for this to be possible. An *ABAB* structure is described as a *hexagonal close-packed* (HCP) arrangement. An ABCABC system is called a *cubic close packed* (CCP) arrangement. A third structure is called a *body-centered cubic* (BCC) arrangement. It has an atom at the corner of each cube and one at the center of the cube. This is a structure that contains more open space than the other two, and is hence less dense.

The packing structure of a metal controls a number of properties, including density, electrical and thermal conductivities, and the way in which layers of atoms slide past each other under tension, pressure, or impact. The sliding planes are called "slip planes" or "shear planes." The existence of many potential slip planes in a structure leads to the property of *ductility*, the ease with which a metal can be extruded into wires. In this sense, BCC and CCP systems, such as Cu, Ag, Au, Al, and Fe, tend to be ductile, but HCP metals, such as Zn or Mn, are generally more brittle because their structure provides fewer slip planes.

The influence of the crystal structure of a metal on thermal or electrical conductivity (see the next section) is less clear-cut. Some of the metals with the highest electrical conductivities (Al, Au, Cu, Ag) have cubic close-packed lattices, which suggests that a high packing density is important for good conductivity. However,

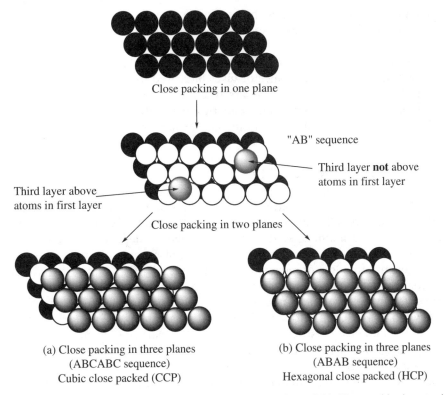

Close packing in one plane

"AB" sequence

Third layer **not** above
atoms in first layer

Third layer above
atoms in first layer

Close packing in two planes

(a) Close packing in three planes
(ABCABC sequence)
Cubic close packed (CCP)

(b) Close packing in three planes
(ABAB sequence)
Hexagonal close packed (HCP)

Figure 8.4. *Illustration of how close packing of atoms in a metal can yield either a cubic close-packed (a) or a hexagonal close-packed (b) structure.*

Pb and Pt have the same packing arrangement but low conductivities. Titanium, which has a low electrical conductivity, exists in a hexagonal close-packed (dense) structure. So there is clearly more to understanding electrical conductivity than knowing the packing density.

Another factor that is sometimes considered to influence metallic properties is the possible existence of metal–metal covalent bonding of the type that is found in small–molecule metallic clusters. As mentioned earlier, covalent bonding may be incompatible with the most efficient packing of spheres. However, it could increase the strength of a material by reducing the influence of slip planes, and it would probably increase the melting point. Because covalent bonding is caused by a pairing of electrons, it might also serve to lower electrical conductivity.

2. Alloys

Alloys are solid mixtures of two or more elements, often produced when a molten mixture is cooled below the melting point. One type of alloy has a structure in which the atoms of an added metal assume positions in the solid lattice formerly occupied by those of the pure host metal. This will normally yield a microscopically homogeneous material, and one in which the properties of the alloy will vary in a predictable way as the ratios of the two elements are varied. However, in other situations the addition of a second element may bring about a change in crystal structure or even a phase separation, in which case the properties will change in a discontinuous way.

Alloys are useful because they allow the introduction of a second or third element that increases corrosion resistance or gives increased toughness. However, the distortion of the lattice that often occurs when atoms with different diameters are mixed together generally lowers the electrical conductivity.

The subject of metal alloys is considered in Chapter 9.

E. ELECTRICAL CONDUCTIVITY

Table 8.3 lists the relative electrical conductivities of a number of common metals.

Why do different metals have different electrical conductivities, and how do metals differ from nonmetals? Electronic conductivity depends on the broad-scale movement of electrons through the bulk material. So, why do the electrons in a metal have this freedom of movement? First, it is easier for a metal atom to release electrons than it is for a nonmetal. Metals readily give up electrons to nonmetals to form salts. On the other hand, nonmetals can either acquire electrons to form anions or they can *share* electrons to form directional covalent bonds. Spin-paired electrons in a covalent bond are restricted to the space of the participating molecular orbitals. By contrast, some of the electrons in a metal are free to migrate throughout the bulk solid. This process appears to require the presence of *unpaired* electrons donated to the lattice from unpaired electrons in the outer shells of the participating atoms.

TABLE 8.3. Electrical and Thermal Conductivity of Selected Metals

Metal	Electrical Conductivity at 20°C Relative to Copper as 100	Thermal Conductivity Relative to Copper as 100
Silver	106	107
Copper	100 ($0.596 \times 10^6 \, \Omega^{-1} \, cm^{-1}$)	100 ($4.01 \, W \, cm^{-1} \, K^{-1}$)
Gold	65	79
Aluminum	59	59
Chromium	55	23.4
Sodium	35.2	35.2
Molybdenum	33.2	34.4
Tungsten	28.9	43.4
Zinc	28.2	28.7
Potassium	23.3	30.9
Lithium	18.1	21.1
Iron	17.7	20
Cobalt	16.3	24.9
Steel (various)	3–15	~20–25
Nickel	12–16	22.6
Tin	13	16.5
Platinum	15	17.9
Lead	7	8.8
Titanium	5	5.5
Mercury	1.66	2.1

In a small molecule the electronic structure can be described as a series of electron pairs that occupy molecular orbitals. As the two electrons that form a covalent bond are brought together, they stabilize each other by forming a spin pair that occupies each orbital. However, the formation of each bonding orbital also generates a higher-level antibonding orbital to which the electrons can jump if provided with sufficient thermal or radiation energy. A bond formed between two atoms thus has an orbital energy structure of the type shown in Figure 8.5.

In a multiatomic small or medium-sized molecule or cluster, there will be bonding molecular orbitals that correspond to each bond that connects an atom to its neighbors, with two spin-paired electrons occupying each bonding orbital, and an equal number of (empty) antibonding orbitals, as shown in Figure 8.5. As more and more atoms are incorporated into the molecule, the number of bonding and antibonding orbitals will grow accordingly and, because of the need to fit many energy levels into the available energy space, they will become closer and closer together. Eventually, in an extended multi-million-atom solid, the energy levels become spaced so closely that they are for all practical purposes part of an energy continuum that we call a "band." The bonding levels, are called the "valence band" and the antibonding levels, the "conduction band."

A characteristic of metals, as distinct from insulators or semiconductors, is that the valence band is only partly filled. This situation comes about because the packing structure described above surrounds each metal atom with more neighbors than it has valence electrons to share with them. For example, for the group I elements, each alkali metal atom has only one valence electron to share with its neighbors. However, it has eight near neighbors in the solid state.

This means that, although the alkali metals can readily form diatomic molecules in the gaseous state, there are not enough electrons available in the solid state to form strong covalent bonds with all of the eight neighbors. So, although those additional energy levels exist, they remain unoccupied until electrons jump into them from the filled levels, or electrons are added from an external source such as an

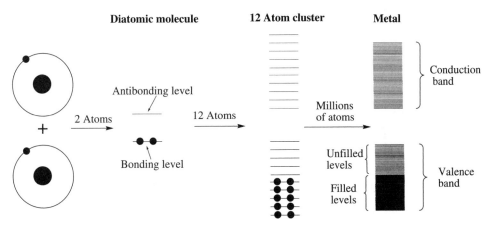

Figure 8.5. *Illustration of how two group I atoms form a covalent bond, but the geometry of metal atom packing in the solid metal forces the electrons into a band structure with a half-filled valence band.*

applied electrical potential. Thus, when electrons are pumped into one end of a metal wire, they populate some of those unfilled levels of the valence band near the point of injection, but they immediately spread out in an attempt to equalize the population throughout the material and are removed at the exit connection. In this way a continuous electric current flows through the material.

Although the alkali metals provide the simplest example of this principle, they are not the most impressive electrical conductors. The transition metals with their *d* orbitals and multiple electrons in the valence shell can contribute numerous electrons per atom to the lattice. This has two effects: (1) the more electrons that are circulating in a metallic lattice, the higher should be the electrical conductivity, and this is generally the case; but (2) the more electrons contributed by an atom, the greater is the chance that they will be able to form covalent bonds to the neighbors through hybrid orbitals, because the number of available electrons will begin to approach the number of near neighbors. When that happens, the strength and toughness of the metal will rise, but the electrical conductivity may decline.

Because an electric current in a metal is carried by unpaired electrons rather than by electron pairs, and because most of the electrons in the valence band will be paired (two spin-paired electrons in each energy level), the question arises of where the unpaired electrons come from. Some arise because they already exist in half-filled energy levels at the top of the populated region of the valence band, a consequence of their origins as unpaired electrons in the isolated atoms. But others can be formed by thermal activation from electron pairs in the filled regions of the valence band (Figure 8.6). The amount of energy needed to promote an electron from the filled levels of the band to the unfilled levels is so small that this must be a significant source of conductive electrons. So, why does the conductivity of a metal *decrease* as the temperature is raised rather than increase as more unpaired electrons are generated by thermal activation? It appears that a more powerful influence on conductivity is the uniformity and rigidity of the lattice. Increased vibration

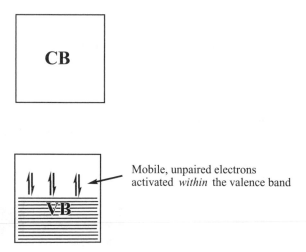

Figure 8.6. *The broad-scale movement of unpaired electrons in a metal is possible because of the ease with which electrons near the top of the filled levels in the partly filled valence band can jump into unfilled levels in the same band.*

of the atoms in the lattice as the temperature is raised interferes with the free flow of electrons, and this overpowers the presence of larger numbers of unpaired electrons.

Thus, although the general principles that underlie electrical conductivity in metals are fairly easy to understand, it is more difficult to explain the behavior of individual metals. What we can say is this:

1. High electrical conductivity is favored by the presence of many electrons in the partially filled levels of the valence band. Thus, high electrical conductivity is associated with transition metals such as Cu, Ag, or Au.

2. Pure metals are usually more highly conducting than alloys because the lattice is more uniform.

3. Impurities impede electron flow by introducing discontinuities into the lattice. This reduces the average free path of electrons.

4. Increases in temperature lower the electronic conduction of metals because lattice vibrations (phonon motions) increase with temperature, and good electron mobility requires a rigid lattice. Liquid metals such as mercury have lower conductivities than might be expected precisely because a rigid lattice no longer exists.

F. THE COLOR OF METALS

Most metals appear grayish or silvery by reflected light. A few, such as copper or gold, reflect a reddish or yellow hue. Bismuth has a faint pink color. Metals such as gold can be beaten into very thin films, so thin that their colors may be viewed by transmission. Gold is blue-green by transmission. What is the source of these colors? And why do polished metals reflect light so well?

The answers lie in the electronic band structure of metals (Figure 8.7). Because the valence band of a metal is only partially filled with electrons, these electrons can absorb light and easily jump to higher, hitherto unpopulated, levels *within the valence band*. This leads to a facile absorption of light. The promoted electrons then release their energy and fall back into the populated section of the band. Light absorbed at the surface of the metal will be re-emitted with the same color as the incident light. This is why most metals appear to be silvery or gray when irradiated with white light.

However, if the valence band is almost full because so many electrons are contributed by a particular metal, as in the case of gold or copper, the higher-energy violet, blue, or blue-green component of the incident white light cannot be absorbed because no empty energy level exists in the valence band that will accommodate such high-energy electrons. In other words, the only way an electron can absorb energy of that wavelength is to attempt to jump the bandgap, and that apparently is not possible. Thus, red and yellow light are absorbed and reemitted (as a yellow or gold color) but the blue-green light will not be absorbed by simple electronic jumps. If the metal is thin enough (as in the case of gold foil) that blue-green light will pass straight through. If the metal is too thick to allow transmission, the short-wavelength light will be converted to heat.

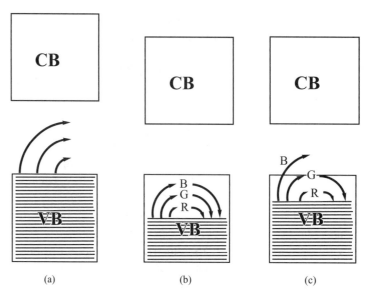

Figure 8.7. Electrons in the valence band of a material can absorb light and be promoted to higher energy levels. An insulator (a) has a bandgap that is so high that electrons activated by visible light do not have sufficient energy to surmount the high bandgap. Thus, the light is not absorbed by electrons, and photons pass unhindered through the solid. A metal has a partly filled valence band. If white light (red plus green plus blue) is absorbed by electrons (b), and if there is sufficient space in the valence band, electrons will absorb the light, be promoted to higher energy levels in the valence band, and then release light of the same frequencies as they fall back to the upper populated levels of that band. Because red, green, and blue light are all absorbed and then reemitted, the metal appears white. However, if the valence band is almost full (c), the higher-energy blue and blue-green light cannot activate electrons into an accessible energy level, and the blue-green light will either be absorbed as heat or pass through a thin sample. The remaining wavelengths will activate electrons that can remain in the valence band, and they will then absorb and reemit reddish-yellow light.

G. THERMAL CONDUCTIVITY OF METALS

A striking characteristic of most metals is their high ability to conduct heat. This is useful for a wide variety of applications from cooking vessels and internal-combustion engines to the dissipation of heat in aerospace vehicles. Heat conduction occurs in a material via one of two mechanisms: (1) through the vibrational motions of the atoms, the mechanism that underlies heat conduction in ceramics and polymers (a relatively inefficient process, which is why these materials are widely used as thermal insulators—think of polystyrene coffee cups); or (2) via the thermal excitation of electrons, where the ease of conduction is dependent on the number and mobility of the electrons. Thus, metals, with their large numbers of mobile electrons, are some of the best thermal conduction materials. Some values for the thermal conductivity of metals are given in Table 8.3. Note that there is a rough parallel between high electrical conductivity and high thermal conduction, for obvious reasons.

H. MAGNETIC PROPERTIES OF METALS

Only a few metals, such as iron, cobalt, and nickel, function as magnets. The rest do not. The magnetic properties of a material arise from the presence of unpaired electrons (paramagnetism). Paramagnetic substances are attracted into a magnetic field because a spinning electron generates its own small magnetic field. Transition metals contain unpaired electrons in their unfilled *d* and *s* shells. Hence, paramagnetism is quite common for these elements and their compounds. However, in the most general form of paramagnetisn the spins of the unpaired electrons are aligned randomly throughout the material *except in the presence of an external magnetic field*. The external field will align the spins so that, as shown in Figure 8.8b, most of them point in the same direction. However, as soon as the external field is removed, the spins become disordered again. Thus, these substances are not permanent magnets.

Contrast this with the special case of *ferromagnetism*, in which the spins are already aligned within a solid-state domain as a result of local mutual interactions, but the alignment in other domains is in differen directions. In such a case the application of an external magnetic field causes alignment of all the spins in all the domains. If the spin-aligned domains can then be stabilized by materials manipulation, the result will be a permanent magnet.

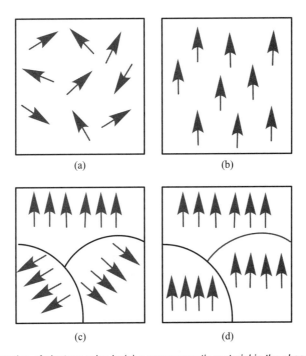

Figure 8.8. *Orientation of electron spins in (a) a paramagnetic material in the absence of an external magnetic field, (b) the presence of a magnetic field, (c) a ferromagnetic material with different domains having different orientations of electron spin, and (d) a ferromagnetic material under the influence of an external magnetic field.*

I. MECHANICAL PROPERTIES OF METALS

The utilization of metals as structural materials depends on their strength, hardness, impact resistance, ductility, and melting temperature. What controls these properties? Strength is clearly a measure of the ability of a metal to resist deformation under load. Metals that are near their melting point at the temperature of use will begin to exhibit flow, deformation, and eventual failure.

Melting point is a measure of the cohesion between atoms in the structure, which, in turn, depends on the packing efficiency of the atoms and the presence of covalent bonds. As shown in the Table 8.2, common metals with melting points over 1000 °C tend to have small atomic radii (Cr, Mn, Ni, Fe, Cu, with radii of 126–128 pm), although there are exceptions (e.g., Ti and Au). Conversely, some of the lower-melting-point metals, such as Hg, Li, Na, Al, Mg, and Zn, have larger atomic radii in the solid state (140–180 pm).

For many uses, strength must be balanced against impact resistance and resistance to metal fatigue. Some of the first commercial passenger jet airliners were involved in disastrous accidents because of metal fatigue. Metal fatigue and failure occur when a solid is subjected to cyclic, fluctuating loads as opposed to continuous stress. The phenomenon is easily demonstrated by clamping a small piece of sheet metal or wire in a vise and bending the material back and forth. This phenomenon is responsible for most of the mechanical failures that involve metals. Cracks appear in a cyclically stressed material, and these eventually propagate through the solid during successive cycles. Metals that are ductile have the ability to retard the propagation of cracks via the migration of slip planes.

In practice, special properties for different applications must be optimized by alloying different metals, and by heat treatment to change the crystal and domain structures. This is a subject beyond the scope of this book, but readers interested in this topic may wish to consult the list of books at the end of the chapter for further reading.

J. FABRICATION OF METALS

The fabrication of metals has been an essential human activity since the beginning of early civilization. The most obvious fabrication method is the casting of molten metal (steel, copper, brass, zinc alloys, etc.) in a mold. Metals such as steel are converted into sheet form by "cold-rolling," or beating them into thin sheets (copper, gold). Blacksmiths use percussion to shape red-hot iron objects by hammering. Rods or tubes are produced by extrusion of molten material through a die.

The coating of one metal on another for corrosion resistance or for improved appearance is accomplished in one of two main ways: (1) by spreading a molten metal on the surface of a sheet of the substrate metal (galvanized steel—a zinc coating on rolled steel—is made in this way), or (2) by electrolytic deposition, a widely used technique for the chromium- or silver-plating of less expensive metals. Electroplating consists of using the base metal as the anode of an electrolytic cell, with the cathode consisting of the metal to be deposited. The electrolyte is an aqueous solution of a salt of that metal. In this way, the corrosion-resistant metal is transferred from the anode to the object at the cathode. Ensuring an even and coherent coating is a challenge that depends on the voltage and current across the cell and the concentration of salt in the solution. Moreover, some electrodeposited metals do not adhere well to other metals, often because of a mismatch in the two crystal structures. In such cases, an intermediate layer of another metal must be deposited. For example, copper may be first electrocoated with nickel, and a layer of silver deposited on top of that. Another method for the deposition of thin metallic films is by chemical reduction of a metallic salt using a reducing agent such as glucose or formaldehyde. Silver mirrors are deposited on glass by this technique. The formation of an even, reflective film depends on the reaction conditions, because the deposition may also yield particles or nanowires rather than a mirror-like film.

K. FUTURE CHALLENGES IN METALLIC MATERIALS

The corrosion of commonly used metals is one of their main weaknesses, and future research will be needed to develop new corrosion-resistant alloys. The weight of many metals is another serious challenge, especially for mobile applications, which means that new alloys based on the lighter metals such as aluminum, magnesium, or lithium will be needed for automobile, aircraft, or marine applications. Composites of metals and nonoxide ceramics with improved compatibility between the two are needed for gas turbines, high-temperature turbines, heat shields, and so on.

L. SUGGESTIONS FOR FURTHER READING

1. Greenwood, N. N.; Earnshaw, A., *Chemistry of the Elements*, 2nd. ed., Butterworth-Heinemann, 1997.
2. Porter, D. A.; Easterling, K. E., *Transformations in Metals and Alloys*, 2nd. ed., Chapman & Hall, 1992.
3. Mizutani, U., *Introduction to the Theory of Metals*, Cambridge University Press, 2001.
4. Crabtree, R. H., *The Organometallic Chemistry of the Transition Metals*, 3rd. ed., Wiley-Interscience, Hoboken, N.J., 2001.
5. Hill, A. F., *Organotransition Metal Chemistry*, Royal Society of Chemistry, Cambridge, U.K., 2003.
6. Russell, A.; Kok, L. L., *Structure-Property Relations in Nonferrous Metals*, Wiley, 2005.
7. Chiusoli, G. P.; Maitlis, P., *Metal Catalysis in Industrial Organic Processes*, Royal Society of Chemistry, U.K., 2006.
8. Thrower, P. A., *Materials in Today's World*, 2nd. ed., McGraw Hill, 1991.
9. Verlinden, B.; Driver, J.; Samajdar, I.; Doherty, R., *Thermomechanical Processing of Metallic Materials*, Pergamon/Elsevier, 2007.

M. STUDY QUESTIONS

1. By constructing models of the different atomic packing arrangements in a metal, or through drawings, convince yourself that the different structures give rise to different opportunities for slip planes to form.

2. By reference to the atomic dimensions listed in Table 8.3, speculate on which alloys of iron, silver, gold, or aluminum with other metallic elements might be homogeneous and which would undergo phase separation. Check your conclusions against data in the literature.

3. The ability of a metal to resist corrosion is an important factor in determining its uses. List a range of applications where resistance to corrosion is essential. Then describe which metals might be used for each of these applications, and why. If the cost/benefit ratio is unacceptable, how would you solve the problem?

4. Specific faces of metal particles are responsible for the ability of that metal to function as a heterogeneous catalyst. Speculate on possible reasons why this should be so. Why does the efficiency of catalysis often depend on particle size?

5. The high cost of a metal usually reflects its rarity in the earth's crust and the difficulty of extraction and purification. Discuss this statement, giving specific examples.

6. Electrical conduction in metals depends on a number of different factors. Review these factors and explain why metals such as silver, copper, and aluminum are good conductors, but titanium and mercury are not.

7. Explain why newly polished copper is reddish in color, brass is yellow, and aluminum is silver-colored. Would an alloy of silver and gold have a color that lies somewhere between those of the pure metals, or would the alloy color resemble that of pure gold or pure silver? Why is the reddish color of copper slowly replaced by a blue-green color following exposure to the atmosphere?

8. The solid-state structure of a metal usually determines its properties and uses. Discuss this statement, giving specific exampled to support your argument.

9. What is the effect of impurities on the electrical and thermal conductivity of metals? Explain, with reference to specific examples. If possible, give literature references.

10. Describe the production of aluminum and its main alloys, and explain how its special properties underlie its utilization as one of the most useful materials.

9

Alloys, Composites, and Defects

A. OVERVIEW

1. Important Mechanical Properties

The underlying theme of this chapter is the relationship between solid-state structure and mechanical properties. To understand this relationship, it is necessary to be clear about the meaning of several key terms.

Toughness is perhaps the most important property that an engineering material can possess. It is the ability to resist breaking when struck by a sharp blow. An excellent example of a tough material is a rubber tire. When hit by a hammer, a tire deforms slightly, absorbs the energy of the impact into molecular motion and heat, and then rebounds to the original shape. In this example toughness is closely associated with *elasticity* and with the laminated construction of the tire (more about this later). Another form of toughness is *malleability*, which is the property shown by many metals to change shape without breaking when hammered or forced through a die.

Brittleness is the opposite of toughness. A sharp impact causes a brittle material to shatter into many pieces. A piece of window glass is an obvious example. However, the failure of metal components in bearings, automobile or jet engines, or bridges can often be attributed to embrittlement that often develops long after the component has been installed. Solids derived from small molecules are also frequently brittle.

Strength in a material is the ability to resist deformation or cold flow. Stone, concrete, and brick are used in building construction because they can withstand enormous pressure without buckling. Unfortunately, strength may be accompanied by brittleness, so that a strong material that withstands considerable static force may

Introduction to Materials Chemistry, by Harry R. Allcock
Copyright © 2008 by John Wiley & Sons, Inc.

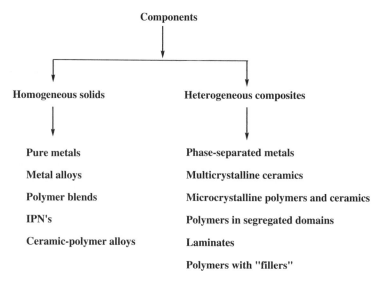

Figure 9.1. *Classification of different types of mixed component materials.*

shatter on impact. Think of the effect of a sledgehammer on a concrete block or a housebrick.

Thus, materials design often revolves around attempts to maximize toughness and/or strength and minimize brittleness. These attempts may involve two distinct approaches. In the first, a solid is manipulated by heating, cooling, or percussion to change its microstructure. This is a method much used for metals and their alloys. In the second approach, two different materials are combined to generate a heterogeneous solid (a *composite* material) in which the useful properties of one component are combined with the advantageous properties of the other. These two approaches to materials design and optimization are discussed in the following sections. Some of the options are summarized in Figure 9.1.

2. Homogeneous versus Heterogeneous Solids

There are several different reasons for the widespread interest in alloys and composites: (1) they provide a means for combining the preferred properties of two or more different materials, while at the same time minimizing their defects; (2) it is sometimes easier to generate a desired combination of properties by mixing two materials together than by synthesizing an entirely new substance with the required combination of properties built in at the molecular level; and (3) if the combination of two different materials leads to phase separation, the possibility exists that the combined material will be stronger than the two constituents alone because one component will reinforce the other and prevent the propagation of cracks and other defects.

With respect to the last factor, an ever-present possibility in the formation of mixed materials is the inability of different metals, ceramics, or polymers to form homogeneous mixtures in the solid state. Phase separation is common, and this means that the properties of the final mixture may not be a simple linear combination of the properties of the constituents. However, as discussed, below, phase separation may provide some advantages not found in a homogeneous system. The difference between homogeneous and phase-separated materials is not always distinct. As the following discussion reveals, some materials that appear to be homogeneous at the macrolevel may in fact be heterogeneous on the micro- or nanoscale.

3. Different Types of Composite Materials

Toughness, strength, and brittleness can be controlled by several different techniques, one of which is the incorporation of two or more different materials into the same solid. Thus, some of the most useful solids, especially those used as structural materials, are mixtures of two or more different substances. The constituents of such mixtures are the types of substances discussed in the preceding chapters on polymers, ceramics, and metals. These mixtures can assume several different forms that range from homogeneous solutions of one element or compound in another (alloys), through crystallites suspended in an amorphous glassy matrix, to solids in which two different components or two different parts of the same polymer molecule segregate into separate domains. Many modern composites consist of whiskers or fibers embedded in a matrix of another material. Other type of composites takes the form of laminates in which layers of different materials, for example, a woven cloth and a crosslinked polymer, are brought together to form multilayer assemblies. The scheme shown in Figure 9.1 sketches some of the possibilities.

4. Defects in Solids

Most solids have strengths that are far less than the values calculated on the basis of chemical bond energies or the sum of all the ionic or van der Waals interactions. In a uniform solid, such as a diamond, failure or shattering of a crystal requires the breaking of millions of C—C covalent bonds—a process that would require an abnormal amount of energy. Yet diamonds can be cleaved fairly easily.

This discrepancy is due to the fact that solid materials frequently "fail" because of the presence of defects. By "defects," we mean discontinuities in the structure that favor crack propagation when pressure or an impact is applied. Three types of defects are particularly important:

1. The existence of *slip planes* in a regular, three-dimensional structure can result in crack propagation. These are the planes that can be drawn between sheets of spherical atoms or molecules in even the most perfect crystalline solids (Figure 9.2). Slippage along these planes provides a natural mechanism for the shape of the solid to change under the influence of pressure. Slip planes can be very useful because they provide a means for a metal to deform slowly under pressure—a property that underlies ductility.

2. If the solid consists of two different, incompatible components segregated into microdomains, crack propagation may occur along interfaces between the two components (Figure 9.3). Thus, in a heterophase solid the strength depends on the *cohesion* between different domains, which can be either chemical or mechanical. Mechanical strength may arise when the domains are locked

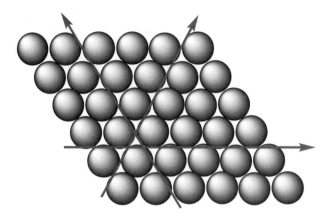

Figure 9.2. *Slip planes in a pure metal along which the solid can undergo distortion.*

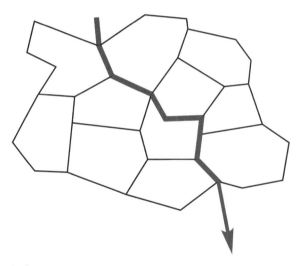

Figure 9.3. *Crack propagation along the boundaries between different domains.*

together physically by shape, even if the chemical affinity of the two surfaces is low. If the domains do slide past each other when pressure is applied, then the material will fail, sometimes suddenly. This concept applies to metal alloys where one phase has crystallized inside another.

3. Crack propagation can be arrested if a spreading defect encounters a separate domain that anchors the two sides of the crack together (Figure 9.4). That second domain may be in the form of crystallites embedded in a polymeric or inorganic glass, or it could be micrometer- or nanometer-sized domains of rubbery polymer embedded in a glassy matrix. The rubbery material absorbs the energy that would otherwise continue to tear the material apart, and converts that energy into harmless molecular motion. The crack-stopping phase can also be a rigid material such as carbon or ceramic powder or a fibrous material such as glass fiber or carbon fiber. Such additives are collectively

Figure 9.4. *Propagation of a crack through a material can be stopped by the presence of a second phase, which can be either an elastomeric domain that absorbs the impact energy or a nonflexible material or crystal that is embedded in the glassy matrix.*

known as "fillers." Ideally, a good filler should form a strong adhesive bond to the matrix material.

4. Defects at the *surface* of a solid may also have an important influence on strength. Even a homogeneous solid, such as an inorganic or polymeric glass, will shatter on impact more easily if scratches exist at the surface. These defects serve as sites for the concentration of stress and impact energy and they initiate cracks, which then propagate rapidly through the solid (Figure 9.5). Thus, an obvious way to strengthen a glass is to ensure that the surface is free from imperfections. This may be accomplished by "flame polishing" to melt the outer surface.

B. PURE MATERIALS AND HOMOGENEOUS SOLID SOLUTIONS

1. Slip Planes, Dislocations, and Grain Boundaries in Metals

Slip planes are an inherent characteristic of metals, even those that have a relatively homogeneous structure. The nature of slip planes is illustrated in Figure 9.2. Different crystal structures give rise to different numbers of slip planes. For example, copper, silver, and gold have cubic close-packed (face-centered cubic) crystal structures, which generate many slip planes. Slip planes are important because the binding forces that hold successive layers of metal atoms together are weak (assuming that covalent bonds are absent), and it requires relatively little force to induce planes of atoms to slide past each other. As mentioned above, this process is the basis of ductility. If a metallic crystal structure that has many slip planes is stretched or bent,

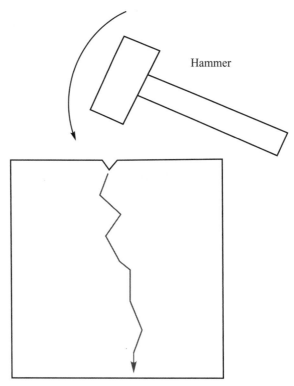

Figure 9.5. *Crack formation is facilitated if scratches or other indentations exist at the surface. These sites act as the focus of impact energy, triggering the separation of grain or domain boundaries and even facilitating the cleavage of covalent bonds.*

the material will deform without shattering as the whole structure readjusts to the new shape by movement along the slip planes. This can be compared to the manner in which geologic forces are relieved by slippage along a geologic fault, except that in a metal it may not occur so abruptly. This property underlies the processes of extrusion of wires and tubes, and fabrication process that depend on metal bending. However, this behavior may be detrimental to the strength of the material under extreme pressure or tension. For example, suspension bridges built entirely from copper might not retain their viability for as long as those constructed from steel or aluminum alloy.

However, the existence of slip planes does not by itself explain the ease with which metal lattices can be deformed or cleaved. Another reason is connected with the fact that crystal lattices are seldom perfect, and the packing of spheres frequently yields discontinuities that result from the random absence of atoms where they should be. This is illustrated in Figure 9.6.

These discontinuities, called *dislocations*, provide a physical pathway for a facile reorganization of the lattice when slow pressure is applied. The existence of dislocations may be the result of errors in the atomic deposition sequence when the crystalline regions are growing from the melt.

Another feature of a pure metal is the existence of *grain boundaries* (Figure 9.7), which reflect the existence of different crystallites growing simultaneously in the

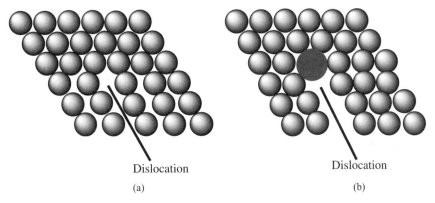

Dislocation   Dislocation

(a)   (b)

Figure 9.6. *(a) Defects in a crystal lattice caused by the absence of an atom, which causes a reorganization of the nearby atoms and the presence of a dislocation. (b) Dislocations may also be generated by the presence of a larger atom that disrupts the local packing arrangement.*

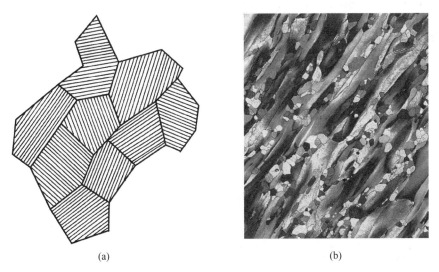

(a)   (b)

Figure 9.7. *(a) Schematic representation of grain boundaries that arise from mismatch between the crystal structures of adjacent domains. The parallel lines represent crystal planes. (b) Optical micrograph of grain boundaries in a metal such as titanium.*

molten metal. When these crystals meet as the metal solidifies completely, the different crystalline domains are mismatched at the boundary in terms of the atomic arrangements. Individual crystallites are physically locked in position with their neighbors. The presence of interlocking crystalline domains can increase both strength and hardness, and possibly brittleness as well. Ductility is correspondingly reduced. The size of the crystallites is important. Many small grains (crystals) provide multiple pathways for crack propagation along grain boundaries and low stress resistance. Fewer, large interlocked grains will give the maximum strength and impact resistance. This type of structure can be induced by "cold-rolling" sheets of the metal, a process that also orients the grains along one axis.

From a practical viewpoint, impact- or temperature-induced stress applied selectively to the *surface* of a metal can harden the surface ("case hardening") without destroying the ductility or impact resistance of the sub-surface material. Moreover, dislocations may be eased out of a metal by heating to just below the melting point, a process known as *annealing*, which allows the internal structure to reorganize and reduces the number of dislocation sites. On the other hand, rapid cooling of a hot metal may harden the surface of a material by generating many interlocked small crystalline regions. This is what happens when a blacksmith hammers an iron horseshoe into shape at red heat, and then plunges it into a bath of cold water. Modern engineering practice also requires that this "quenching" process is carried out routinely on metal components.

Discontinuities may also arise from stresses (impact, stretching, or bending) applied to the metal after it has cooled below its melting point. They are responsible for the failure of metals by "metal fatigue," a process in which repeated bending or stressing of a metal leads to catastrophic failure.

Impurities in a metal also have a profound effect on strength and ductility. The presence of small amounts of impurity atoms, especially those with atomic radii that are markedly different from the host material, will cause discontinuities and dislocations, and this could bring about an increase in brittleness. On the other hand, larger amounts of a second element may lower the melting point by disrupting the overall crystal structure.

Note that the imperfections described here also emerge at the *surface* of metals and their alloys, features that, together with "steps" and "ledges," play an important role in catalysis (see Chapter 15).

2. Homogeneous Metallic Alloys

Metallic alloys were mentioned briefly in Chapter 8, and much of the foregoing comments in this chapter apply to metallic alloys as well as to other materials. Metal alloys are produced by melting the components together and cooling. The speed of cooling often determines the properties of the final material. A characteristic of a true alloy is that the different atoms are distributed uniformly throughout the volume of the solid rather than occurring as separate domains or phases. However, this is composition-dependent, and phase separation often occurs beyond some concentration limits. Typical alloys are those formed by small amounts of carbon in iron (steel); iron, chromium, and nickel in stainless steel; silver and mercury in dental "silver amalgam"; copper and gold in jewelry; and copper, gold, and nickel in coins. A key requirement for the formation of a homogeneous mixture of metals is that the atomic radii of the participating metals should be similar, so that one type of metal atom can replace another in the solid without serious disruption of the crystal structure. However, if a second metal has a larger atomic radius, it will create discontinuities at the slip planes and prevent movement along those planes. Hence the alloy will become harder.

3. Polymer Alloys—Blends

Homogeneous mixtures of two or more different polymers are important because they, too, provide a method for combining the diverse properties of two different materials, especially when the attributes of one compensate for the deficiencies of the other. These materials are also known as *polymer blends*. One advantage of homogeneous polymer alloys is that the properties of the blend change in a predictable way with the ratio of the two constituents. Alloys of macromolecules may be produced by dissolving two or more different polymers in an organic liquid, followed by evaporation of the solvent. Alternatively, the two molten polymers can be mixed mechanically, and the mixture cooled. As with metals, the final morphology may depend on the speed of cooling and the subsequent thermal history.

Tentative evidence that the product is a homogeneous blend is its transparent appearance. This means that segregation into two phases has not occurred (there is no light scattering from separate domains with different refractive indices). By contrast, if the solid mixture is opaque, this suggests that phase separation has occurred. A fairly common situation is the initial formation of a homogeneous, transparent material, which slowly becomes opaque over time as a result of phase separation. Transmission and scanning electron microscopy (Chapter 4) are tools that are widely used to probe the morphology of polymer alloys.

In addition, the detection of only one glass transition temperature (by DSC analysis) intermediate between those of the two constituent polymers is evidence for the existence of compatible mixtures. Whether this homogeneity persists at the nanostructure or molecular level is a matter for debate.

Some scientists believe that true homogeneous alloys of two or more polymers may not exist. Nevertheless, many examples exist of mixtures of two different polymers that appear to be homogeneous at the level of optical microscopy or even scanning electron microscopy. For two polymers to be miscible, the free energy of mixing must be less than zero. This can be accomplished in one of three ways: (1) the *entropy* of mixing should be high, (2) the *enthalpy* of mixing should be negative, or (3) there should be some favorable balance between these two factors. If the two polymers are both high-molecular-weight species, the entropy of mixing will be relatively small, because the two separated polymers will already be in a highly disordered state and have little to gain by mixing with another polymer. Thus, unless there is a strong interaction between the two polymers that lowers the enthalpy, they will be incompatible. A significant decrease in enthalpy can occur if the polymers have complementary functional groups. For example, homogeneous mixture may be formed if one polymer bears OH, COOH, or NH_2 groups that can form hydrogen bonds with the oxygen or nitrogen atoms in the second polymer.

An alternative way to improve the compatibility of two different macromolecules is to add a polymer compatibilizer, which is another polymeric component that interacts well with the two main constituents. An example is the use of a block copolymer of polyisobutylene and poly(dimethylsiloxane) to convert heterogeneous mixtures of the two highly incompatible polymers to a homogeneous polymer blend. Presumably the two components in the compatibilizer associate with their counterparts in the individual polymers to lower the enthalpy of the final elastomer.

4. Interpenetrating Polymer Networks

An important type of polymer alloy is an *interpenetrating polymer network* (IPN) (Figure 9.8).

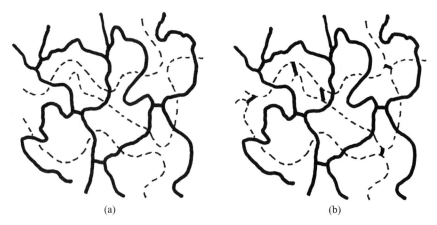

(a)   (b)

Figure 9.8. Schematic representation of (a) a partial and (b) a full interpenetrating polymer network (IPN). The partial IPN consists of an uncrosslinked polymer that permeates the crosslinked matrix of a second macromolecule. In a full IPN, both polymers are crosslinked within the matrix of each other.

Two polymers that would normally be incompatible with each other and would undergo phase separation in the solid state may be prevented from doing so by the following procedure. First, a crosslinked version of polymer A is prepared. This crosslinked network is then swollen by absorption of the liquid monomer for polymer B, and this monomer is polymerized. The resultant structure then consists of the crosslinked network of polymer A randomly penetrated by polymer B, an arrangement that prevents the two polymers from separating to form segregated domains. This is called a *partial IPN*. In theory, polymer B could reptate from the matrix of A and form a separate phase, although this would take a long time to occur. However, if polymer B is crosslinked after it is formed in the matrix of A, the whole structure is locked together and phase separation is not possible. This is called a *full IPN*, and homogeneity is strongly favored, at least at the microstructure level. In theory, some heterogeneity may exist at the nanostrcture level because a number of chains of polymer B may be located adjacent to each other and, in principle, could constitute a separate nanophase. The importance of this factor would depend of the amount of polymer B formed within the matrix of A.

5. Ceramic–Polymer "Alloys" (Ceramers)

Homogeneous mixtures of different silicate ceramics are a well-known aspect of glass technology. They differ from polymer alloys because the temperatures required to melt a mixture of two or more ceramics are sufficiently high that chemical bond interchange occurs, so that the final materials have a different composition at the molecular level and would resemble copolymers rather than alloys. Nevertheless, this is an effective method to maximize strength, toughness, and impact resistance, and is practiced widely in the production of borosilicate and other glasses (see Chapter 7).

An alternative approach is to employ the sol–gel process for the preparation of ceramic alloys. As discussed in Chapter 7, this is a method for the preparation of crosslinked silicate or aluminosilicate polymers at moderate temperatures. Alloys are accessible when water- or alcohol-soluble organic or inorganic polymers are incorporated into the initial sol–gel reaction mixture and become trapped within the crosslinked silicate or aluminate matrix as sol–gel crosslinking occurs. Thus, these alloys are partial interpenetrating polymer networks (IPNs) in which the ceramic matrix provides strength and the trapped polymer imparts impact resistance. Reactions can be devised to crosslink the organic polymer inside the silicate matrix to give a full IPN.

As with polymer blends, the degree of homogeneity might be estimated from the transparency of the final product, but the nature of the sol–gel process, and its ability to generate a wide range of particulate intermediates, makes transparency or the lack of it an uncertain criterion.

C. HETEROPHASE MATERIALS

1. General Observations

The incorporation of two or more incompatible substances into one composite material can have benefits not found in the individual components. The hard tissues in mammals, such as bone and tooth enamel, are excellent examples of composite ceramic/polymer materials that are both strong and tough. They are combinations of a polymer (collagen) and various calcium phosphates known as *hydroxyapatites*. The mineral component provides the strength and load-bearing qualities, while the collagen gives impact resistance. This subject is revisited in Chapter 16 on biomaterials. Another example is the use of laminated carbon fiber–epoxy composites for aircraft construction, where the carbon fiber cloth increases the strength and reduces the brittleness of the crosslinked organic polymer. Other examples include the growth of crystals in an inorganic glass to toughen and provide thermal shock resistance, and the behavior of block copolymers that contain an elastomeric segment linked to a "hard" glassy segment, which phase-separate to give rubbery and glassy domains for a combination of strength and impact resistance.

Certain general factors underlie the utility of phase-separated materials:

1. The properties of a heterophase solid depend on the *size* of the component domains. To take an obvious example, very little improvement in properties would be anticipated from incorporating 1-cm-diameter spheres of silica into silicone rubber. The surface area of contact between the two components is too small to seriously counteract the low strength of the rubber. However, if the silica spheres have a diameter in the micrometer or nanometer range, the area of interfacial contact will be many times larger, and a striking improvement in strength will be evident because the cohesive forces between the two components will be increased dramatically. This is the same reason why carbon powder is an essential component of automobile tires and the shock-absorbing components in aircraft landing gear.

2. As implied by the above examples, the degree of *adhesion* between the two components of a composite is an important aspect of materials design. If, for example, the surface of an insoluble filler has poor adhesion to a matrix material, the interface

between the two components may break down as the material is flexed or subjected to repeated impact. This can be the cause of the catastrophic failure sometimes seen in composite materials, which, when new, appear to have excellent mechanical properties.

3. The *shape* of at least one of the components in a heterophase composite is also important. Although spherical micro- or nanoparticles are frequently used as reinforcing fillers in polymers, this shape is relatively ineffective at providing enhanced strength if the forces applied to the composite material are concentrated in only one direction. Separation between the matrix and the spheres may occur readily. In this case, *fibers* are more effective reinforcement additives for metals, polymers, and ceramics. It is more difficult to separate a fiber from a matrix by repeated stress than it is in the case of a sphere. In fact, fiber reinforcement is preferred for most high-technology applications; for example, fibers of glass, carbon, or synthesized nonoxide ceramics are the favored additives for aircraft construction, turbine blades, high-performance automobiles, and racing yacht hulls. Note, however that the smaller the diameter of the fiber, the more effective will be its influence on mechanical properties. This is why carbon nanotubes (see Chapter 17) are being investigated intensively as additives. The shape and size influence also applies to metals, where the growth of nanofibers of a second phase may stabilize dislocations and hence lower brittleness even in seemingly homogeneous metallic alloys.

2. Reasons for Phase Segregation

Thermodynamic factors that explain why different polymers undergo phase separation have already been discussed. An additional reason for the insolubility of one material in another is that one of the substances is highly crosslinked. Examples are carbon fiber, glass, or silica. Crosslinked "fillers," as these materials are known, do not allow the uncrosslinked phase to penetrate into the surface layers of the insoluble component. Hence, such materials are held in suspension in the solid matrix mainly by the high viscosity of the matrix material and by whatever surface forces exist to promote adhesion.

Phase separation occurs in metals, ceramics, and glasses when one component crystallizes from the molten material as it is being cooled. This often results from the complex metathetical exchange reactions that occur in molten inorganic materials. One or more of the products of these reactions may be insoluble in the molten matrix and will crystallize out, leaving behind a slightly altered composition of the molten material from which they came. This is a plausible mechanism for the formation of heterophase metamorphic rocks, as well as synthesized heterophase glass–ceramics like Corning Ware[R].

An equally interesting situation from a chemical perspective is the behavior of solids derived from block copolymers, especially those consisting of blocks derived from normally incompatible macromolecules. Block copolymers of this type mimic, at the molecular level, the behavior or the two individual incompatible, polymers at the macrolevel. Specifically, one block prefers to associate with the similar blocks in neighboring macromolecules rather than with the block to which it is attached. Thus, phase separation occurs, but with the two phases connected through covalent bonds. This gives rise to some extraordinary solid-state arrangements, discussed later.

3. Phase-Separated Metals

The crystallization of *pure* metals into multicrystalline structures is described above. However, heterogeneous structures that involve the segregation of different elements or compounds within an alloy when two or more different metals are melted together and cooled are quite common. Various forms of steel are formed by the precipitation of iron carbide (Fe_3C) from homogeneous alloys of iron and carbon when the carbon concentration exceeds the solubility limit. Alloys formed from aluminum and lithium are lightweight metals that are important in aerospace engineering. The structure and properties of the metal depend on the composition. Starting with pure aluminum, the addition of lithium gives a homogeneous alloy in which the lithium atoms are randomly distributed among the aluminum atoms, and the original cubic close-packed crystal structure is maintained. However, beyond a certain point, the addition of more lithium produces a second phase that is insoluble in and separates from the first. It has a composition of Al_3Li and, when cooled, it crystallizes in a face-centered cubic structure with lithium atoms at the corners of a cube and aluminum atoms in the centers of the faces. Thus, the solid composite consists of a true alloy, with a second phase that has a different crystal structure. The presence of the domains of the second phase hardens the metal by interrupting the slip planes and dislocations within the true alloy. Hence, crack propagation is inhibited, and the material is hardened. It is useful to compare this process of phase separation with the formation of grain boundaries in a pure metal. The effects caused by these two processes are quite similar.

4. Heterophase Mineralogical Materials

Many minerals are composites of different inorganic phases. Granites are naturally occurring composites of silica with the oxides of aluminum and several other elements crystallized from molten rock. The various components crystallize to give a range of phases strongly bonded to each other by amorphous domains, an arrangement that accounts for the strength of these minerals. Concrete is an example of a synthesized composite of sand and Portland cement (a crystalline reinforcing phase and a binder).

5. Microcrystalline Polymers

As discussed in Chapter 6, except in very rare cases, crystalline polymers consist of microcrystalline domains embedded in an amorphous matrix. Part of a single polymer chain may pass through a microcrystallite while the rest of the chain forms part of the amorphous matrix. The amorphous regions of the solid provide impact resistance while the microcrystallites serve as "anchors" that connect different chains and provide strength by preventing the chains from sliding past each other.

A characteristic of microcrystalline polymers is that they are opalescent or opaque, due to the different refractive indices of the crystalline and amorphous regions. The presence of crystallites may raise the glass transition temperature by reducing internal reorientation. Melt fabrication of the polymer is facilitated by the fluidity of the polymer when heated above the crystallite melting temperature, while allowing the rigidity to return when the material is cooled to room temperature.

6. Heterogeneous Ceramic–Polymer Composites

Some of the most useful modern composite materials are, like bone, obtained by a combination of ceramics and polymers. Synthetic tooth restoration resins and bone repair materials are hydroxyapatite/synthetic polymer composites. The sol–gel process allows insoluble organic polymers to be incorporated into the solid-state silicate or aluminate structure with a marked increase in the toughness of the material. Another type of ceramic–polymer composite is formed from a polymer that contains particles of a ceramic incorporated as a "filler." These are mentioned at the end of this chapter.

7. Phase-Separated Polymer–Polymer Composites

The rapid growth in the use of polymers in engineering technology is due largely to the development of composites, especially phase-separated composite materials. The reasons for this are fairly obvious given the strengths and weaknesses of polymers, as discussed in earlier chapters. Two of these weaknesses include the relatively low glass transition temperatures, which leads to materials creep when the polymer is subjected to prolonged pressure or tension, and the brittleness of many glassy polymers.

As discussed earlier in this chapter, different polymers are often incompatible with each other and will phase-separate if given a chance. It will be obvious from the foregoing discussion that an ideal structural material is one that has enough elasticity to absorb impact without cracking but is rigid enough to maintain its shape over a wide temperature range. This challenge is often solved by mixing two or more polymers together to form an alloy or polymer blend. For example, a rubbery polymer such as polybutadiene might be blended with a glassy polymer such as polystyrene. In practice, the two polymers will probably become segregated into different domains, but, provided the domains are small enough, the system will provide engineering advantages compared to either of the two separate components.

A phase-separated composite of two different polymers can be identified in several ways:

1. As mentioned earlier, the opacity of the combined system will probably be an indicator of phase separation, assuming that the domain sizes are larger than the wavelength of light.
2. The glass transition temperature, typically determined by DSC, will also provide evidence. Two completely incompatible polymers when mixed will show two T_g values that correspond to those of the individual components. Partial miscibility will be indicated if these two T_g values are now accompanied by a third T_g at a temperature between the other two. Total miscibility will result in the appearance of only one T_g at some intermediate temperature.

3. Scanning or transmission electron microscopy may reveal the presence of separate domains in a phase-separated system.

It is important to recognize that phase separation in polymer mixtures, as in metal alloys, is often concentration-dependent. Thus, a small amount of polymer A in polymer B or a small amount of B in A may form homogeneous blends, whereas a 50/50 mixture is phase separated. Moreover, the exact behavior may depend on the temperature. Increasing temperatures often reduce the tendency for phase separation, with the temperature at which only one phase is formed being designated as an *upper critical solution temperature* (UCST). Other systems exhibit a *lower critical solution temperature* (LCST), below which only one phase exists. Clearly, these factors must be taken into account when considering the processing of mixed polymer systems.

8. Phase-Separated Block Copolymers

A method to retard the gross segregation of different polymers in a mixture is to connect them to each other by covalent bonds. Block copolymers are one type of structure that accomplishes this end. A block copolymer may give rise to a homogeneous solid-state structure, but more frequently it allows phase separation to occur but on a smaller than normal dimensional scale. Whereas the domains in a mixture of two unconnected polymers may be micron size or larger, the heterogeneity of a block copolymer can be at the level of nanometers. An example of a flexible polymer chain linked to a rigid chain is provided by butadiene–styrene block copolymers. These usually undergo phase separation in the solid state such that the butadiene blocks from different polymer molecules aggregate as a separate phase, as do the styrene blocks. Because the length of each block is small, the size of the domains is also small and the texture of the phase-separated material can be very fine, leading to an efficient combination of the properties of the two blocks. Block copolymers are widely used for engineering applications where rigidity and toughness must be combined, and for this reason these materials are used extensively in automobile manufacture.

The various solid-state arrangements generated by block copolymers depend not only on the incompatibility of the component blocks but also on the block lengths. This leads to considerably structural complexity and a remarkable range of accessible properties. Figure 9.9 illustrates some of the structures that have been identified in solids derived from block copolymers.

The principle that underlies the formation of different morphologies is illustrated by the following example. Consider a block copolymer that contains a hydrophilic block linked to a hydrophobic block. The hydrophilic segments of each chain will tend to associate into one domain. The hydrophobic blocks will aggregate to form a separate domain. However, the precise way in which the domains are organized depends on the strength of the interdomain interactions and the block lengths. The simplest arrangement is the formation of classical spherical micelles when the block copolymer is suspended in water. In these, the hydrophobic blocks assemble into a spherical hydrophobic core while the hydrophilic blocks point out as a "corona" toward the water. Now imagine a similar situation, but in the absence of the water. The hydrophobic blocks still self-assemble to form spherical cores while the hydro-

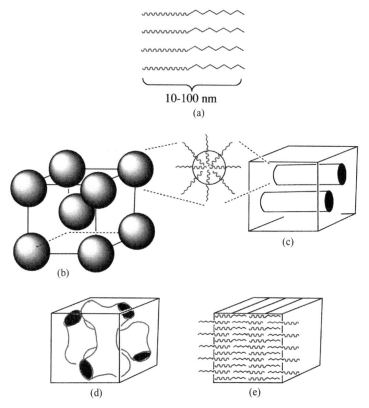

Figure 9.9. Diblock copolymers (a) undergo phase separation in a number of different ways depending on block length and composition. The basic motif is the spherical micelle [the structure shown between (b) and (c)]. Four example morphologies are as follows: (b) spherical domains derived one block distributed in a matrix formed from the other block, (c) cylinders of one block in a matrix of the second block, (d) bicontinuous system of gyroids of one block in a matrix of the second block, and (e) laminar domains generated from macromolecular interleafing of the blocks.

philic blocks mingle with their counterparts from other micelles to form the matrix. This would be an example of the situation shown in Figure 9.9b. The organization of the cores may be body-centered cubic or face-centered cubic (see Chapter 8). The cylindrical morphology illustrated in Figure 9.9c arises when the micellar cores become elongated into "worms" instead of spheres, and the coronas from different worms intermix to form the matrix. The cylinders often form hexagonal packing arrays. The complex gyroid pattern shown in Figure 9.9d is a further development from the cylindrical morphology. The laminar structure in Figure 9.9e arises by an *interleafing* of similar blocks to form a slab-like arrangement.

Which block type forms the matrix and which forms the coronas? This depends on the relative lengths of the two blocks. The spherical micellar structure tends to appear when one block length (call it block A) is appreciably longer than the other. As the block length of A becomes shorter, the morphology may change progressively to cylindrical, to gyroidal, and then for nearly equal block lengths, to laminar. If the length of the other block (B) is now increased, the reverse sequence of phases will be formed—gyroidal, cyclindrical, to spherical, with block B now constituting the matrix phase. These phase components have dimensions that are in the nanometer range, because they depend on the 10–100 nm lengths of the blocks.

9. Laminates

Laminates are composites in which layers of two or more different and incompatible materials are bonded together to yield a material that is stronger than the individual components. Two types of laminates are of particular interest. First, laminates that consist of two outer layers of a glass cemented by a central layer of an impact-resistant adhesive polymer are used in automobile windshields and related applications. The glass provides the strength and transparency, while the thin central layer holds the broken structure together if an impact occurs. Poly(vinyl butyrate) is often used as the central, impact-absorbing layer.

A second type of laminate consists of layers of a reinforcing woven cloth, typically of glass fiber or carbon fiber, infused with a thermosetting monomer (Figure 9.10). In a laminated assembly of this type, multiple layers of the cloth are infused one by one with the liquid monomer. The assembly is then compressed and heated to crosslink the polymer, thus trapping the reinforcing agent within the polymer matrix. The advantage of a woven cloth or mat of fibers over chopped fibers is that the cloth or mat can provide reinforcement in two dimensions over a large area. Carbon fiber–epoxy composites of this type are now widely used in aircraft construction. Glass fiber–epoxy laminates are the basis of boat manufacture and printed circuit boards. Glass fiber– or nylon fiber/rubber laminates are used in automobile tires. The ultimate strength of laminates of this type depends on the thermal stability of the fiber and the adhesion between the fibers and the resin. Failure of a composite is usually the result of a breakdown of the adhesive bond between the components.

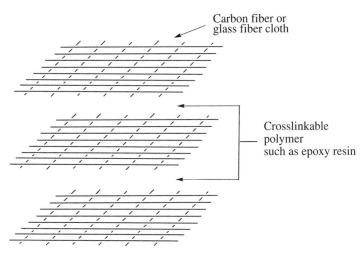

Figure 9.10. *Construction of a typical laminated composite.*

10. Filled Thermoplastics and Thermosetting Materials

The addition of fine particulate matter such as carbon or silica to a polymer has been mentioned throughout this chapter. Such composites can bring about a major improvement in the engineering properties of the material. This is why carbon black is incorporated into automobile tires. If the filler becomes chemically bonded to the polymer, the reinforcement effect is likely to be stronger. The smaller the particle size, the greater is the effect. This is one of the least-heralded but most practical uses of nantechnology. The use of carbon nanotubes as reinforcement fillers and electrical conductors for electrostatic dissipation is a subject of growing interest.

D. SUGGESTIONS FOR FURTHER READING

1. Thrower, P. A., *Materials in Today's World*, 2nd ed., McGraw-Hill, New York, 1996.
2. Ellis, A. B.; Geselbracht, M. J.; Johnson, B. J.; Lisensky, G. C.; Robinson, W. R., *Teaching General Chemistry—a Materials Science Companion*, American Chemical Society, Washington, DC, 1993.
3. Klempner, D.; Sperling, L. H.; Utracki, L. A., eds., *Interpenetrating Polymer Networks*, American Chemical Society Advances in Chemistry Series Vol. 239, Washington, DC, 1994.
4. Solc, K., ed., *Polymer Compatibility and Incompatibility*, Harwood Academic Publishers, 1982.
5. Huang, H.-H.; Orler, B.; Wilkes, G. L., Ceramics: Hybrid materials incorporating polymeric/oligomeric species with inorganic classes by a sol-gel process, Polym. *Bull.*, **14**:170–839 (1985).

E. STUDY QUESTIONS

1. Compile a list of different polymer pairs that you would guess intuitively should form compatible polymer blends, and another list of pairs that you think might be incompatible. Provide the reasons for your selections. Now discuss methods to convert the incompatible pairs to compatible solids and vice versa.

2. Why might micro- or nanolevel composites of two different materials have better properties than do the two pure materials used separately or the two materials cemented side by side?

3. Why is it easier to produce an alloy by melting together two metals than it is to obtain a homogeneous blend by melting together two different polymers?

4. Explain why alloys formed from two or more different metals often have superior properties compared to the pure metals.

5. Heterogeneous composites of two different polymers, or a polymer plus a powdered ceramic or chopped carbon fibers, are widely used in technology. Explain why such heterogeneous mixtures have desirable properties, and what special features of the polymer and the additive enhance the performance of the final material.

6. How would you determine experimentally whether a mixture of two metals or two polymers is homogeneous or heterogeneous?

7. Can two different ceramics form homogeneous alloys? Discuss.

8. Describe why many block copolymers form heterophase materials.

9. You have been engaged as a consultant to design a new material for helicopter rotor blades. The blades must be strong; impact-resistant; easy to fabricate; resistant to rain, seawater, and strong sunlight; and remain viable at temperatures as low as 50 °C below freezing. From what you have learned in this and earlier chapters, suggest three different composite materials and describe the advantages and possible weaknesses of each.

10. The hulls and other components of yachts and power boats are often constructed from composite materials. Discuss why this is so, give specific examples of the composites used, and review view how such materials are fabricated.

Part III

Materials in Advanced Technology

10

Semiconductors and Related Materials

A. IMPORTANCE OF SEMICONDUCTORS

Semiconductors are found in virtually every technological device, from computers, telephones, radios, light-emitting diodes, lasers, and television screens to medical sensors, smoke detectors, cameras, and automobile and aircraft controls. Crude semiconductor devices were first introduced in the 1950s, but the subsequent incorporation of hundreds to millions of semiconductor units onto a small area the size of a fingernail has been one of the major triumphs of modern science and technology. Probably no single material has had a greater influence on our way of life in the last hundred years than the semiconductor.

Semiconductors are solids that conduct electricity at a level that falls between those of an insulator and a metal—typically between about 10^{-8} S/cm (siemens per centimeter) to around 1 S/cm. Many different solid materials fall into this category. Examples include not only well-known substances such as silicon or germanium, and compound semiconductors such as gallium arsenide or cadmium sulfide, but also certain unsaturated organic polymers such as polyacetylene and poly(phenyleevinylenes). Although the inorganic semiconductors (such as silicon) have traditionally dominated the electronics field in the past, polymers are beginning to be used for special applications where materials flexibility and relatively low cost of production are important.

Most semiconductor materials need to be modified by the introduction of a "dopant" before their electronic behavior can be tailored for useful applications. A dopant is typically a small-molecule compound or element that donates electrons to the host material (an n-type semiconductor) or removes electrons from the host (a p-type semiconductor). Typical dopants are minute quantities of elements such as boron or phosphorus added to silicon, or larger quantities of compounds such as iodine or arsenic pentafluoride that are added to organic polymeric semiconductors. The role of dopants is discussed in the following pages. In subsequent sections we consider semiconductor theory, the preparation of semiconductor-grade materials, the fabrication of integrated circuits, and some applications of semiconductors.

B. SEMICONDUCTOR THEORY

The most important defining feature of a semiconductor is its bandgap. As discussed earlier, the electronic structure of a conductive material is best described in terms of energy bands rather than individual orbitals. The result of combining millions of orbitals into a band structure is the formation of a lower-energy valence band derived from all the bonding orbitals of the participating atoms, and a conduction band derived from all the antibonding orbitals. These are separated by a bandgap, which varies from substance to substance (Figure 10.1). As discussed in Chapter 8, metals have a partly filled valence band and an empty conduction band. Electronic conduction occurs by excitation of electrons to levels in the unfilled region of the valence band. Insulators have a filled valence band, an empty conduction band, and a high-energy gap between them. On the other hand, semiconductors have a filled valence band and an empty conduction band, but the two bands are separated by a relatively small energy gap in the range of 1 eV. The bandgaps of some different solids are listed in Table 10.1.

Electrical conduction in an inorganic semiconductor occurs through the movement of either unpaired electrons and/or holes. A "hole" is the positively charged entity that is left when an electron is removed from an electron pair. Unpaired electrons can be formed in two ways. First, exposure of the material to heat or light can cause electrons in the upper levels of the filled valence band to jump the bandgap into the lowest levels of the conduction band. Once promoted in this way, they are free to move through the solid under the pressure of an external negative electrical potential. The positive holes that are left behind in the valence band are

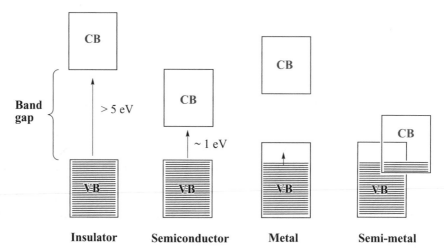

Figure 10.1. *Illustration of the band structures and bandgaps found in different materials.*

TABLE 10.1. Bandgaps of Different Solids (eV)

Material	Bandgap	Material	Bandgap
Te	0.3	GaP	2.3
ZnSb	0.6	InN	2.4
Ge	0.7	AlP	2.5
GaSb	0.8	CdS	2.6
Si	1.1	ZnSe	2.8
InP	1.3	GaP	2.3
GaAs	1.5	InN	2.4
CdTe	1.6	AlP	2.5
Se	1.8	CdS	2.6
Cu_2O	2.2	ZnSe	2.8
ZnTe	2.2	BP	6.0

also free to move, but in the opposite direction from the electrons. Thus, current passes via these counterflowing streams of charged units. However, the flow of holes and electrons may not be equal. Some semiconductors conduct primarily through the movement of electrons and others, mainly through hole migration. The situation that prevails depends on the solid structure and the elements involved. For example, the ratio for the relative mobility of electrons versus holes in silicon is 3:1. In indium phosphide it is 33:1.

The second mechanism that underlies the behavior of semiconductors involves the dopants or impurities that are distributed throughout the lattice. Silicon may be doped with either an element that lies in group III (13) of the periodic table, such as boron, or with an element such as phosphorus or arsenic in group V (15). The addition of each group V element to the silicon lattice adds an extra electron because the group V elements have one more electron in the valence shell than does silicon. Where does this electron go? It goes into a new energy level in the bandgap (Figure 10.2b), from which it can be promoted easily by the absorption of heat or light energy into the conduction band. Conversely, the presence of a group III element impurity creates a new energy level in the gap *into* which an electron can jump from the valence band (Figure 10.2a). Thus, the impurities serve to lower the actual bandgap and provide additional electrons or holes.

What would happen if equal amounts of *both* boron and phosphorus were added as impurities to the lattice? Very little, because the two would cancel each other out with respect to adding electrons or holes to the system. However, both n and p dopants may be used in adjacent areas of the same semiconductor for ease of integrated circuit fabrication. It is sometimes easier to start with, say, an n- or p-doped material and then add a smaller or larger amount of the opposite dopant to gain a close control over the amount of impurity introduced and the exact level of conduction.

This raises another question. How does this apply to *compound* semiconductors, such as gallium arsenide or gallium phosphide, in which the whole solid lattice is composed of equal amounts of group III and group V elements held together by a combination of ionic and covalent bonds? Unlike silicon, these materials are *direct-bandgap* semiconductors, which are the favored materials for light emission from light-emitting diodes or semiconductor lasers. Do these materials need to be doped, or do they already have their own built-in dopant in the lattice? The answer is that

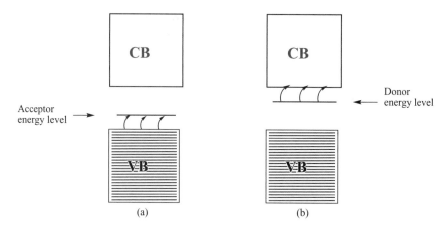

Figure 10.2. *(a) Incorporation of B or Ga into silicon generates a new energy level in the lower region of the bandgap into which mobile electron from the top of the valence band can readily jump. (b) Doping of silicon with P or As inserts electrons into a new energy level in the upper region of the bandgap, from which electrons can easily jump to the lower levels of the conduction band. Both processes lower the effective bandgap and lead to increased conduction.*

a "dopant" is still needed. It may take two forms: (1) GaAs, for example, can be p-doped by the addition of a group II element such as berylium or zinc, or n-doped by a group IV element such as carbon or silicon, or an excess of gallium or arsenic will serve the same purpose; or (2) the doping can be in the form of a third element that gives rise to variable-composition alloys. This third element controls the bandgap. This is important because these materials are the mainstay of the light-emitting diode and laser diode industries, and the added element controls the color of the light emitted when an electric current is passed through the solid. These materials are described by a formula such as $GaAs_{1-x}E_x$, where E is the dopant element. Thus, $GaAs_{1-x}P_x$ is widely used in LEDs, with the amount of phosphorus controlling the color of the light emitted. For example, GaAs itself emits in the infrared but, as an increasing amount of phosphorus replaces the arsenic, the emitted color changes to red and eventually to green. Other colors, such as blue or violet are generated when nitrogen or tellurium are introduced.

C. PREPARATION OF SEMICONDUCTOR-GRADE SILICON AND COMPOUND SEMICONDUCTORS

1. Semiconductor-Grade Silicon

As mentioned in Chapter 3, the manufacture of pure silicon is an involved process that requires first a reduction of sand by carbon to give a mixture of silicon and silicon carbide, followed by reaction with chlorine to give $SiCl_4$ and CCl_4, which are then separated by fractional distillation. Treatment of the $SiCl_4$ with a reactive metal such as zinc or magnesium or with hydrogen gives partially pure silicon. However, this is not nearly pure enough for semiconductor use. A final purification is effected by the process of pulling an ingot from a bath of molten silicon and/or zone refining (Figure 10.3).

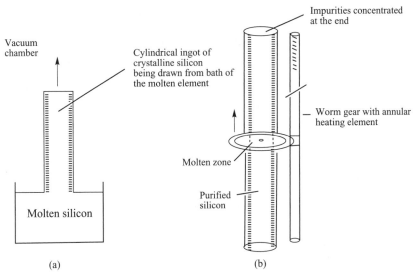

Figure 10.3. *Purification of silicon for semiconductor applications. (a) A cylindrical ingot of crystalline silicon is drawn from a molten silicon bath. (b) Zone refining by movement of a molten zone in which impurities are concentrated and deposited at one end of an ingot.*

Ingot growth and zone refining are both based on the principle that the solid that crystallizes from an impure molten material is purer that the molten material. This occurs because the mechanism of crystallization and the growth of a crystal lattice tend to exclude any atoms or compounds that do not fit efficiently into the solid lattice. Thus, if a crystal of pure silicon is suspended at the surface of molten silicon, it will stimulate the deposition of additional pure material on the original crystal. As the solid material is pulled very slowly from the surface of the melt, an ingot of very pure silicon will be obtained. Zone refining involves taking an ingot of semi-pure silicon and moving a circular heating element vertically from one end to the other. As the hot zone (~1420 °C) of the heating element passes up the rod, the silicon melts and then recrystallizes, a process that concentrates the impurities in the moving molten zone. After a number of passes of the hot zone, the impurities will be concentrated at one end of the ingot, while the rest of the silicon will be highly pure.

Purified ingots are then cut into thin wafers by means of a diamond saw, are polished to reduce surface roughness to less than a ten-millionth of a meter, and are then ready for microlithography. A typical modern ingot will have a diameter of 300 mm (~12 inches), which will give wafers that allow the fabrication of up to 2000 integrated circuits per wafer.

2. Amorphous Semiconductor Silicon

The process described above for the preparation of pure crystalline silicon is expensive—too expensive for the large-scale production of large-area devices such as solar cells. An alternative approach is the use of amorphous silicon produced by chemical vapor deposition (CVD) on a flat surface. When silane (SiH_4) is decomposed in a radiofrequency plasma, it yields fragments that range from silicon atoms to SiH, SiH_2, and SiH_3. If these are allowed to impinge on a flat surface such as glass or a polymer, amorphous silicon contaminated by Si—H species is deposited as a coherent film. *Amorphous silicon*, as its name implies, consists of a random, three-dimensional network of Si—Si bonds interrupted by dangling Si—H units. The hydrogen serves as a dopant, and the films are semiconductive. Their efficiency as semiconductors is quite low. For example, when used in solar cells, amorphous silicon devices have an energy conversion efficiency of only 4% compared to about 10% for crystalline silicon.

3. Preparation of Compound Semiconductors

Some of the most important semiconductor materials are variable-composition alloys that contain combinations of three elements such as gallium and arsenic with phosphorus, or gallium and nitrogen, with aluminum or indium. Compound semiconductors such as $GaAs_{1-x}P_x$ can be prepared by heating the ingredients together in a sealed, evacuated tube. However, chemical vapor deposition of a mixture of organometallic derivatives of the elements is a more reproducible approach using, for example, a single crystal of GaAs as a "template" to control the crystal structure of the depositing material. As these materials are used mainly for LEDs, quite small crystals ($\sim 1\,mm^3$) are satisfactory. A major additional use for these species is in diode or "double heterostructure" lasers, again manufactured by chemical vapor deposition, as discussed later in this chapter.

D. ORGANIC POLYMER SEMICONDUCTORS

1. Background—Polyacetylene

Until the 1970s polymers were considered to be electrical insulators. An unstable inorganic macromolecule, poly(sulfur nitride) $[(SN)_x]$ was known to conduct electricity, but this was considered to be a bizarre aberration. Several events served to change that picture. First, early attempts to polymerize acetylene by free-radical techniques yielded black, insoluble, but electrically conductive materials (reaction 1). But these reactions can be explosive, and there was no way to fabricate films or fibers from these materials. Then in 1967, a method was found to polymerize acetylene to a new form of polyacetylene, $(CH=CH)_n$ (reaction 2) a material that looks more like a metal than a conventional polymer. This polymer forms silvery or gold-colored films that were shown to conduct electricity at a level typical of a weak semiconductor ($10^{-8}\,S/cm$). However, when treated with reagents such as iodine. arsenic pentafluoride, or antimony pentafluoride, or when oxidized with sodium, lithium, or radical anions, or reduced electrochemically, polyacetylene becomes a modest metal-level conductor ($10^5\,S/cm$) in the same range as mercury. These dopants add or remove electrons from the solid polymer, and facilitate conduction by a still controversial mechanism. Other methods to produce polyacetylene are shown in reactions 3 and 4.

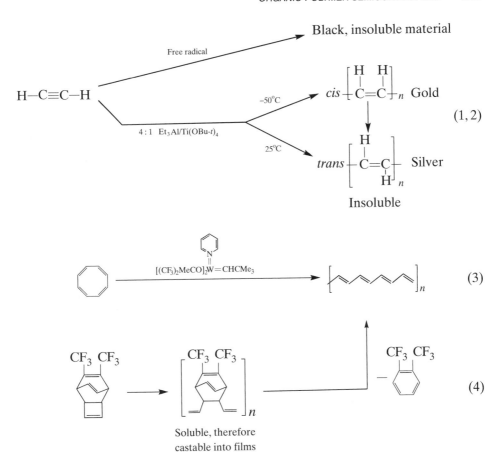

$$(1, 2)$$

$$(3)$$

$$(4)$$

Soluble, therefore
castable into films

Doped polyacetylene films are flexible, and have an almost unique morphology. They consist of polymer chains organized into microfibrils, clustered into larger fibrils that are intertwined into a fiber mat (Figure 10.4).

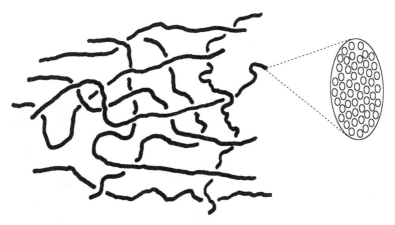

Figure 10.4. *Representation of the morphology of polyacetylene films. The high-surface-area porous structure is valuable for uses as electrodes. It also facilitates the entry of dopants into the material.*

Since the 1970s a number of other unsaturated polymers have been shown to yield electronic semiconductors when doped. These include poly(phenylene vinylene), polyphenylenes, polypyrrole, polythiophene, and polyaniline (reactions 5–9). They are discussed briefly in the following sections.

2. Poly(phenylene vinylene)

Although the electrical behavior of polyacetylene proved to be a groundbreaking discovery, uses for this polymer are severely limited by its instability to oxidation in the atmosphere and its insolubility. This restricts the options for device design and fabrication. Thus, the emphasis has shifted to studying other polyunsaturated polymers such as poly(phenylene vinylenes) or polythiophene, which are more stable in the atmosphere and are soluble in some organic solvents. Like polyacetylene, poly(phenylenevinylene) is an insoluble polymer and therefore cannot be fabricated by solution-casting techniques. Instead, a soluble polymeric precursor in first synthesized and fabricated, and a final reaction converts this to the required polymer. A typical process is illustrated by reaction 5.

Although poly(phenylene vinylene) itself is insoluble, the linkage of organic side chains to the main skeleton improves the solubility considerably, but at the expense of causing twisting of the chains and reduction of electron delocalization. Moreover, changes to the repeating unit structure bring about a change in the bandgap and the wavelength of the light emitted in electroluminescent devices. This is a crucial requirement for semiconductors used for display applications. A major use for poly(phenylene vinylenes) is in flat panel light-emitting devices, as described later in this chapter.

3. Poly(*p*-phenylene)

In principle, this polymer (reaction 6) when doped should be an excellent polymeric semiconductor due to the opportunities that exist for long-range electron delocalization and the stability of the polymer in air. In practice, certain constraints apply, including the difficulty of controlling the chain length through condensation reactions and the insolubility of the polymer when more that four or five phenyl rings are linked together. A more serious problem is the lack of coplanarity of the phenyl rings due to the steric interactions between the hydrogen atoms in the ortho positions. This raises the band gap and, in applications such as light-emitting diodes (see Chapter 14), limits the emission color to the blue region of the spectrum.

4. Polypyrrole and Polythiophene

Pyrrole, thiophene, and their substituted derivatives can be polymerized electrochemically as shown in reactions 7 and 8. The polymers form as films on the cathode, and can even be produced as a continuous strip by use of a revolving, metal drum electrode. Undoped polypyrrole and polythiophene have very low conductivities, in the region of 10^{-12} S/cm. When doped chemically or electrochemically, the conductivity rises to the 10^2 or 10^3 S/cm level, which is in the same range as silver. Polypyrrole is a black material that has been considered for use as an electrode component in batteries, while polythiophene is of interest for its ability to change color in the presence of different compounds, which makes it potentially useful as a sensor.

5. Polyaniline

This is a polymer that has been known almost from the beginning of organic chemistry, although not recognized as such. It is formed by the oxidation of aniline either chemically (using ammonium peroxydisulfate) or electrochemically (reaction 9). The oxidation process goes through several stages, each of which gives rise to a different level of conductivity. The polymer structure shown in reaction 9, where $x + y = 1$, is a good starting point for understanding the different forms of this polymer. *Leucoemeraldine*, with $x = 1$ and $y = 0$, is the least oxidized form, whereas *pernigraniline*, with $x = 0$ and $y = 1$, is the most oxidized state. The most useful form for practical applications is the green-colored "emeraldine base" state where both x and $y = 0.5$, which is a semiconductor when doped with an acid such as HCl.

Emeraldine base is the most useful form because it is less prone to degradation in the atmosphere than the other two modifications. Moreover, it can be synthesized in the form of nanofibers, is used for corrosion protection and antistatic coatings, and is employed as the thin, hole-injection layer in light-emitting diodes. A demonstration integrated circuit has been built based on polyaniline transistors imaged by photolithography. The imaging process is based on the fact that irradiation of the polymer by strong ultraviolet light reduces the conductivity by a factor of 10.

6. Mechanism of Conduction in Unsaturated Organic Polymers

Although it was originally suspected that semiconduction in doped organic unsaturated polymers might occur via a similar mechanism to that found in doped silicon, it was soon realized that significant differences existed. First, much larger amounts of dopant are needed than in conventional inorganic semiconductors, and the dopants are oxidizing or reducing *molecules* rather than atoms. Further, attempts to detect significant amounts of unpaired electrons in the doped polymers by electron spin resonance spectroscopy were unsuccessful. Hence it appeared that conduction may not occur simply via free electrons moving down the π-conjugated backbone (Figure 10.5a). It was also noted that the polymer chains are shorter than the length of the conductive sample, and so the question arose as to how electrons, if they are the charge carriers, jump from the end of one polymer chain to another.

The initial theory was based on polyacetylene conduction. The most widely accepted explanation is that pure polyacetylene is a wide-bandgap semiconductor (Figure 10.5b). However, the pristine polymer shows evidence of an e.s.r. signal that would be consistent with unpaired electrons (Figure 10.5a). The addition of an electron to a conjugated polymer chain initially generates a radical anion, called a *polaron*, in which three electrons occupy two energy levels within the bandgap. Addition of a second electron yields a dianion or *bipolaron*, which is unstable because of the proximity of the two negative charges. Thus, the two electron pairs migrate away from each other to a safe distance that equalizes their energies in the band gap (a soliton pair), and conduction occurs as these two electron pairs migrate in tandem along the polymer chain or transfer from chain to chain. From a chemist's perspective it would be expected that the movement of a soliton or bipolaron along an unsaturated polymer chain would require a lengthening and shortening of skeletal bonds as the charge carrying unit moves through each site on the polymer—a relatively slow process. This is the same problem that has been used to argue against conduction by isolated electrons (solitons). Removal of an electron from a polymer would generate a radical cation, and similar arguments apply. It appears that hole conduction via radical cations may be more important in unsaturated polymers than conduction by electrons or dianions. Explanations for electronic conduction in polyacetylene and the other conductive polymers are still evolving.

E. PHOTOLITHOGRAPHY AND MICROLITHOGRAPHY

1. Principles of Semiconductor Fabrication

Photolithography is based on an old method for printing photographs known variously as the *dye transfer process* or the "washoff relief process." In this process (no longer employed for everyday photography) a sheet of transparent film is coated with a thin layer of a photocrosslinkable polymer that was often gelatin containing dichromate as a sensitizer. Light passing through a negative to the sensitized film caused crosslinking of the gelatin, but only in those areas exposed to light. After a

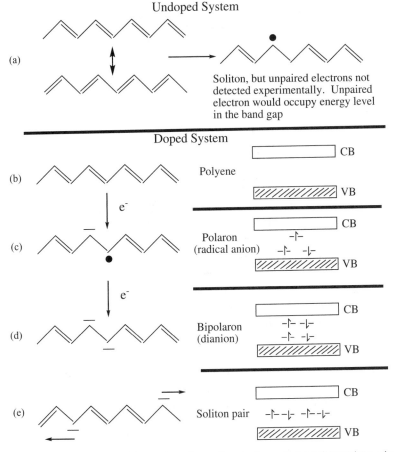

Undoped System

Doped System

Figure 10.5. *Electronic conduction in a polymeric semiconductor such as polyacetylene, showing the formation of a conductive soliton pair.*

development process to amplify the crosslinking, the unexposed, uncrosslinked gelatin was removed by washing with warm water to leave a raised relief image that corresponded to the transparent parts of the negative. This raised relief image was then allowed to absorb a water-soluble dye, and the printing "plate" or matrix was pressed in contact with a sheet of paper to allow the dye to diffuse to the paper in proportion to the thickness of the relief image. Thus, a single printing plate could be used to make many prints by repeating the dye absorption–printing cycle. Full-color prints were produced by superimposing dyes from three separate matrices to transfer yellow, magenta, and cyan colors to the print.

A similar process is used today to make printing plates for the production of newspapers, magazines, and books. Here, a sheet of aluminum is coated with a photosensitive polymer, which is exposed to ultraviolet light through a transparent negative or positive of the original image. This image will typically be in letterpress form (i.e., with only all-black or all-white areas), or a "halftone," in which an image is broken into larger or smaller dots to simulate a continuous tone original. Exposure of the printing plate to ultraviolet light causes either crosslinking of the polymer

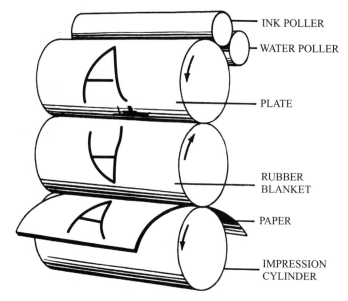

Figure 10.6. *Offset lithography, the method used for nearly all large-scale printing, depends on the adsorption of a hydrophobic ink on hydrophobic regions of the printing plate coupled with the adsorption of water on the hydrophilic regions.*

(an older process) or a change in chemical composition that leads to a change in hydrophilicity or hydrophobicity. The crosslinked polymer plate is then washed with a suitable solvent to remove the uncrosslinked polymer, and the raised relief image is then inked and the ink transferred to paper in a rotary printing press.

If exposure of the printing plate causes a change in hydrophobicity, a different process is employed. The printing plate is attached to a rotating drum and is treated first with water from a roller and then with a hydrophobic ink from a second roller. The ink adheres only to the hydrophobic areas of the plate and is repelled by the water on the hydrophilic areas. The ink is then transferred by rotary action to the paper (see Figure 10.6). This is called "offset printing," which is the method used for virtually all printing these days.

A variant of the photolithography process was and still is used to make printed circuit boards, although silk-screen printing is an alternative approach. However, microlithography to produce integrated circuits with millions of transistors has the closest resemblance to the production of letterpress images with their raised relief profile. The difference is that microlithography is carried out using extreme miniaturization to produce features on the surface that can be as small as 45 nm in size.

2. Overview of the Semiconductor Manufacturing Process

a. Microlithography Principles. Figure 10.7 illustrates the two main approaches to microlithography—specifically, the role of the polymer resist and the differences between negative tone and positive tone resists. A "resist" is a thin layer of polymer that adheres to and covers the silicon oxide coating on the surface of a silicon wafer. The resist must be photosensitive to ultraviolet light and be resistant to the chemi-

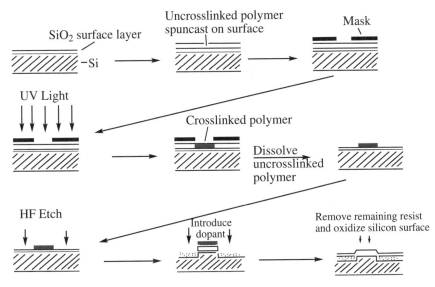

Figure 10.7. *Sequence of events in photolithography through use of a negative tone resist (see color insert).*

cals (such as HF) that can remove the silicon oxide layer in regions of the wafer that are not protected by the polymer. The elemental silicon regions exposed after removal of the oxide layer are then doped either by vapor deposition or by exposure to solutions of the dopant.

A negative tone resist is a polymer that is crosslinked and therefore insolubilized by exposure to ultraviolet light. A positive tone resist is one that is solubilized by irradiation by ultraviolet light. The significance of this difference is illustrated in Figures 10.7 and 10.8.

b. The Overall Sequence of Steps in Microlithography. The sequence of steps in the manufacture of integrated circuits (microprocessors) is as follows:

1. Wafers of pure, crystalline silicon are exposed to hot oxygen or water to form a thin coating of silica on the surface.
2. A very thin film of a photosensitive polymer (the "resist) is spun-cast onto the surface.
3. The photosensitive film is exposed to ultraviolet light through a miniature mask (a letterpress-type negative—opaque and transparent regions only) that bears one or several identical images of one layer of the integrated circuit. This is usually accomplished by *projection* of the image through a lens system. The areas exposed to ultraviolet light are either crosslinked to make them insoluble (a negative tone resist) or are changed chemically to make then more soluble (a positive tone resist). Multiple printing of this image on a 6-, 9-, or 12-inch wafer is accomplished by moving the wafer for a new exposure by means of a "stepper"—a machine that can move and register an image with great precision. A 6-inch wafer will accommodate about 100 integrated circuits, a 9-inch disk about 200, and a 12-inch wafer up to 2000. The

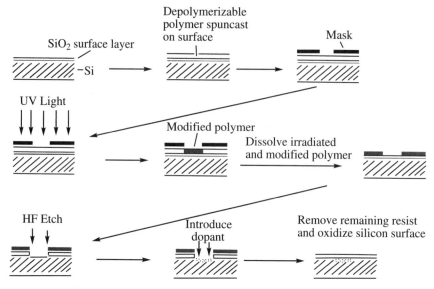

Figure 10.8. *Use of a positive tone resist (see color insert).*

size of circuit features (e.g., transistors) is now less than 100 nm, and is decreasing year by year,

4. The silicon wafer is removed from the equipment and washed to remove the unexposed and therefore uncrosslinked or photosolubilized resist.

5. The silicon oxide surface film exposed by dissolution of areas of the resist is then removed by exposure to hydrofluoric acid or a reactive plasma. The remaining resist protects the silicon oxide that lies beneath it. The wafer is then washed again.

6. The pure silicon exposed by the etching process is exposed to the dopant either by immersion in a solution of a dopant precursor or by chemical vapor deposition. In either case the wafer must be heated to allow the dopant to diffuse into the outer layer of the silicon.

7. Exposure to oxygen then covers the doped area with a layer of silicon dioxide.

8. Separate microlithography steps are needed to introduce the wiring connections within each chip.

9. The whole process is then repeated to superimpose a second imaged, etched, and doped layer on top of the first. As many as 25–50 cycles of this procedure may be needed, each with a separate mask, so that as many as 500 total steps may be required to produce each finished wafer.

10. An incorrect step in any of these procedures has the potential to ruin a chip or the whole wafer, and this becomes more challenging with each advance in miniaturization. With miniaturization now reaching the level of more than one million transistors per chip, the yield of satisfactory chips may be as low as 50%, due to microlithography defects (dust particles, poor registration of successive images, equipment malfunctions, etc.).

11. Individual chips are then tested, cut from the wafer, packaged, and shipped.

It will be obvious from this description why a modern fabrication facility for the microlithography of 300-mm (12-inch) wafers to produce 256-megabit (Mb) dynamic random-access memories (DRAMs) costs well over 3 to 5 billion US dollars.

Special features of each of these steps are described in the following sections.

3. Equipment

a. Microlithography Masks. The miniaturization required for the process is accomplished in two stages: first in the fabrication of each mask, and second in the projection of the image of the mask onto the resist. Masks are the "negatives" used to print an image onto the photoresist. A separate mask is required for each cycle of the photolithography process. One of the major challenges in microlithography is the design of the mask to control the location of transistors, resistors, wiring, and other components. An initial step is to draft the circuit at a "macro" level and then reduce it in size. This reduction can be accomplished optically in a projection unit that resembles a photographic enlarger operated in reverse.

In other words, the large-scale original "negative" is projected through a lens onto a quartz plate that has been coated with a photoresist. Those parts of the photoresist that receive light become crosslinked or made more soluble. The resist is then developed to remove unwanted polymer, and an opaque layer of chromium is sputtered onto the areas of quartz so exposed. Removal of the crosslinked polymer then yields the finished mask. The complexity of mask manufacture is such that many things can go wrong. There may be only one perfect mask in existence for each layer of a given chip, and that mask may be used for making from 100 to 1 million chips. The final reduction in image size is accomplished in the equipment used for chip manufacture, as described in the following section.

b. Microlithography Equipment. The equipment used to generate each level of the chip consists of an intense ultraviolet laser that projects light through the mask and through a complex series of large quartz lenses on to the resist layer on a silicon wafer. The final reduction in the size of the image is achieved at this stage. A schematic representation of the equipment is shown in Figure 10.9.

An intense ultraviolet laser is needed as the light source because the exposure of the resist must be sufficiently rapid to allow multiple images to be transferred in sequence to the wafer by "stepping" the image from one site to another. As feature sizes become smaller, shorter and shorter wavelength ultraviolet light must be used because the irradiation wavelength must be shorter than the smallest feature size on the chip to maintain resolution. For example, a 1-Gb DRAM chip needs features that can be as small as 180 nm. Thus, the crowding of more and more transistors into smaller and smaller space has required a progressive reduction in wavelength from 435 nm (mercury vapor g-line), to 365 nm (mercury vapor *I*-line), 248 nm (krypton–fluorine excimer laser), 193 nm argon–fluorine excimer laser), to the use of 157 nm (fluorine laser) radiation.

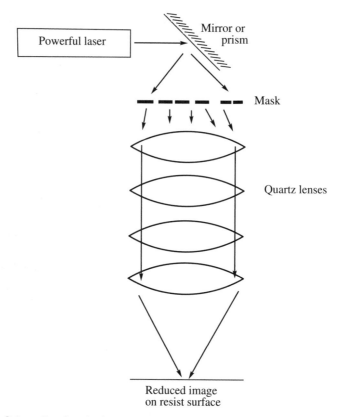

Figure 10.9. Schematic of projection system used to reduce the size of the mask image during microlithography.

The lenses are quartz because this is virtually the only glass that is transparent into the short-wavelength ultraviolet. Moreover, the lenses must have a wide diameter to allow the maximum amount of light to pass through, again to shorten the exposure. Wide-aperture single lenses usually give rise to optical aberations that would blur the pattern projected on the resist. Thus, multiple lenses are needed in the light path to correct for these aberations.

c. Pellicles. As mentioned above, the yield of chips is often no more than 50%. One reason for this is the presence of minute dust particles that could create imperfections in the projected image. Some of these contaminants settle on the resist or the mask during photolithography, so a UV-transparent polymer film known as a "pellicle" is interposed between the mask and the wafer to catch small particles in an out-of-focus region of the light path. Teflon has been the main material used for pellicle films for 248 nm and 193 nm radiation. Pellicles for 157-nm microlithography are hard to find. Few materials except quartz transmit at this wavelength, and this problem has proved to be a handicap to using this wavelength.

d. Steppers. Sequential printing of, say, 200–2000 integrated circuits on one wafer and registration of each successive lithography cycle requires a machine that will move the wafer in two dimensions with a precision that is remarkable. A substantial part of the cost of a fabrication plant is in the steppers.

F. PHOTORESISTS

1. General Features of Resists

The earliest resists used for printed circuit and 435-nm microlithography were taken almost directly from the printing industry. The progressive use of shorter and shorter wavelengths in the microlithography process, and the need for both positive tone and negative tone resists, has stimulated an enormous amount of research and development to produce photosensitive polymers with a wide range of properties.

As discussed, photoresists may function by becoming insoluble when exposed to light ("negative tone" resists) or by becoming soluble after light exposure ("positive tone" resists). The terms "positive" or "negative" tone refer to the fate of the image as it is created on the resist. Negative tone resists generate crosslinked regions that correspond to the transparent areas of the mask. Positive tone resists yield images where the transparent regions of the mask are reproduced as polymer-free regions in the resist.

It might be imagined that any polymer that either crosslinks or decomposes when irradiated with ultraviolet light could be used as a photoresist. In practice, special chemistry is needed to ensure adequate photosensitivity and satisfy several other requirements. A good resist must have the following characteristics:

1. There must be an ease of conversion from a soluble to an insoluble modification or vice versa.
2. The resist must be partially transparent to ultraviolet radiation. If the resist is opaque to the radiation, only the surface layer will respond to UV light, and the underlying material will be unaffected.
3. The resist, as its name implies, must be stable to the aggressive conditions generated during the subsequent reactive ion etching process to remove silicon oxide from the exposed surface.

A very large number of resist systems have been described, and it is beyond the scope of this book to deal with this subject in detail. The following is a brief illustration of some well-known examples.

2. Novolac Positive Tone Resists

Novolac-type polymers have been known almost from the beginning of the polymer industry in 1872. They are based on phenol–formaldehyde compositions that were originally employed to produce rigid thermoset materials such as ashtrays and electrical insulators. Treatment of a phenol with formaldehyde leads to the process shown in reaction 10.

(10)

Hydrophobic photosensitizer;
insoluble in base

Hydrophilic, acidic;
soluble in base

Soluble polymer

The application of heat drives off water and eventually yields a highly crosslinked material. However, the condensation process can be arrested at an earlier stage to yield either water-soluble polymers or species that are on the borderline between soluble and insoluble. These are the polymers used widely in microlithography. Photosensitization and insolubilization in aqueous base of a Novolac polymer is accomplished by the addition of 20–50% of a water-insoluble diazoquinone (reaction 10). Apparently this is a *physical* insolubilization of the polymer due to the presence of the large amounts of the hydrophobic additive. Diazaquinones are photosensitive. When exposed to ultraviolet light, they decompose to yield carboxylic acids (reaction 10), which are soluble in aqueous base, and this tips the balance between insolubility and solubility of the polymer. Thus, polymer in the exposed area of the resist becomes soluble in aqueous base and is removed, leaving the unexposed regions still insoluble. When these are baked, the polymer undergoes further crosslinking to give an insoluble, raised relief, *positive* tone image, which protects the underlying silicon oxide from the subsequent plasma-etching process. Novolac resists are still widely used for printed circuit and low resolution microlithography that employs 350–450-nm mercury vapor radiation, but they are too absorbing of short-wavelength ultraviolet to be used for the highest resolution images. However, it is amazing that such an ancient polymer has been used as the basis for one of the most advanced of all modern technologies.

3. Chemical Amplification

The high sensitivity of the resist to ultraviolet light reduces the time of exposure and allows more circuits to be printed in a given time. This must be accomplished without, as discussed earlier, total absorption of the light at the polymer surface. A solution to this problem is to design a system in which a *catalyst* is generated by the action of light, and this then amplifies the effect of polymer solubilization or insolubilization. One example of this principle is a negative tone resist produced by a Novolac-type polymer combined with a photochemical acid generator that induces crosslinking through a third component: a melamine compound. This system is developed by treatment with aqueous base, which removes the unexposed portions of the image. Features with 250 nm dimensions can be produced by this system. Acid-catalyzed systems are also used to amplify polymer depolymerization reactions that yield positive resist.

Photochemical acid generators fall into several categories. A few examples are triphenylsulfonium hexafluoroantimonate $Ph_3S^+SbF_6^-$, diphenyliodinium triflate $Ph_2I^+CF_3SO_3^-$, and dialkylphenacylsulfonium hexafluoroarsenate $Ph_2C(O)CH_2(Alk_2)S^+AsF_6^-$. When exposed to 200–300 nm ultraviolet light, the first two compounds undergo free-radical cleavage of the carbon–iodine or carbon–sulfur bonds. Subsequent heating at >200 °C induces a series of reactions that generate a strong acid, which reacts with the resist polymer to change its solubility and regenerate additional acid. This provides an amplification mechanism. Phenacyl sulfonium salts yield ylids and acid when irradiated.

4. Poly(4-hydroxystyrene) Resists

This polymer dissolves too rapidly in aqueous base to be useful as a resist. However, the *tert*-butoxycarbonyloxy (*t*-BOC) derivative is useful because the *t*-BOC group can be removed by photochemically generated acid to re-form the hydroxyl group. The *t*-BOC-protected polymer is hydrophobic. The hydroxy form is hydrophilic. Exposure of the exposed portions of the imaged resist to aqueous base or an alcohol removes the hydroxyl form of the polymer to give a positive tone image. Development with a hydrophobic organic solvent such as methylphenyl ether removes the unchanged *t*-BOC regions to form a negative tone resist (reaction 11).

$$\tag{11}$$

Hydrophobic and insoluble → Soluble in aqueous base or alcohol

5. Multilayer Lithography

Integrated circuits for 1 Gb or larger applications require features that are organized in three- rather than two dimensions on the chip. Thus, the repeated process of lithography, etching, doping, and oxidation of the exposed silicon for many cycles generates a profile on the surface that begins to resemble a complex series of mountain ranges and valleys. This causes focusing problems that result from the very shallow depth of focus of the image, which itself is a consequence of the wide lens apertures needed for short exposures. A solution to this problem is to introduce a "planarizing" layer of resist that buries the valleys and provided just one plane on which the image is tightly focused. Subsequent steps then involve cutting the exposed or unexposed regions down to the base layer by means of plasma-etching techniques without losing the edge sharpness needed for the tight tolerances need for the closely spaced features. Because polymers that contain silicon are more resistant to oxygen reactive-ion etching than are completely organic polymers, considerable research has been directed toward using resists made from these materials to prevent loss of image integrity during multilayer procedures.

6. All-Dry Resists

Wet development of a resist to remove the soluble sections is a step that takes time and runs the risk of distortion of the image. Thus, methods to remove the affected parts of the resist by other means are an attractive alternative. One alternative is to treat the exposed resist to a gaseous reagent that silylates phenolic groups in the altered regions of the chip. Subsequent oxygen reactive-ion etching then removes regions of the resist that contain no silylated polymer while, at the same time, converting the silylated areas to a plasma-resistant form. This gives a negative tone resist.

G. ELECTRON BEAM LITHOGRAPHY

Use of an electron beam for lithography in place of ultraviolet light has the advantage that the wavelength is shorter (~13 nm) so that the feature size on the chip can be reduced. Moreover, no mask is needed because electron beams can be focused and moved magnetically and can therefore be used to write directly on the resist. However, this is a slow process, and each chip must be written separately. Thus, chips produced by this method are expensive, and the process is restricted to specialized applications such as the production of masks for ultraviolet lithography. Poly(methyl methacrylate), which depolymerizes when exposed to an electron beam, is a resist material that is appropriate for this technique.

H. X-RAY LITHOGRAPHY

Short-wavelength X rays (down to 1 nm) offer the possibility for producing lithographic feature sizes in the range of atomic dimensions. However, X rays from an X-ray tube cannot be focused (they can only be collimated or reflected from a mirror), and so radiation from a synchrotron is needed to take advantage of the technique. The masks needed for X-ray lithography must use heavy metals for the opaque regions to generate sufficient contrast from the penetrating radiation. Overall, this is a very expensive process, which is restricted to research activities and specialized applications.

I. CIRCUIT WIRING

The "wiring" that connects the transistor regions on a microchip was traditionally made from aluminum, using standard microlithographic and vapor deposition techniques. The most modern chips use copper connections because of its higher conductivity. Wires can be produced that are thinner than the wavelength of the light being used to etch them by generating interference patterns. A typical width is about 90 nm. At the research level it is now possible to etch lines that are only 30 nm in width, and features 65 nm in width are already produced in manufacturing. The subject of nanowires is addressed briefly in the nanoscience chapter (Chapter 17).

J. SEMICONDUCTOR DEVICES

1. Devices Based on Presence of a Single Semiconductor

a. The Thermistor. One of the earliest applications of semiconductors is a simple device for the measurement of temperature, known as a *thermistor*. Opposite ends of a piece of doped semiconductor are connected to a source of direct electric current. The amount of current passing through the semiconductor depends on the temperature. This phenomenon arises because the conductivity depends on the number of electrons that have jumped the bandgap and the number of holes that are left behind (Figure 10.10a). As the temperature rises, more and more electrons jump into the conduction band and more current flows. Calibration of the device against, for example, a thermometer allows a direct correlation to be made between current flow and temperature.

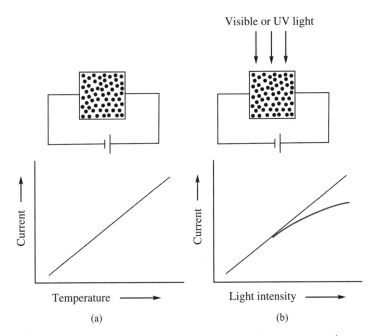

Figure 10.10. (a) Principle of the thermistor. Increased temperature causes an increase in current flow due to the higher population of electrons in the conduction band. (b) The photocell. Increasing intensity of light falling on the semiconductor increases the number of electrons promoted from the valence to the conduction band, thereby providing more current carriers. The falloff in conductivity at the highest intensity levels is due to saturation of the accessible energy levels of the conduction band.

b. The Photocell. A photocell uses the same principles as the thermistor except that electron jumps to the conduction band are stimulated by the absorption of light, so that the current flowing depends on the light intensity (Figure 10.10b). This device is used commonly in cameras to match the shutter speed or lens aperture to the light intensity. Earlier camera photocells used cadmium sulfide, a system that was prone to inaccuracies due to insensitivity at higher light intensities and memory effects. Modern photocells use doped silicon semiconductors. Note that a photocell requires the presence of an external electric current. On the other hand, a *photovoltaic cell* generates electric current through use of a p–n junction.

2. The Transistor and the Metal Oxide Integrated Circuit

An integrated circuit consists of a complex array of transistors, resistors, rectifiers, connecting wires, and insulators. The transistors are the key components that control the flow of current and the logic of the device. They can be viewed either as on/off switches, or as devices for amplifying an electric signal. They function as miniature analogs of the thermionic tubes or valves that once dominated radio or television sets.

The principles of a metal oxide semiconductor transistor are explained in Figure 10.11. It consists of a p-type semiconductor into which have been fabricated (by overdoping) two closely spaced regions of n-doped material (the opposite configuration is also possible). Two electrodes (the "source" and the "drain") are then deposited in contact with the n-type areas. The surface of the silicon is now covered by a thin layer of silicon dioxide on top of which is deposited the third electrode (the "gate"). The gate electrode serves the same role as the grid in a thermionic tube. Positive potential applied to the gate causes holes in the p-type material to retreat from the surface, and this allows electrons to move beneath the silica layer from one of the other two electrodes to the other. The important point is that this control via the gate electrode can either shut off the electron movement or allow it to take place (the on/off function), or it can control the *amount* of current passing. A small potential on the gate can control a larger current beneath the gate. These devices are given the general name of field-effect transistors.

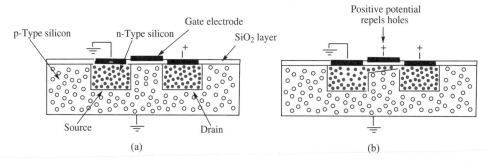

Figure 10.11. Cross section of a metal oxide semiconductor transistor. (a) With no potential applied to the gate electrode, the source and drain n-type regions are insulated from each other by the p-type region. However, application of a positive potential to the gate electrode (b) causes the p-type region to retreat, thus allowing an "n channel" to form beneath that electrode. The current that flows through the n channel can be controlled by the positive potential applied to the gate (see color insert).

The fabrication of millions of these devices on one chip, together with all the other circuit components, is undoubtedy one of the major triumphs of the combination of chemistry, physics, and engineering.

3. Phenomena Based on a p–n Junction

a. Consequences of a p-n Junction. A crucial characteristic of semiconductors is the location of the Fermi level in the bandgap. The Fermi level is the energy level below which the probability of finding an electron is $\frac{1}{2}$. For an intrinsic (i.e., undoped) semiconductor or insulator at room temperature, the Fermi level will be near the center of the bandgap. For a p-type semiconductor, this level will be closer to the valence band (fewer electrons are in the conduction band). For an n-type semiconductor (with more electrons in the conduction band), the Fermi level will be closer to the conduction band. Contact between a p-doped semiconductor and an n-doped counterpart generates an important set of phenomena. As shown in Figure 10.12, because the dopant energy levels in the bandgap have different energies, the Fermi levels are also different. When two materials with different Fermi levels are brought together, the two levels need to assume the same energy. A consequence is that the valence and conduction bands at the junction must bend as shown in Figure 10.13.

A second consequence of a p–n junction is that electrons flow from the n-type region at the junction to form a thin, electron-rich layer on the p side in an attempt to neutralize charge, and the holes do the same thing on the n side (Figure 10.13). This has consequences for the use of junctions in photovoltaic cells.

In practice, p–n junctions are not made by simply pressing the two materials together (the contact would be inefficient and variable from sample to sample) but utilize chemical vapor deposition (CVD), molecular beam epitaxy (MBE)

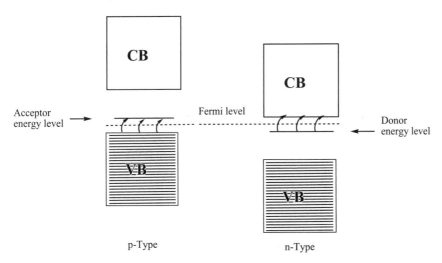

Figure 10.12. *The Fermi levels of the p-doped and n-doped components must be equal at a p–n junction. This means that the linkage of the valence and conductance bands requires band bending at the junction, as shown in Figure 10.13.*

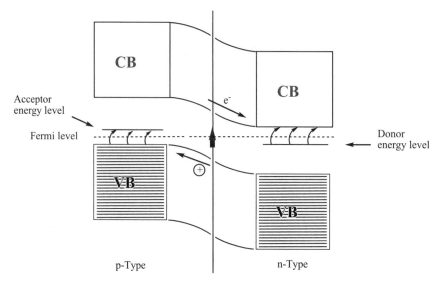

Figure 10.13. *Bond bending at a semiconductor junction caused by the need to equalize the two Fermi levels. Absorption of light causes electrons to jump the bandgap and flow ("dowhill") to the right. Holes flow in the opposite direction.*

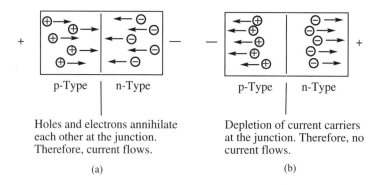

Figure 10.14. *Illustration of how a p–n junction can function as a rectifier. In (a) electrons injected from the right, and holes from the left, annihilate each other at the junction, leading to current flow. A reversal of the current, as in (b), causes both electrons and holes to retreat from the junction, preventing current flow.*

techniques or, especially for integrated circuits, n and p doping of adjacent regions on a chip.

b. The Rectifier. A *rectifier* is a device that converts alternating electric current to some semblance of direct current. A semiconductor rectifier uses a p–n junction to achieve this end. Consider the charge-carrying species near a p–n junction (Figure 10.14). As a negative electric potential is applied to the n-doped component (called *forward bias*), electrons will move toward the junction where they meet holes moving in from the p-doped side. The electrons and holes anihilate each other at the junction as current flows. However, if the potential is reversed, electrons move away from the junction attracted by the positive potential. At the same time, holes move from the junction region toward the now-negative potential. This depletes both electrons and holes from the junction region, and current ceases to flow. Thus, only one component of the alternating current can pass, and this converts an alternating current into a (chopped) direct current. Smoothing circuits are needed to make this resemble a normal direct current.

c. Photovoltaic Cells. A semiconductor photovoltaic (solar) cell generates electricity when a p–n junction is exposed to light. The principles involved are illustrated in Figure 10.15. If the light has energy that exceeds the bandgap, electrons will be promoted into the conduction band while holes will remain behind. The promoted electrons are attracted toward the p-type side of the junction, while the holes will move to the n-type region. This happens because (1) the electrons are attracted toward the thin layer of positive charge on the n side of the interface and (2) the electrons can lower their energy by moving in that direction. One of the practical problems with silicon photovoltaic cells is that the conversion of light to electricity is quite inefficient, partly because electrons and holes may recombine without generating an electric current. The cofiguration of a silicon photovoltaic cell is shown in Figure 10.15.

Most current photovoltaic cells use crystalline or amorphous silicon as the semiconductor material. Polymeric semiconductors have many potential advantages for solar cell applications, including impact resistance, lightness of weight, and ease of fabrication. However, their lifetimes are currently rather limited because of the sensitivity of conjugated organic polymers to photodecomposition.

Note that the photosensors in digital cameras may contain 10 million or more diode p–n junctions arrayed across the ~6-cm^2 surface of the device, itself a remarkable technological achievement.

Another form of photovoltaic cell is based on oxide semiconductors, liquid electrolytes, and dyes. The underlying principle behind these cells is summarized in Figure 10.16. A thin coating of a transparent titanium dioxide semiconductor, in the

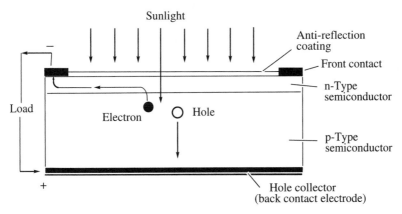

Figure 10.15. *Cross section of a semiconductor solar cell illustrating how sunlight induces separation of electrons and holes in the region of the junction, and how this leads to the generation of an electric current.*

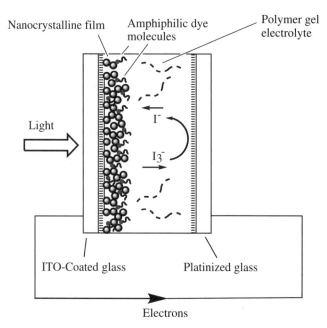

Figure 10.16. Schematic cross section of a dye-based solar cell. The nanocrystalline TiO_2 semiconductor layer receives electrons from the adsorbed dye molecules, which are reactivated via the cycling of an I^-/I_3^- couple.

form of nanoparticles or nanorod surface features, is deposited or etched on a conductive electrode. Dye molecules are then deposited on the high-surface-area titanium dioxide layer. A conductive counter electrode of platinum on glass is spaced a short distance away from first electrode and is separated from it by a liquid or gel ion conductive electrolyte (Chapter 12) that contains a redox-active (reduction–oxidative-active) salt such as lithium triiodide. Light photons impinging on the cell are captured by the dye molecules, and this energy is transferred to the titanium dioxide and thence into an external circuit. This process depletes electrons from the dye/semiconductor layer, and these are replenished by the I^-/I_3^- species in solution via electrons arriving at the ITO electrode. In principle, dye-based solar cells can be produced on a larger scale and less expensively than can cells based on crystalline silicon, or even on amorphous semiconductor silicon. Challenges to be overcome include the photosensitivity of the dye and the efficiency of electron transfer between the dye and the titanium dioxide layer. Figure 15.5 (in Chapter 15) shows nanostructured surfaces of TiO_2 designed to improve the efficiency of dye-based solar cells.

d. Inorganic Light-Emitting Diodes. These are devices based on a p-n junction that generates light from an electric current. Their operation can be understood in terms of the illustrations in Figure 10.13, 10.14, and 10.15. Assume the reverse of the scenario just illustrated in Figure 10.15 for a solar cell. Instead of using light photons to generate an electric current, an electric current is forced across the semiconductor junction. Current continues to flow as electrons combine with holes at the junction. However, for this to occur, electrons and holes must be concentrated in the same region to allow the electrons to fall from the conduction band to the valence band and, in so doing, combine with holes to emit energy. That energy can be in the form of light, the wavelength of which depends on the bandgap. Thus, alterations to the effective bandgap by changes in the composition of the semiconductor materials will generate light of different colors. This is why the color of light emitted from a gallium arsenide semiconductor depends on the amount of phosphorus present in the crystal.

e. The Semiconductor Laser. A *diode laser* is a device that works on the same principle as does an inorganic light-emitting diode in the sense that light is emitted when electrons in the conduction band fall through the bandgap to combine with holes in the valence band. However, a two-layer device that consists of thin slabs of p- and n-doped semiconductor is a relatively inefficient means for generating light. Instead, a preferred design involves three layers to favor light amplification by lasing. A typical laser consists of a thin sandwich of a low-bandgap material between two higher-bandgap layers, one of which is n-doped and the other p-doped, to give a so-called double heterostructure (DH) device. The central thin layer allows electrons and holes to become concentrated in this layer. The structure of such a laser is summarized in Figure 10.17. Two materials that are commonly used are GaAs and AlGaAs.

Electrons (from the n-doped layer) and holes (from the p-doped layer) injected separately into the top and bottom layers via a "forward" electric potential, will meet, combine, and generate light within the middle layer. The color of the light for a relatively thick middle layer will be defined by its bandgap. Moreover, because *recombination of electrons and holes is assisted by the presence of photons*, this becomes an amplifying process. Light travels back and forth along the middle layer, being reflected at the edges, until it bursts through as a laser beam. Alternatively, if the middle layer is thin enough, it serves as a "quantum well," which allows the emitted wavelength to be controlled by the layer thickness. A feature of triple-layer lasers is that the emergent light is not collimated—it spreads out and must be focused or collimated by a lens. Red lasers are widely used in devices like lecture pointers or barcode readers. Blue lasers use GaN (bandgap = 3.4 eV) and alloys such as $Al_xGa_{1-x}N$ or $In_xGa_{1-x}N$. These lasers emit light in the blue or green regions

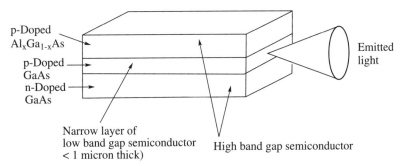

Figure 10.17. *Schematic diagram of a semiconductor diode laser.*

of the spectrum. The blue emitters are now widely used in high-density optical storage devices such as CD and DVD players. The fabrication of semiconductor lasers is accomplished via chemical vapor deposition of organometallic compounds such as trimethylgallium.

f. Polymeric Electroluminescent Devices. Polymeric semiconductors, such as poly(phenylene vinylenes), are used increasingly in light-emitting devices, mainly because they are relatively inexpensive to produce, and because they can be formed on robust flexible films. The operation principle for these devices is slightly different from that for the inorganic light-emitting diodes described above, as shown in Figure 10.18.

g. Quantum Dots. A *quantum dot* is a nanometer-sized semiconductor particle that contains about 100–1000 atoms. It is composed of silicon, germanium, or a compound semiconductor such as cadmium selenide. Quantum dots are typically produced by either molecular beam epitaxy or colloidal chemical precipitation or crystal growth. The significance of quantum dots is that they can absorb white light and reemit light at a specific wavelength that is determined by the *particle size* rather than the composition. The smaller the particle, the shorter is the wavelength. This means that, provided control can be exercised over the particle size, the color of the emitted light may be varied across the whole visible spectrum from red to blue without any change to the composition of the material. Moreover, the purity of the color (the narrowness of the emitted line) is unusually high. This phenomenon arises because the small size of the particles confines both electrons and holes to a small region and limits the mode of recombination.

Quantum dots have been widely studied as tracers in biological processes. For example, quantum dots can be linked to a biomolecule, and the movement of that molecule through a tissue or cell can be monitored through the emission of a specific color during irradiation with light. The photolytic stability of the particles is a significant improvement over the fluorescent organic dyes that are normally used for

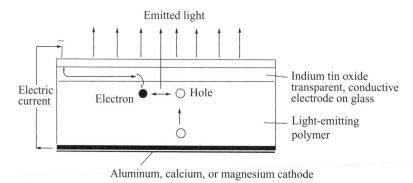

Figure 10.18. *Emission of light from a polymeric electroluminescent device. Electrons and holes from an anode and cathode are forced through a semiconducting polymer. Combination of the two charged species generates energy that is emitted in the form of light.*

this purpose. Applications in light-emitting panels are possible by distributing the dots within a polymer film. Because quantum dots can be "addressed" by a very small electric current, they may also be the basis of nanometer-scale transistors for use in fast-acting integrated circuits or, for example, in nanoscale blue lasers.

K. UNSOLVED PROBLEMS IN SEMICONDUCTOR MATERIALS SCIENCE

Undoubtedly the single greatest challenge in the semiconductor industry is further miniaturization beyond what has already been achieved. Approaches that are under development have been mentioned throughout this chapter, including electron beam and X-ray lithography and interference-based reduction in the wavelength of ultraviolet light. At the other end of the scale, progress can be anticipated in the *printing* of integrated circuits from soluble organic semiconductive materials using silicone rubber printing plates. Field effect transistors based on this principle are already under development. For many applications this type of process offers the promise of mass production of computer and television screens, or of integrated circuits for automobiles and appliances at a fraction of the cost of devices made by photolithography. However, for the finest features, it seems clear that one of the various forms of photolithography or a method based on assembling molecules on a surface will be needed to give the ultimate level of miniaturization—which is down to the molecular level.

L. SUGGESTIONS FOR FURTHER READING

1. Streetman, B. G.; Banerjee, S. K., *Solid State Electronic Devices*, 6th ed., Pearson Prentice Hall, Upper Saddle River, NJ, 2006.
2. Yu, P. Y.; Cardona, M., *Fundamentals of Semiconductors*, 3rd eds., Springer, 2005.
3. Numai T., *Fundamentals of Semiconductor Lasers*, Springer, 2004.
4. Bredas, J. L.; Silbey, R., (eds.), *Conjugated Polymers*, Kluwer/Springer, 1991. (This book contains articles by some of the leading early contributors to the field of conductive polymers.)
5. Chiang, C. K.; Finncher, C. R.; Park, Y. W.; Heeger, A. J.; Shirakawa, H.; Louis, E. J.; Gau, S. C.; MacDiarmid, A. G., Electrical conductivity in doped polyacetlyene, *Phys. Rev. Lett*, **39**:1098–1101 (1977).
6. Burroughs, J. H.; Friend, R. *et al*, Light-emitting diodes based on conjugated polymers, *Nature*, **347**:539–541 (1990).
7. Zheng, J.; Swager, T. M., Poly(arylene ethynylene) in chemosensing and biosensing, *Adv. Polym. Sci.* **177**:151–179 (2005).
8. Grätzel, M.; O'Regan, B., "A low-cost, high-efficiency solar cell based on dye-sensitized colloidal TiO_2 films", *Nature* **353**:737–740 (1991).
9. Hagfeldt, A.; Grätzel, M., "Molecular Photovoltaics", *Accounts Chem. Res.* **33**:269–277 (2000).
10. Günes, S.; Neugebauer, H.; Sariciftci, N. S., "Conjugated Polymer-Based Organic Solar Cells", *Chem. Rev.* **107**:1324–1338 (2007).

M. STUDY QUESTIONS (for class discussions or essays)

1. Discuss the reasons why silicon is the dominant semiconductor material in present-day devices. Discuss which other semiconductors are candidates for use on a similar broad scale and speculate on the devices that might accelerate their introduction.

2. Explain the challenges faced by the developers of integrated circuit chips when the feature size falls below 50 nm.

3. Why have poly(phenylenevinylene) and its derivatives emerged as a preferred organic semiconductor? Describe its advantages, and some of the disadvantages that exist to the use of other polymers.

4. Describe the problems that must be overcome before X rays or electron beams can be routinely used to image a photoresist. What chemistry could be used to produce negative or positive tone resists using these types of radiation? What would be the advantages?

5. How are p–n juctions fabricated for inorganic and organic semiconductors? What are the limits to miniaturization of these junctions?

6. You have been asked to design an integrated electronic circuit that will withstand temperatures above 500 °C for long periods of time. How would you approach this challenge?

7. Discuss the following statement. "The bringing together of polymer chemistry and semiconductor science has played a major role in the advances in computer technology during the past 20 years."

8. Dicuss the staement: "Photolithography is chemistry in action."

9. The doping of an inorganic semiconductor is a key step that underlies virtually all modern electronic science and technology. Explain the theoretical and experimental principles that are involved.

10. What do a semiconductor rectifier, photovoltaic cell, and a light emitting diode have in common? Describe how this common feature accounts for their properties and uses.

11. How might an integrated circuit be constructed that uses tetracene molecules (Chapter 5) instead of silicon or a conductive polymer as the semiconductor material?

12. Discuss the advantages and disadvantages of photovoltaic cells based on silicon versus those that use dyes to capture photonics.

13. Numerous semiconductors exist that contain two, three, or more elements. What are the advantages of these types of materials over silicon or germanium. Describe devices in which they are the preferred materials.

11

Superconductors

A. OVERVIEW

Superconductors are materials that offer no resistance to the passage of an electric current. The significance of this phenomenon is profound. Without electrical resistance, a current will continue to flow indefinitely within the material, magnetic fields generated by the current will be stronger and need very little current to maintain them, and superconducting objects placed in an external magnetic field will levitate the magnet—specifically, the magnet will become suspended in the air. Applications that utilize these phenomena include strong magnets for use in NMR spectroscopy, trains that move above the track but do not touch it, high-performance electric generators and motors, and power lines that transmit much larger amounts of electric current than do conventional wires. More fanciful proposed applications include shields against strong magnetic pulses, noncontact bearings, rocket launchers, energy storage rings, and rapid-switching electronic circuits and detectors. The catch to all these applications is that no materials are known that are superconductors at room temperature. All the known examples must be cooled to temperatures that vary from near zero degrees Kelvin to 138 K, depending on the type of material.

The superconductivity of mercury at a temperature near 4 K was first reported by Onnes in 1911. Between then and 1986, roughly 30 metals, a number of simple alloys, some lead molybdenum sulfides such as $PbMo_6S_8$ known as *Chevrel phases*, and an unstable polymer, poly(sulfur nitride), were found to become superconductors when cooled to temperatures close to absolute zero. The maximum temperature (the transition temperature) at which superconduction could be generated during this period was 23 K. In 1986 a far-reaching discovery was made by Bednorz and Muller in Switzerland that certain ceramic derivatives produced from copper oxide plus rare-earth oxides became superconducting when cooled below 30 K. The

Introduction to Materials Chemistry, by Harry R. Allcock
Copyright © 2008 by John Wiley & Sons, Inc.

significance of this discovery lay less in the 7 K increase in transition temperature and more in the recognition that copper oxide ceramics, with all the opportunities for synthetic variations that they provided, offered the promise of dramatic increases in the temperatures at which superconduction could be achieved. Indeed, subsequent developments in rapid succession raised the superconducting transition temperature to 138 K, a temperature that can be reached by cooling with liquid nitrogen rather than liquid helium. This "highest" temperature material contains mercury, thallium, calcium, and barium as well as copper and oxygen. Altogether more than 6000 different materials are now known to have superconducting properties, including more than 50 different copper oxide–based ceramics. Thus, these high-temperature species have been the focus of most of the recent research in this field. Some classical metallic superconductors and their superconducting transition temperatures are listed in Table 11.1.

The plot of electrical resistance versus temperature shown in Figure 11.1 illustrates the behavior of most superconductors. At room temperature they are metal-level conductors or semiconductors. The resistivity decreases only slightly as the temperature is lowered until the superconducting transition temperature T_c is reached, at which point the resistance falls sharply to zero. Provided the temperature is maintained below this point, current continues to flow through the material indefinitely, generating a strong magnetic field that can be utilized in a wide range of electrical devices.

In addition to metal alloys and cuprate ceramic superconductors, superconduction has also been detected for a polymer, poly(sulfur nitride) (polythiazyl), near absolute zero. Carbon nanotubes and fullerenes also superconduct under special conditions. Moreover, a few conventional polymers such as oxidized polypropylene show high conductuvities at room temperature—not quite superconducting but higher than metals. These are termed "ultraconductors."

In this chapter we focus on high-temperature cuprate superconductor species in terms of (1) the nomenclature, (2) how they are synthesized, (3) solid-state structure, (4) possible mechanisms of superconduction, and (5) current and emerging applications. The other types will be mentioned briefly in the appropriate sections.

TABLE 11.1. Examples of Low-Temperature (Type 1) Metallosuperconductors

Metal	T_c (K) at Ambient Pressure
Titanium	0.4
Cadmium	0.517
Osmium	0.66
Zinc	0.85
Molybdenum	0.915
Aluminum	1.175
Thallium	2.38
Chromium	3.00
Indium	3.41
Tin	3.72
Mercury	4.15
Lanthanum	4.88
Lead	7.196
Niobium	9.5

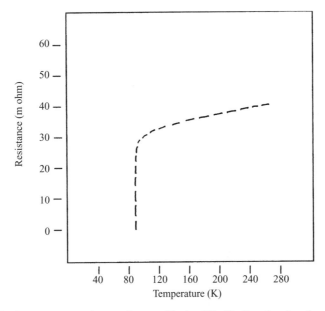

Figure 11.1. *Resistance versus temperature profile for $YBa_2Cu_3O_7$, showing the onset of super-conductivity as the material is cooled to 90 K.*

TABLE 11.2. Formulas of Several High-Temperature (Type 2) Superconductors and Their Superconducting Transition Temperatures

Formula	T_c (K)
$La_{1.8}Sr_{0.2}CuO_4$	35
$TlSrLaCuO_5$	40
$Bi_2(CaSr)_3Cu_2O_8$	90
$(Ba,Sr)CuO_2$	90
$YBa_2Cu_3O_7$	92
$Bi_2(CaSr)_4Cu_3O_{10}$	110
$Tl_2Ca_2Ba_2Cu_3O_{10}$	125

B. NOMENCLATURE

High-temperature superconductors have a number of different formulas, as summarized by the examples in Table 11.2.

There are two different name–abbreviation systems in use. The simplest system lists the number of elements of a certain type in the formula. For example, $YBa_2Cu_3O_{7-z}$ is a "1,2,3-type superconductor," with the numbers simply denoting the number of metal atoms in the formula unit. The number of oxygen atoms is not specified because that number varies according to processing conditions. Using the same system, $Tl_2Ba_2CuO_6$ is a 2,2,1 material, $Tl_2CaBa_2Cu_2O_8$ would be a 2,1,2,2 material, and $Tl_2Ca_2Ba_2Cu_3O_{10}$ would be a 2,2,2,3 superconductor. This system can soon become unmanageable unless the types of metal atoms are arrayed in some previously agreed on sequence.

An alternative nomenclature is based on the crystal structure and on the number of *planes* within a conductive unit or block. All copper oxide superconductors contain sheets of conductive copper oxide planes that alternate with electrically insulating and spacer sheets. Thus, this naming system specifies four numbers. The first is the number of insulating layers between conducting sheets. For example, TlO layers are insulating. The second number denotes the number of spacer sheets between copper oxide sheets: BaO layers are spacers. The third number indicates the number of separating layers: CaO layers are separators. The fourth number refers to the number of CuO planes within a conductive unit or block. Thus, $TlBa_2Ca_2Cu_3O_9$ is a 1,2,2,3 system. Clearly, it is necessary to know the crystal structure before the material can be named. The safest system for a chemist is to simply specify the full formula.

C. SYNTHESIS OF HIGH-TEMPERATURE SUPERCONDUCTORS

The synthesis of ceramic superconductors is, in principle, a process that requires a complicated sequence of simple steps. The equipment needed is found in most chemistry, chemical engineering, or ceramics laboratories. A typical procedure for the synthesis of $YBa_2Cu_3O_{7-z}$ is as follows:

1. A finely ground mixture of yttrium oxide, barium carbonate, and copper oxide in a ratio of Y:Ba:Cu of 1:2:3 is heated in air in a furnace to ~950 °C for 18–24 h in an alumina or porcelain dish. This process generates the fundamental 1,2,3 crystal structure, but with a deficiency of oxygen. The conglomerate is then cooled to room temperature and pulverized.

2. The material is heated again for several more hours with the temperature rising from 500 °C to ~950 °C under a flow of oxygen, and then held at ~950 °C for 18 or more hours before the temperature is allowed to fall slowly (~100 °C/h), again all the time under a flow of oxygen. This step increases the amount of oxygen in the solid, a process known as *oxygen annealing*. The material is then reground to a powder.

3. Finally, the black powder, or a pellet pressed from it, is placed in an alumina dish and is then heated to 950–1000 °C for about 18 h before being allowed to cool at a rate of less than 100 °C/h under oxygen. The slow rate of cooling in this step and in step 2 is critical for obtaining the best superconducting material. A crucial aspect of this procedure is providing an opportunity for oxygen to enter or leave the lattice at temperatures above 400 °C. A typical heating/cooling sequence is shown in Figure 11.2.

Following the synthesis process, a good test for superconductivity is the Meissner effect. With the superconductor pellet in a container, a small rare-earth magnet is placed on top of it and liquid nitrogen as added *slowly*. When the temperature of the superconductor falls below its transition temperature, the magnet should rise and levitate above the superconductor. Another test is a direct measurement of conductivity. The conductivity is measured by means of a four-point probe (see Chapter 4) as the temperature is lowered by adding liquid nitrogen. As the super-

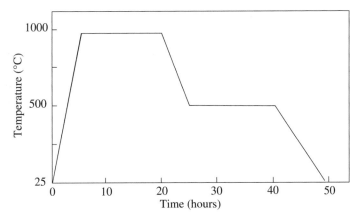

Figure 11.2. *Temperature profile used for the preparation of YBa₂Cu₃O₇.*

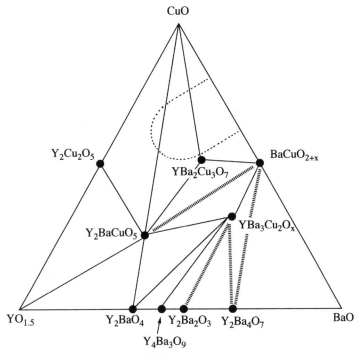

Figure 11.3. *Phase diagram of the species formed by varying the ratios of CuO, BaO, and YO₁.₅, followed by heating at 950–1000 °C. The region bounded by the dotted lines represents compositions that undergo partial melting. YBa₂Cu₃O₇ lies close to this boundary.*

conducting transition is reached, the current flow suddenly increases dramatically and the resistance falls to zero.

The composition of a superconductor depends on the ratio of the starting materials and on the way in which different phases separate at different temperatures. The ternary phase diagram shown in Figure 11.3 illustrates the various compositions that can be achieved by changes to the amounts of the three starting materials. Note both the composition ratios that lead to partial melting and the location of $YBa_2Cu_3O_7$ in the diagram, which is very close to the melting contour. Removal of oxygen, initially to give the $YBa_2Cu_3O_{7-z}$ species, is an essential part of the heating–cooling cycle.

Each species shown in Table 11.2 requires a different set of conditions to optimize the superconductive behavior. For example, the thallium-based superconductors are different in the sense that oxygen annealing is not required. Moreover, Tl_2O melts at the low temperature of 300–400 °C, allowing better contact between the components during the sysnthesis process. However, Tl_2O is volatile and toxic, and any volatilization will change the stoichiomentry so that the composition of the starting mixture is not necessarily related to the composition of the final material. For example, the 2:1:2:2 125 K superconductor is produced from a 2:2:2:3 starting mixture.

D. SOLID-STATE STRUCTURE

An understanding of why superconducting ceramics behave as superconductors requires a knowledge of the crystal structures, which can be determined by X-ray diffraction techniques. The starting point for understanding these crystal structures is the "ABX_3" perovskite unit cell, which is shown in Figure 11.4.

A typical perovskite structure, based on the mineral $CaTiO_3$, has a titanium atom near the center of the unit call, eight calcium atoms at the corners, and six oxygen atoms in the centers of the edges. The relationship of this structure to the more complex arrangement in $YBa_2Cu_3O_7$ will be evident from Figures 11.5 and 11.6.

Two obvious differences from the classical perovskite structure are the absence of oxygen at certain sites and the presence of a fourth element, in this case yttrium,

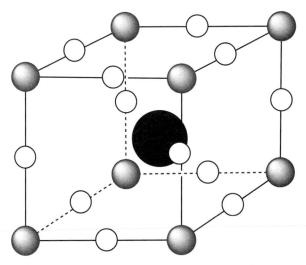

Figure 11.4. *The classical perovskite unit cell, with a metallic cation (black circle) at the center, 8 smaller metallic cations (partially shaded circles) at the corners, and 12 nonmetallic anions (open circles) in the center of the edges.*

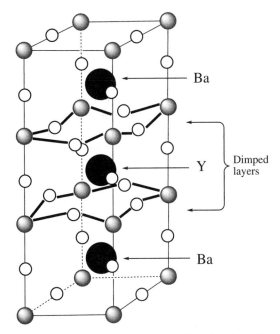

Figure 11.5. *Unit cell of the superconductor YBa₂Cu₃O₇. Note the dimped copper (partially shaded circles) oxide layers that are believed to provide the superconducting pathway and the deficiency of oxygen atoms in the layers that contain the barium ions.*

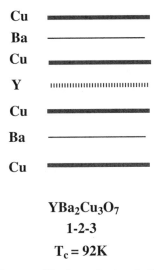

$$YBa_2Cu_3O_7$$
$$1\text{-}2\text{-}3$$
$$T_c = 92K$$

Figure 11.6. *Schematic diagram of the layer structure in YBa₂Cu₃O₇ (see color insert).*

which means that the unit-repeating distance along one axis is much longer than in the classical perovskite structure. Note especially the location of those copper oxide layers that lie on the planes that flank the yttrium ions. These layers are "dimped," that is, the oxygen atoms in those planes are slightly displaced from the copper planes toward the adjacent copper oxide layer. The oxidation state of the copper ions is higher than 2, which is unusual. Superconduction of electrons may take place along the layers occupied by the dimped copper oxide units, whereas the planes that contain the yttrium atoms appear to *insulate* the adjacent copper oxide planes from each other. The planes occupied by the barium oxide are viewed as *spacers* because they separate the two adjacent, dimped copper oxide layers by a considerable distance (two classical perovskite cell lengths). The copper oxide layers in the spacer planes are not dimped and may not participate in superconduction. However, some superconductors contain no copper oxide layers at all (barium bismuth oxides with potassium or rubidium ions), and so the significance of these structural features is a subject for debate.

The structures of other cuprate superconductors are more complex, particularly when more than four different elements are present, but the basic structural principles described above are reproduced in the more complex species. For example, the five-element thallium superconductors such as $Tl_2CaBa_2Cu_2O_8$ or $Tl_2Ca_2Ba_2Cu_3O_{10}$ have structures that contain double thallium–oxygen sheets (Figure 11.7).

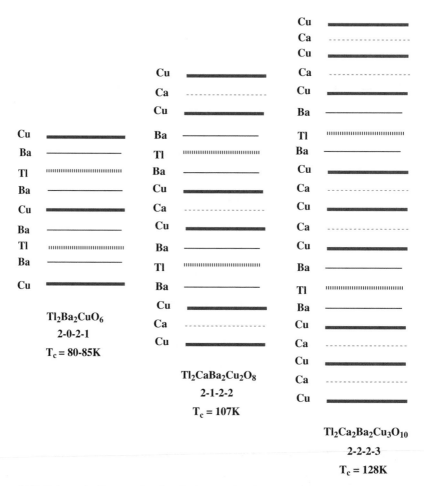

Figure 11.7. Schematic diagram showing the layers of thallium, calcium, barium, and copper oxides in the repeating sequences of three cuprate superconductors. The superconducting transition temperatures (T_c) rise with increasing numbers of copper oxide layers (see color insert).

In these systems the presence of adjacent copper oxide planes separated by calcium ions give higher T_c values as more copper oxide layers are added. Cuprate-type superconductors conduct electric current more easily along the direction of the copper oxide planes than in other directions. This is not true for bismuth-based systems, in which the conduction is anisotropic.

E. THEORIES OF SUPERCONDUCTION

The reasons for superconduction in metals and metal alloys (type 1 systems) is fairly well understood. These species show metal level conductivity at room temperature and must be cooled to temperatures close to absolute zero before superconduction is detected. Their behavior can be understood in terms of the *Bardeen–Cooper–Schreiffer* (BCS) *theory*, which asserts that electrons passing through a crystalline lattice cause the lattice to deform toward the electrons, a process that generates a "phonon trough" with a positive charge. Specifically, as an electron passes a site in the lattice, the nearby metal cations move toward the moving electron (the lattice distortion). However, as the electron moves on, the massive cations, with their large inertia, do not immediately return to their original location, but provide a strong attraction for a second electron. Thus, the electron that follows the first will be accelerated toward the deformation (Figure 11.8). In a sense, this resembles a synchrotron accelerator. Superconductivity will occur, but only if the normal lattice vibrations are reduced by cooling to near absolute zero. At higher temperatures the lattice vibrations will quench any coordinated quantized lattice distortion and thus prevent superconduction. This theory has also been applied, with some variations, to high-temperature, type 2, superconductors.

An aspect that plays a key part in the theory is the existence of "Cooper pairs" of electrons. Normally, electrons are repelled by other electrons, but in superconductivity it is postulated that two electrons will become spin-paired through their interactions with the lattice. This process will facilitate their movement through the material. An analogy is the way in which two cations in water may become paired because their water solvation shells bind them together. Thus, superconductivity arises when the transport of the Cooper pairs becomes coordinated with the lattice vibrations in such a way that their movement is not hindered by attractions to the

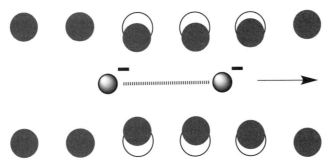

Figure 11.8. Representation of the lattice distortion that could accelerate a Cooper pair of electrons and lead to superconductivity.

nuclei in the lattice. Cooper pairs may function as *s* waves, having the characteristic spherical shape of *s* orbitals, or as *d* waves, which resemble the four-leaf-clover shape of *d* orbitals.

Another explanation invokes a static distortion of the lattice in which the valence of the copper atoms changes from site to site, resulting in a synchrotron-type effect to accelerate the electrons. An alternative explanation of superconductivity in type 2 materials is that they facilitate the formation of rivers of charge flowing through the materials, for example, along the copper oxide planes. These rivers are called "stripes." Stripes have been proposed to explain the behavior of ultraconductors.

The explanations of high-temperature superconduction are still in flux. With so many superconductors now known, it is possible that slightly different theories apply to different materials that coincidentally behave in similar ways. This is clearly an area in which the development of theory lags behind the synthesis and discovery aspects.

F. OTHER SUPERCONDUCTING SYSTEMS

a. Poly(sulfur nitride) (Polythiazyl). The discovery that this polymer becomes a superconductor when cooled to 0.3 K was made in the 1970's. It remains today the only example of a superconducting polymeric material. Gold-colored poly(sulfur nitride) [$(SN)_x$] is synthesized by polymerization of the unstable cyclic dimer [$(SN)_2$] (see Chapter 6). At room temperature it is a metal-level conductor that is oxidatively unstable and has a reputation for spontaneous detonation. Early claims that it might be used for power transmission were clearly exaggerated. The electrical behavior probably depends on the solid-state structure, which is shown in Figure 11.9.

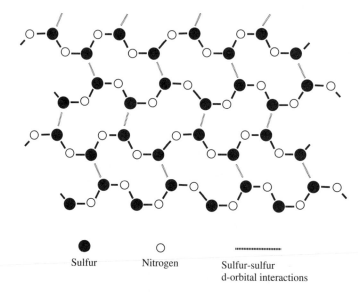

●	○	
Sulfur	Nitrogen	Sulfur-sulfur d-orbital interactions

Figure 11.9. *Crystal structure of poly(sulfur nitride) showing the cis-trans planar chain conformation and the close interchain sulfur–sulfur distances that may allow electronic transmission between chains and perhaps explain the superconductivity.*

The most important feature is that the chains are organized in a cis–trans planar conformation, with opportunities for sulfur–sulfur interchain electronic interactions as well as electron or hole transmission along the delocalized orbitals of the skeleton. At temperatures above $0.3\,K$ the conductivity is mainly along the polymer chain axis, but below the superconducting transition the conduction is anistropic. Reasons for the superconductivity are not clear, but an understanding of the mechanism might point the way toward other polymeric superconductors.

Other superconductors that do not fall into the established categories include fullerenes doped with metals (Chapter 5) and doped carbon nanotubes.

G. CURRENT AND PROPOSED USES FOR SUPERCONDUCTORS

Numerous applications for superconductors have been proposed, but few have been developed beyond the laboratory or conceptual stage. The two applications that have progressed furthest are the use of superconductors in powerful magnets and in the levitation of trains. Superconducting magnets are now widely used in NMR equipment. The coils in such devices are fabricated from low-temperature superconducting metallic materials, which need to be cooled in liquid helium. Electric generators, transformers, and fault limiters that contain superconducting windings are much more efficient than those that use copper wire, and these are just coming into use in stationary power generation and manipulation. An advantage that these stationary uses have over mobile applications such as in trains is the ease with which liquid nitrogen or helium can be stored, condensed, and reused in a stationary structure.

However, superconducting magnetic levitation trains have reached a high level of development in Britain, China, Japan, and the United States. These work on the principle of the Meissner effect, using a high magnetic field generated on a train, which induces a current and a comparable magnetic field in the rails. Thus, repulsion between the train and the wheels dramatically reduces friction. As early as 1999, an experimental Japanese train using superconductors reached a speed of 361 miles per hour.

The "superconducting supercollider" accelerator in Europe is another development that uses superconductor magnets, and proton–antiproton and electron–proton colliders have been developed in the United States.

The promise of low-cost electric power transmission is being realized in demonstration projects in Denmark and the United States. The Danish test involved a 30-m power line that used high-temperature superconducting wires cooled in liquid nitrogen. A 122-m length of cable has been tested in Detroit, and a 350-m cable is in the planning stage for Albany, New York. The major challenges for such cable systems are that they need to be heavily insulated to conserve the coolant. This leads to complexity of construction. They must also be buried in the ground for protection. The production and use of very large volumes of liquid nitrogen is probably one of the smallest drawbacks. Energy storage equipment to stabilize the existing electric power grid uses superconducting magnet technology.

H. CHALLENGES FOR THE FUTURE

The use of superconductors in large-scale power generation is possibly the application that could be of the greatest benefit worldwide, because it would improve the efficiency of electrical power production and also has the capacity to significantly reduce the amount of carbon dioxide released by power plants. The improved efficiency of electric motors would be a supplementary benefit. Wide-scale use of superconductors in both power generation and power utilization will, in turn, depend on improvements to the technology for the fabrication of wires and tapes from high-temperature superconductors that are inherently brittle. Much progress has been made in the fabrication area, but more needs to be done. Fabrication would be less of a problem if superconducting *polymers* with high T_c values could be found.

Obviously, a major challenge for the future is the design or discovery of materials that become superconducting at even higher temperatures than existing examples. Room-temperature superconductivity is the ultimate objective, and this is a major incentive for research in this field.

I. SUGGESTIONS FOR FURTHER READING

1. Onnes, K. H., "On the Sudden Rate at which the Resistivity of Mercury Disappears", *Comm. Phys. Lab.* Nos **122** and **124**, 1911.
2. Burns, G., *High-Temperature Superconductivity—An Introduction*, Academic Press, 1992.
3. Bednorz, J. G.; Müller, K. A., *Z. Phys B.*—"New class of T'-structure Cuprate Superconductors", *Condensed Matter*, **64**:189–193 (1986).
4. Bednorz, J. G.; Müller, K. A., "Perovskite-type Oxides—The New Approach to High-*Tc* Superconducitivty", *Reviews of Modern Physics*, **60**:585–600 (1988).
5. Tinkman, Michael, *Introduction to Superconductivity*, 2nd Ed., Dover Books on Physics. Dover, Mineola, NY, 2004.
6. Lebed, A. G., The Physics of Organic Superconductors and Conductors 1st Ed., Springer Series in Material Science Vol **110**, 2008.
7. Labes, M. M.; Love, P.; Nichols, L. F., "Polysulfur Nitride—a Metallic, Superconducting Polymer", *Chem. Rev.* **79**:1–15 (1979).
8. MacDiarmid, A. G.; Heeger, A. J.; Garito, A. F., in *Yearbook of Science and Technology*, McGraw-Hill, New York, 1977.

J. STUDY QUESTIONS (for class discussions or essays)

1. What factors have so far limited the development of applications for high-temperature superconductors? What developments are needed for the widespread use of these materials?

2. To what extent do the theories that explain the superconductivity of some metals at tmperatures near absolute zero also apply to high-temperature copper oxide–type superconductors? Why is it important to understand the underlying theory? Speculate on how a *room-temperature* superconductor might be developed.

3. Why is the processing sequence so crucial in the preparation of many high-temperature superconductors?

4. Explain why the phenomenon of levitation is an important property of superconductors, and discuss the various ways in which this property can be utilized in technology. Stress the advantages and disadvantages of the uses described.

5. The discovery of new phenomena such as high-temperature superconductivity is often a complete surprise to most scientists because it would not have been predicted by a logical analysis of existing data. How can you, as a scientist, increase the chances that you will make a breakthrough discovery in any area of materials science?

6. Explain the main differences between metals, semiconductors, and superconductors in terms of their solid-state structures and electronic characteristics.

7. Discuss the conductivity–temperature characteristics of poly(sulfur nitride) (polythiazyl), and explain why this material is not a viable candidate for superconducting applications.

8. By means of a literature search, write a survey of the ways in which traditional low-temperature supercomputers have been utilized in technology, and speculate on possible uses for the future.

9. How is superconductivity detected and measured?

10. Why is copper the key element in most high temperature superconductors? Which other elements might replace copper in these materials, and what would be their advantages or disadvantages?

12

Solid Ionic Conductors: Advanced Materials for Energy Generation and Energy Storage

A. GENERAL OBSERVATIONS

The generation and storage of energy is one of the most important challenges of our time. The reasons include the looming crisis in the supply of oil and natural gas and the effects of burning hydrocarbon fuels on the climate of the whole planet. Solutions to these problems include electricity generated from solar cells, windmills, tides, or nuclear fission, together with the replacement of hydrocarbon fuels by hydrogen, methanol, or ethanol.

One of the most appealing solutions for both stationary and mobile power generation is the fuel cell (Figure 12.1)—a device that converts fuel directly into electricity with a higher efficiency than do methods that depend on the combustion of coal or hydrocarbons. However, fuel cells are still in a developmental stage, with hurdles to be overcome that involve mainly the limitations of existing materials. Thus, the search for new materials for fuel cells provides an excellent example of the way in which materials science holds the key to solving a crucially important problem. A critical component of a fuel cell is the material that conducts ions from

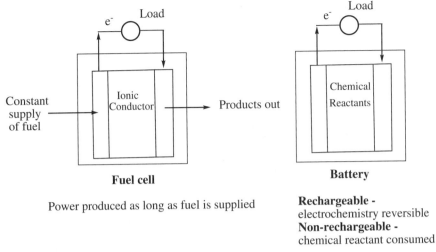

Figure 12.1. *Essential differences between a fuel cell and a battery. A fuel cell provides electric current as long as the fuel is supplied. A battery can yield current only until its chemical fuel is exhausted. After that it must be recharged or discarded.*

one electrode to the other. The performance of most current fuel cells is severely restricted by the limitations of the available ion-conducting materials.

Power generation is not the only challenge facing the modern world. A related problem is the *storage* of power, particularly for use in mobile applications such as electric automobiles, laptop computers, cellular phones, and medical devices. The main method for the storage of electric power is the use of primary and secondary batteries. A primary battery yields its energy by consuming a reactive metal, while a secondary battery uses a reversible electrochemical reaction that utilizes electricity for the recharge cycle. For practical reasons, the storage of significant amounts of power requires the use of *rechargeable* batteries. Various rechargeable battery technologies are summarized in Figure 12.2. Three key requirements are to reduce the weight of batteries, increase the amount of energy stored per unit weight (energy density), and increase the amount of power that can be released in a given time. For some applications, resistance to failure under harsh conditions is also an important requirement. Conventional batteries use liquid electrolytes such as sulfuric acid or flammable organic solvents. This adds weight, makes them unnecessarily fragile, and in some cases creates a fire hazard. Batteries that use solid membrane electrolytes offer many advantages for the future, but here again the materials properties of the available solid electrolytes are inadequate for widespread use. In this chapter we examine some of the challenges and possible solutions to battery materials and to fuel cell membranes.

Finally, electrical energy can also be stored in capacitors. Traditionally, small capacitors are key components of electric circuits in radios, television sets, and microcomputers. However, very large capacitors could, in principle, store significant amounts of energy. The development of new materials for this purpose is another subject of growing importance.

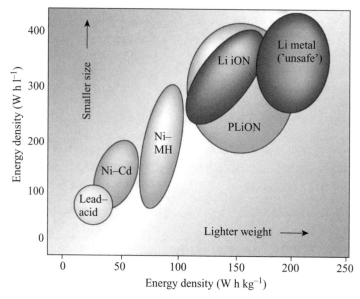

Figure 12.2. Energy density of different energy storage systems (from Scrosati, B.).

B. FUEL CELL MATERIALS

1. Background

Fuel cells are an alternative to both heat engines and batteries for numerous applications. A comparison of the principles of batteries and fuel cells is shown in Figure 12.1. The key characteristic of a fuel cell is that it generates electrical power as long as a chemical fuel is pumped into the device. Moreover, compared to an internal-combustion engine or a steam-driven power station, a fuel cell has fewer moving parts, generates less noise and far less pollution, and can utilize a variety of fuels. The fuels may be hydrogen, methanol, or hydrocarbons that are converted by catalytic reactors to water and/or carbon dioxide, depending on the type of cell.

2. General Principles

A fuel cell is an electrochemical device that takes a reaction that would normally be highly exothermic, and moderates that process to generate electricity as well as heat. Example processes are the oxidation of hydrogen, methane, gasoline, or methanol, all of which, under normal oxidative chemical reaction conditions, would generate a fire or explosion. Note that a single fuel cell is only rarely sufficient to generate the required electrical power, and stacks of multiple cells are normally needed for practical applications.

There are many different designs for fuel cells. However, five main types of fuel cell are already in use or are under development: (1) the polymer electrolyte membrane (PEM) fuel cell, (2) the phosphoric acid electrolyte cell, (3) the alkaline fuel cell, (4) the molten carbonate fuel cell, and (5) the solid oxide system. The first two function through the presence of a *proton-conducting* material that separates the anode from the cathode. The remaining three use chemistry that involves the transport of *anions* from the cathode to the anode. The main differences between these units are (1) the type of electrolyte, (2) the temperature of operation, (3) the type of metal catalysts that are needed to facilitate the reactions that occur at the electrodes, and (4) the degree to which the different designs can tolerate impurities in the fuel supply. These five systems are summarized in Table 12.1, and are discussed in turn.

3. Polymer Electrolyte Membrane (PEM) Fuel Cells

The main driving force for the development of polymer electrolyte fuel cells is their suitability for operation at moderately low temperatures and for mobile applications such as in automobiles and portable devices such as cellular phones or laptop computers. Unlike internal combustion engines, they produce power without the formation of nitrogen oxide pollutants. A cross section of a polymer electrolyte membrane fuel cell, together with the reactions involved, is shown in Figure 12.3.

A PEM fuel cell consists of two electrodes that contain finely divided precious-metal catalyst particles separated by a proton conducting polymer membrane. The anode typically is constructed from carbon fiber cloth and/or powdered graphite,

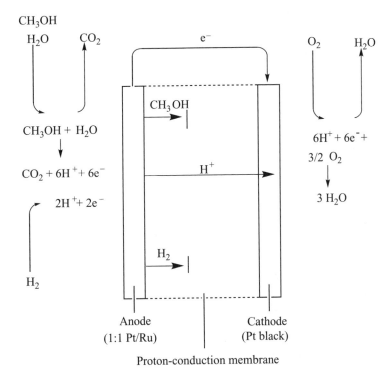

Figure 12.3. Schematic cross section of a direct hydrogen or direct methanol polymer electrolyte membrane (PEM) fuel cell.

TABLE 12.1. Characteristics of Different Types of Fuel Cells

Type of Fuel Cell	Fuel[a]	Temperature of Operation (°C)	Anode Catalyst	Cathode Catalyst	Chemical Emissions	Electrical Efficiency (%)	Overall Efficiency (Electrical plus Heat)	Advantages	Disadvantages
Polymer electrolyte	Hydrogen, methanol, or ethanol	80	Pt/Ru	Pt	H_2O or $H_2O + CO_2$	35–40	—	Robust, lightweight, portable	Lack of suitable membranes (especially for direct methanol)
Alkaline	Hydrogen, oxygen	50–250	Nonprecious metals	Nonprecious metals	H_2O	60	—	Simplicity (spaceflight)	System poisoned by even traces of CO_2
Phosphoric acid	Hydrogen, methane	150–200	Pt	Pt	H_2O or $H_2O + CO_2$	37–42	80	Efficiency	Heavy, nonportable; corrosive electrolyte
Molten carbonate	Methane	650	Ni	? Ni	$H_2O + CO_2$	50	85	Efficiency	Startup delay, heavy, nonportable
Solid oxide	Hydrocarbons, hydrogen, carbon monoxide	700–1000	None	None	$H_2O + CO_2$	50–60	—	Efficiency, no catalysts needed, uses variety of fuels	Startup delay to reach operating temperature; heavy, nonportable

[a]Fuel actually entering the device. Many fuel cells require a "preconverter" to change methane or gasoline to hydrogen and carbon dioxide. Fuel cells that use line hydrogen are dependent on off-site conversion of methane or higher hydrocarbons to hydrogen and carbon dioxide, or the reaction of carbon (coke) with water (the water–gas shift reaction) to a mixture of hydrogen, carbon monoxide, and carbon dioxide.

with platinum or platinum–ruthenium alloy particles covering the surface of the carbon. The purpose of the anode is to convert the fuel to protons and electrons at a temperature of 80 °C or higher. When the fuel is a hydrocarbon such as methane, a catalytic reactor is positioned ahead of the fuel cell to convert the hydrocarbon to hydrogen, carbon dioxide, and perhaps carbon monoxide. If hydrogen is the fuel entering the fuel cell anode, the only products are protons and electrons. When methanol is the fuel, a catalyst at the anode converts it to carbon dioxide, protons, and electrons. In both cases the electrons pass through the external circuit and eventually reach the cathode.

The cathode is also typically constructed from carbon fiber cloth or particles, with both infused with platinum particles. At the cathode, the protons that have traversed the membrane react with both oxygen from air and the incoming electrons to form water, again at 80 °C or higher temperatures.

Not shown in Figure 12.3 are the metal or graphite gas distribution channels that funnel the fuel and air evenly into the porous anode and cathode.

The most critical part of a polymer electrolyte fuel cell is the proton-conducting membrane, which serves two purposes; it must (1) conduct protons from the anode to the cathode and (2) not allow the fuel (hydrogen or methanol) to pass through to the cathode. If the fuel does reach the cathode, it will be oxidized without generating an electric current. Proton conduction in a polymer electrolyte membrane is achieved by the linkage of acidic functional groups, such as sulfonic acid (SO_3H) groups, to a polymer. The acidic groups provide "way stations" for protons to be transferred from one acidic site to another. A cartoon showing how this may occur is shown in Figure 12.4.

Sulfonic acid groups are the most commonly used acidic units, although experimental membranes with phosphonic acid or sulfonimide units are also being studied. These membranes are called "single-ion conductors" because only the protons are able to move. The anions, such as the sulfonate, or phosphonate ions are covalently linked to the polymer chains of the membrane, and are thus immobilized.

The design, synthesis, and fabrication of a suitable proton-conducting membrane is a major challenge in this technology. Part of the problem is the need for membranes that function effectively at temperatures above 80 °C, because higher temperatures allow the use of less expensive electrode catalysts that are less sensitive to catalyst poisoning. High temperatures also accelerate the movement of protons from anode to cathode. Unfortunately, most of the existing polymer electrolyte materials make use of water trapped in the membrane to facilitate the transport of protons, and this water is lost at temperatures above 80–90 °C. Example polymers that have been or are being investigated for fuel cell membranes are shown as species **12.1–12.8**.

12.1   **12.2**

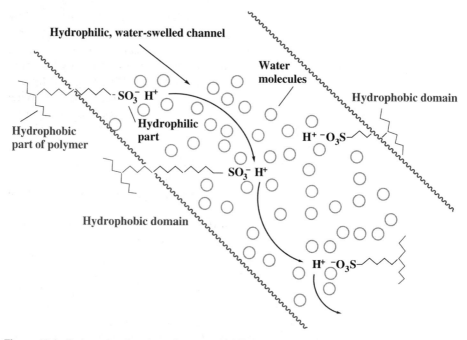

Figure 12.4. *Proton migration through an amphiphilic membrane is believed to occur through water-rich channels by transfer from one acidic site to another. Each proton moves accompanied by a solvent sheath of water molecules. The higher the loading of SO_3H groups, the higher will be the proton conductivity, but the more the membrane will swell by absorbing water. This hydrophilicity must be balanced by hydrophobic regions and/or by crosslinking the chains to give mechanical strength to the membrane.*

12.3

12.4

12.5

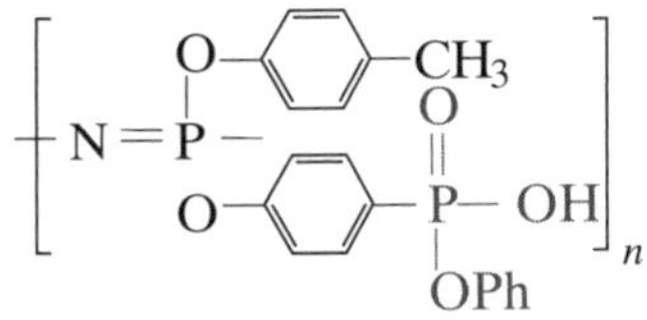

12.6

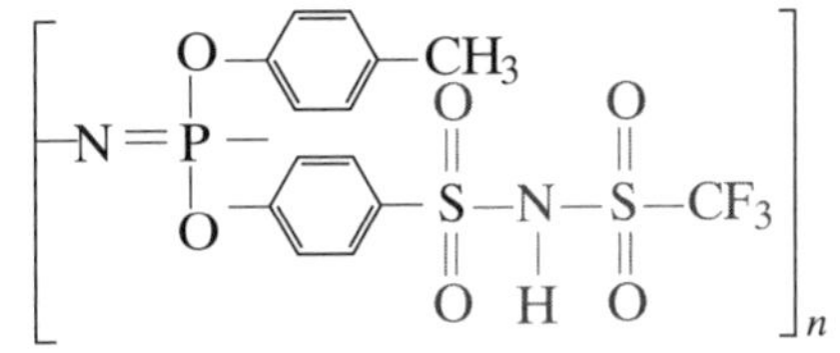

12.7

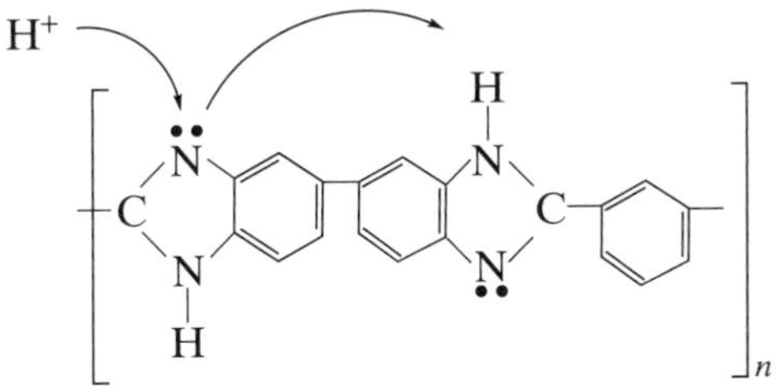

12.8

a. Nafion. Nafion 117® is the classical proton-conductive fuel cell membrane. It is a fluorocarbon polymer with sulfonic acid groups connected to the structure. A representation of the structure of Nafion is shown as structure **12.1**. This polymer consists of a hydrophobic part (the fluorocarbon) and a water-absorbing hydrophilic part (the sulfonic acid groups), which phase-separate to generate water-containing hydrophilic channels through which the protons diffuse. The exact mechanism of proton conduction is still a subject for debate. However, it seems clear that the protons generated at the anode move through the membrane from one acidic site to another, assisted by the shell of water molecules that accompanies each proton. Thus, when the water is lost by evaporation at temperatures above ~80°C, the proton conduction ceases. The hydrophobic fluorocarbon domains are probably needed to provide strength to the membrane since they cannot absorb water. It is possible to operate Nafion-based fuel cells using elevated pressures to retard water loss, but this increases the engineering complexity of the system.

Although Nafion prevents transport of molecular hydrogen across the membrane, it has a high affinity for and is highly permeable to methanol. Thus, methanol crosses the membrane and is oxidized at the cathode without the production of electric power. For this rerason, Nafion is unsuitable for the small portable direct methanol fuel cells that are needed to power laptop conputers, cellular phones, or even automobiles that use methanol as a fuel. Use of dilute solutions of methanol in water reduces methanol permeation but seriously decreases the efficiency of the process.

Nafion is an example of a single-ion conductor, in which protons are free to move but the anionic sulfonate groups are immobilized on the polymer.

b. Sulfonated Polystyrene (Structure 12.2). Sulfonated polystyrene was one of the earliest and simplest fuel cell membranes. It was used in the Gemini space mission. This material was employed in a pure hydrogen/oxygen fuel cell. It had a satisfactory performance over the short time-period of the mission, but suffered from oxidative degradation of the polymer during extended operation.

c. Sulfonated Polyimides. These are complex condensation polymers with some combination of the repeating units shown in structure **12.3**. They are thermally stable, their mechanical properties can be varied by changes to the ratio of different repeating units and sulfonic acid groups, they can be used for both hydrogen and direct methanol fuel cells, and they can be crosslinked to provide membrane strength. The disadvantage is the complexity of synthesis.

d. Sulfonated Poly(phenylene oxide) (Structure 12.4). Poly(phenylene oxide) is an inexpensive, highly stable condensation polymer that can be sulfonated readily. The ratio of hydrophilic sulfonate groups to hydrophobic units is controlled by the degree of sulfonation. The methanol permeability is low.

d. Polyphosphazenes with Acidic Functional Groups (Structures 12.5–12.7). The electrochemical stability of the polyphosphazene backbone offers some advantages for fuel cell applications. Most research with these polymers has utilized macromolecules that bear aryloxy side groups to which are linked acidic units such as sulfonic acid, phosphonic acid, or sulfonimide functional groups. Typical examples are shown in structures **12.5–12.7**. Because membranes derived from these polymers show resistance to methanol crossover, the promise for them lies mainly for uses in direct methanol fuel cells.

F. Poly(benzimidazoles) (Structure 12.8). Attempts are being made to combine the attributes of proton donors such as phosphoric acid (see the following section) with the solid properties of polymer membranes. For example, poly(benzimidazole) copolymers such as structure **12.8** can be "doped" with phosphoric acid or other acids to allow proton conduction at elevated temperatures *in the absence of water*. The mechanism of proton transfer is believed to be via a hopping of protons from the lone-pair electrons of one skeletal nitrogen atom to another, a process that eliminates the need for water. However, the oxidation reaction at the cathode generates water, and some concern exists that this water may extract the phosphoric acid over long periods of fuel cell operation.

Numerous other polymers have been synthesized and evaluated for fuel cell applications. Some of these, such as structures **12.3–12.8**, are better than Nafion at resisting methanol crossover. However, few of the water-containing fuel cell membranes function well above 80–100 °C except when the device is operated under sufficient pressure to prevent the volatilization of water. This is clearly an area of keen interest for future energy generation devices.

4. Phosphoric Acid Fuel Cells

One of the most highly developed fuel cells for stationary applications, such as in office buildings, hospitals, and small power stations, is the phosphoric acid fuel cell. Its design and mode of operation are shown in Figure 12.5.

The fuels are hydrogen and oxygen. The electrodes are typically carbon in the form of carbon fiber "paper" or some other porous modification of this material.

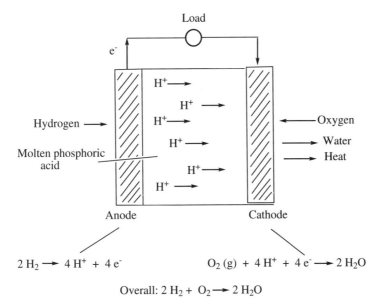

Figure 12.5. *Configuration and reactions in a phosphoric acid fuel cell.*

Like the PEM fuel cell described above, molecular hydrogen enters the device at the anode, where it is converted to protons and electrons by a catalyst such as platinum. However, the transport of protons to the cathode takes place through a *liquid* electrolyte of molten phosphoric acid. At the cathode, protons and the electrons coming in through the external circuit react with oxygen from the air to produce water.

The overall efficiency of a phosphoric acid fuel cell is high (>40%). Moreover, the system is relatively insensitive to impurities such as CO_2 in the fuel. Molten phosphoric acid is a preferred electrolyte because of its low volatility and stability. However, the proton mobility in phosphoric acid at room temperature is low, and the electrolyte must be heated to temperatures near 200 °C before satisfactory conductivities can be achieved. A deterrent for the use of these devices in mobile applications used by the general public is the corrosive characteristic of the molten acid, which would constitute a serious hazard. For stationary applications, a significant advantage is the production of high-temperature steam as a byproduct. This can be used for heating buildings or driving a turbine for supplementary electricity generation. This supplementary energy production raises the overall energy efficiency into the range of 80%.

5. Alkaline Fuel Cells

The principle of operation for alkaline fuel cells is different from that for the two examples described above in the sense that they depend on the transport of *anions* from the cathode to the anode. The electrolyte is an aqueous solution of potassium hydroxide in an inert porous matrix instead of an ion conductive polymer or molten acid. The electrodes are constructed from a chemically inert material such as carbon

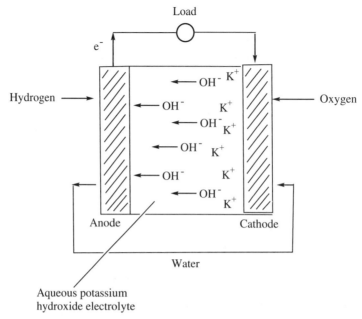

Figure 12.6. An alkaline fuel cell, which has many advantages for spaceflight, uses a liquid electrolyte of potassium hydroxide in water, but requires the use of highly purified hydrogen and oxygen as fuels.

fiber (Chapter 7) or porous carbon. The fuels are hydrogen and oxygen, and the products are water and electrical power. A schematic of an alkaline fuel cell is shown in Figure 12.6.

Pure hydrogen gas enters the device at the anode and combines there with hydroxide ions to give electrons under the influence of an inexpensive catalyst such as microparticulate silver or the more expensive platinum. The water formed in this process is transferred to the cathode. At the cathode, pure oxygen gas reacts with potassium ions and water to regenerate potassium hydroxide.

Alkaline fuel cells have been used to provide power and water during the Apollo space mission and space shuttle programs. The temperatures of operation range from ~65 °C to 220 °C, which is acceptable for mobile applications. However, although these fuel cells have a high efficiency (~70%), they are easily poisoned by carbon dioxide and other impurities and, for this reason, they are currently unsuitable for more general power generation where inexpensive and possible impure fuels must be used.

6. Molten Carbonate Fuel Cells

Like the alkaline fuel cell just described, molten carbonate fuel cells function through the transport of *anions* from the cathode to the anode. The cell consists of an anode of porous, sintered nickel powder and chromium, while the cathode is porous nickel oxide doped with lithium. The nickel serves as a relatively inexpensive catalyst, which can be used because of the high operating temperatures. The

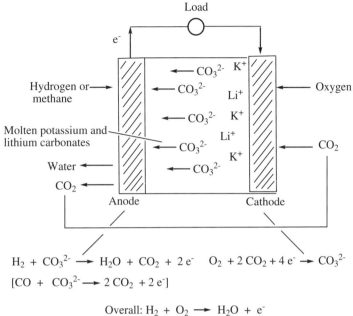

Figure 12.7. *Molten carbonate fuel cell.*

electrolyte between the electrodes is a molten mixture of potassium and lithium carbonates embedded in a porous, inert ceramic matrix. The reaction chemistry and general structure of the cell are shown in Figure 12.7.

Several features of molten carbonate fuel cells are important: (1) they operate at high temperatures in the range of 650 °C; (2) because of the high temperature, nickel can be used as a catalyst in place of platinum, and this lowers the cost of the device; and (3) although the fuel could be hydrogen and oxygen, in practice the combination of the high temperature and the catalyst allows the direct use of methane, which, in the presence of carbonate ions, is converted at the anode to hydrogen, water, and CO_2. The water and CO_2 are transferred to the cathode, where they participate in the formation of carbonate anions, which traverse the electrolyte and are discharged at the anode.

The efficiency of these cells is very high (~50%) in terms of direct electricity production; however, if this is combined with the use of the generated steam to power a turbine, the overall efficiency rises to ~85%. Because of the high temperatures involved, these units are suitable for stationary power generation rather than portable applications.

7. Solid Oxide Fuel Cells

As the name implies, these cells utilize a solid ceramic oxide electrolyte, such as yttrium oxide–stabilized zirconium oxide, which is capable of transporting oxygen dianions from the cathode to the anode and, at the same time, preventing the diffusion of fuel from the anode to the cathode. The cathode is typically lanthanum manganese oxide (LaMnO$_3$) doped with strontium, a perovskite-type structure (see Chapter 11) that is an electronic conductor. The anode is usually porous nickel, which provides both electronic conductivity and catalytic properties. A simplified layout of such as cell and the reactions taking place in it is shown in Figure 12.8.

A major difference from the other types of fuel cells is the extremely high temperature of operation, which may exceed 1000 °C. This has advantages and disadvantages. The elevated temperature allows the cell to operate at high efficiency (40–70% for direct electricity generation, and up to 80–85% if the byproduct heat is utilized).

Moreover, these cells do not require expensive catalysts, and they can use hydrocarbon fuels directly and may even utilize the products from sulfur-containing coal gasification plants. On the other hand, the thermal and chemical instability of metals and even ceramics at this temperature is a major problem. Thus, expensive materials such as lanthanum chromite or yttrium–chromium alloys may be required just to provide the structural components and interconnects of the cell. The high-temperature operation also means that the time required for startup from room temperature is too long to allow for intermittent use in mobile applications. Thus, these units are suitable for stationary power generation where continuous operation is required and under conditions where the excess heat generated by the chemistry can be utilized to heat buildings or drive steam turbines. A major challenge for scientists and engineers is to find materials and designs that allow solid oxide fuel cells to operate efficiently at temperatures well below 1000 °C.

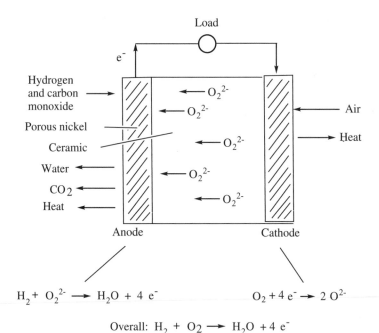

$$H_2 + O_2^{2-} \longrightarrow H_2O + 4\,e^-$$

$$O_2 + 4\,e^- \longrightarrow 2\,O^{2-}$$

Overall: $H_2 + O_2 \longrightarrow H_2O + 4\,e^-$

Figure 12.8. *Schematic representation of a solid oxide fuel cell. The anode and cathode are separated by a ceramic layer that conducts oxygen anions.*

C. BATTERY ELECTROLYTE MATERIALS

1. Background

Energy storage is an ever-present component of modern life, with examples that range from batteries for flashlights, radios, and cell phones to those in laptop computers, automobiles, cameras, and satellites. Batteries fall into two categories— primary (non-rechargeable) and secondary (rechargeable). Primary batteries, such as the common alkaline zinc-manganese dioxide cell or the alkaline lithium–manganese dioxide cell consume a high-energy chemical, such as metallic zinc or lithium, to generate power. When that chemical is used up, the battery is finished. On the other hand, secondary batteries utilize a reversible electrochemical reaction, one cycle of which consumes a high energy metal and produces an external electric current, while the recharge cycle reverses the reaction. Well-known examples include the lead–lead sulfate automobile battery, the nickel–cadmium battery, the nickel–metal hydride battery, and the lithium ion battery.

Secondary batteries have obvious advantages in applications such as cell phones, laptop computers, satellites, and electric automobiles. However, the disadvantages of many existing types of rechargeable batteries are their low power densities (power:weight ratio) because they use heavy metals, their use of environmentally unfriendly metals (lead, cadmium), and the fact that they contain a liquid electrolyte that may be corrosive (sulfuric acid) or flammable (volatile organic solvents).

For these reasons considerable research is underway to develop lightweight, safe, environmentally friendly, liquid-free secondary batteries that can be fabricated into intricate shapes to facilitate their accommodation into the housings of computers or cell phones, into the shell of automobiles, or into the flat-panel structure of solar cell arrays for terrestrial or space applications. For reasons of weight, stored energy, and environmental safety, lithium is the anode material of choice, especially when incorporated into a cell that employs a solid or gel polymer electrolyte. Thus, the focus in the following sections is on lithium batteries.

2. Lithium Ion ("Rocking Chair") Batteries

Two types of rechargeable lithium batteries are known: (1) "rocking chair" batteries, which employ lithium intercalated into graphite as both the anode and cathode; and (2) batteries that employ metallic lithium as the anode and an intercalation solid as the cathode. Hybrids of these two types are also known. These designs and their materials characteristics are described in turn.

From a theoretical viewpoint, the use of metallic lithium as the anode in an electrochemical cell has several advantages, including high power density. However, the high reactivity of this metal, and the possibility of an ignition if the battery is damaged, is a concern. A less sensitive alternative is the use of a graphite anode with lithium atoms intercalated in the galleries (see Chapter 7), and a graphite cathode to receive lithium atoms as the cell undergoes discharge. The layout of a typical battery of this type is shown in Figure 12.9a. Such a device produces electric current in the following way. During discharge, lithium atoms give up an electron first to the graphite anode and then to the external circuit. The resultant lithium ions leave the galleries of the anode and diffuse toward the cathode through a liquid

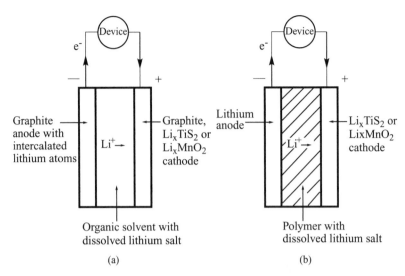

Figure 12.9. *Schematic comparison of (a) a "lithium ion" or "rocking chair" rechargeable battery and (b) a metallic lithium anode rechargeable battery with a polymer membrane electrolyte.*

or gel electrolyte. At the cathode the lithium ions combine with the incoming electrons to form lithium atoms. The electrolyte usually consists of a lithium salt such as lithium triflate (structure **12.11** later, in Section C.5.b) dissolved in a coordinative organic solvent such as propylene carbonate. Recharging involves pulling electrons from the cathode, thereby converting lithium atoms to lithium cations, and simultaneously feeding electrons into the anode, which attracts lithium ions from the cathode, across the electrolyte, and back into the galleries of the anode. These are called *lithium ion* or "rocking chair" batteries for obvious reasons. Many current lithium batteries use this technology, with the electrolyte being a solution of a lithium salt in an organic solvent.

Although cells of this type are believed to be safe because they contain no metallic lithium, in fact they suffer from two problems: (1) the amount of lithium that can be intercalated in a graphite anode is far less than exists in a metallic lithium counterpart, which clearly limits the amount of energy that can be stored in the battery; and (2) the use of a volatile, flammable solvent in the electrolyte raises the danger of ignition if the battery suffers damage that leads to a short circuit or it overheats. One way to overcome these problems is to use a polymeric membrane in place of an organic solvent

3. Principles behind Ion Transport Membranes

Although polymeric *electronic* conductors are well known (see Chapter 17), other polymers exist that are electrically conductive because of the movement of *metal cations* through the solid material. Some of the best-known polymeric ionic conductors allow the transport of lithium ions from one side of a membrane to the other.

Ion transport membranes allow an electric current to pass through a solid via the diffusion of ions from one side of the membrane to the other. The movement of ions through a polymeric solid requires a special set of membrane characteristics. In a lithium battery the ions are provided by a dissolved salt. Normally, salts are insoluble in conventional polymers. A salt will dissolve only if the polymer has coordination sites for one or both ions. For example, the salts lithium perchlorate or lithium trifluoromethylsulfonate (triflate) will not dissolve in polystyrene or polyethylene, but they are soluble in poly(ethylene oxide) (structure **12.9**) or the polyphosphazene MEEP (structure **12.10**), both of which are polymers with etheric oxygen atoms that coordinate to cations.

$$\left[-CH_2-CH_2-O-\right]_n$$

12.9

$$\left[\begin{array}{c} OCH_2CH_2OCH_2CH_2OCH_3 \\ | \\ -N=P- \\ | \\ OCH_2CH_2OCH_2CH_2OCH_3 \end{array}\right]_n$$

12.10

However, if the ions are too strongly coordinated to the polymer molecules, they will be unable to diffuse freely through the polymer matrix and will remain bound to one fixed point on the polymer. Even if the polymer molecules themselves are slowly reptating through the matrix, the mobility of the bound ions will be restricted and will be insufficient to allow passage of a useful electric current. Thus, *weak* coordination between the polymer and one or both ions is a prerequisite for ion diffusion. In this way the thermal motions of the polymer molecules will allow the ions to transfer from one coordination site to another until they reach the opposite electrode (Figure 12.10).

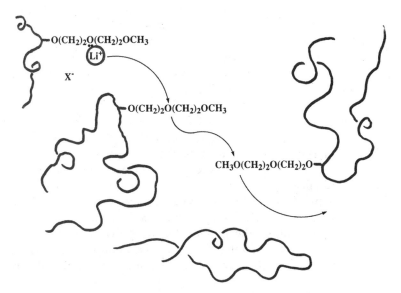

Figure 12.10. *Transport of lithium cations across a polymer membrane via hopping of the cations from one oxygen coordination site to another assisted by the flexural and reptation motions of the polymer chains.*

If the lithium ions are free to move but the counteranion is firmly attached to the polymer, electric current can still flow through what is termed "single-ion conduction." This situation would correspond to the one described earlier for proton-conducting fuel cell membranes.

Ion-conducting membranes are also affected by the surface sorption and desorption processes at the electrodes. For example, in a rechargeable lithium battery, lithium ions generated from lithium metal at the anode must leave the metal and penetrate through the membrane surface. Similarly, lithium ions leaving the membrane at the other side must find their way into a graphite, TiS_2, or MnO_2 intercalation cathode material. The process is reversed when the cell is recharged. Thus, surface effects play an important role in these systems.

4. Metallic Lithium/Solid Polymer or Gel Electrolyte Batteries

A solid polymer rechargeable lithium metal battery has the cross section shown in Figure 12.9b. The metallic lithium anode, in the form of a thin sheet of the metal, is backed by a "current collector" of carbon fiber or nickel mesh, and is in close contact with a thin membrane of the polymer electrolyte. The membrane is a suitable polymer in which a lithium salt has been dissolved. The cathode consists of a powdered solid such as manganese dioxide, titanium disulfide, or lithium cobalt oxide ($LiCoO_2$) in close contact with the cathodic side of the membrane. A carbon cloth or metallic current collector, intimately incorporated into the cathode, connects to the external circuit.

The operation of the cell is as follows. During discharge, lithium metal dissolves from the anode to form lithium cations, and they migrate through the polymer toward the cathode. The electrons formed simultaneously at the anode enter the external circuit and pass through the load before arriving at the cathode. Movement of the lithium ions through the polymer membrane is probably via a "hopping" mechanism, by which each ion jumps from one coordination site on a polymer molecule to a similar site on adjacent repeating units or nearby macromolecules. This process is assisted by the broad-scale motion of polymer chains and/or side groups at temperatures well above the glass transition. At the cathode each lithium ion can enter the galleries of the MnO_2, TiS_2, or $LiCoO_2$ particles, concurrently receiving electrons that are arriving through the external circuit. Recharging involves pumping electrons into the cathode to regenerate lithium cations, which migrate back across the membrane to meet electrons that are being arriving at the anode. As each cation reaches the anode, it receives an electron and plates out on the lithium metal of the anode.

This is the principle. In practice, the power produced by the battery depends on the number of current carriers (i.e., lithium cations) in the membrane, the behavior of the counteranions, the mobility of the polymer molecules, the ease with which lithium ions are generated from the anode during discharge, and the facility with which they can move from the membrane into the galleries of the cathode.

An increase in the lithium salt concentration in the membrane above a certain concentration actually lowers the conductivity because cations will coordinate to several polymer chains simultaneously and lower both ion transport and polymer chain mobility through ionic crosslinking. Thus, they raise the glass transition temperature to the point where ion transport is severely restricted.

The anions dissolved in the polymer exert a significant role through their tendency to bind to the cations. Ion pairs will not conduct an electric current. Hence, the salts used must be chosen with care to ensure that the anions are large enough to prevent tight anion–cation binding and do not compete with cation–polymer coordination. Some experimental polymer electrolytes are designed to bind the anions tightly to generate a "single ion" conductor so that the anions cannot concentrate at the cathode and inhibit transfer of the cations to the cathode.

Polymer chain mobility is crucial for the effective operation of the cell. Hence, polymers with very flexible chains and/or side groups are needed to allow good cation mobility. Even so, no polymer has yet been found that allows lithium ion conductivity higher than about 5×10^{-4} S/cm. A value of 1×10^{-3} S/cm is considered to be the minimum needed for commercial battery development. For this reason, organic solvents are added to the polymer electrolytes to plasticize them and raise the ion mobility to achieve conductivities of 10^{-3} S/cm or higher. This may seem like a step backward, since the plasicizers are often the same volatile solvents as those used in liquid electrolyte batteries. However, the amounts of solvent needed are far less that in liquid electrolytes, and the "gel electrolytes" so formed are more robust and sometimes less flammable than liquid electrolytes. Moreover, the battery container can be lighter than in liquid electrolytes because the solvent vapor pressure is much less.

5. Example Polymers for Lithium Battery Applications

a. Poly(ethylene oxide) (PEO) (Structure 12.9). This is the classical polymer for electrolyte applications. The etheric oxygen atoms in its backbone give molecular flexibility and provide weak coordination sites for lithium cations. Unfortunately, the maximum room temperature conductivities attainable with this polymer range from 10^{-8} to 10^{-7} S/cm with the use of lithium triflate or lithium trifluoromethylsulfonimide as added salts. These low values are attributed to the fact that poly(ethylene oxide) is a 75–80% crystalline polymer, which means that only a small fraction of the polymer matrix provides a pathway for ionic conduction. Copolymers of poly(ethylene oxide) with poly(propylene oxide) are less crystalline, and these materials have room-temperature conductivities in the range of 10^{-4} S/cm. However, commercially usable conductivities can be generated only after appreciable amounts of solvents such as propylene carbonate have been added to yield gel electrolytes.

b. Polyphosphazenes. The polymer known as "MEEP" {poly[bis(methoxyethoxyethoxy)phosphazene]} (structure **12.10**) is one of the most flexible macromolecules known. It has a glass transition temperature of $-80\,°C$ and bears flexible side groups that have 6 times as many oxygen coordination sites per repeating unit as PEO. It is also noncrystalline even at low temperatures. With dissolved lithium triflate (structure **12.11**) or lithium bis(trifluoromethyl)sulfonimide (structure **12.12**) as salts, it yields conductivities in the range of $2–4 \times 10^{-4}$ S/cm at $25\,°C$. Experimental batteries using MEEP as the electrolyte have survived more that 600 charge–discharge cycles. The addition of small amounts (~10%) of propylene carbonate raise the conductivity to well above above 10^{-3} S/cm.

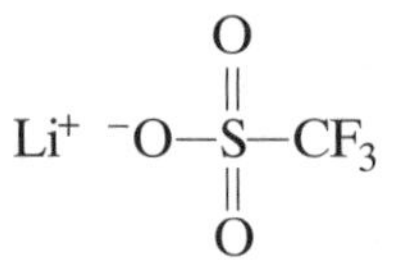

$$Li^+ \ ^-O-\overset{\overset{\displaystyle O}{\|}}{\underset{\underset{\displaystyle O}{\|}}{S}}-CF_3$$

12.11

$$CF_3-\overset{\overset{\displaystyle O}{\|}}{\underset{\underset{\displaystyle O}{\|}}{S}}-\overset{\displaystyle Li^+}{N^-} \ ^-\overset{\overset{\displaystyle O}{\|}}{\underset{\underset{\displaystyle O}{\|}}{S}}-CF_3$$

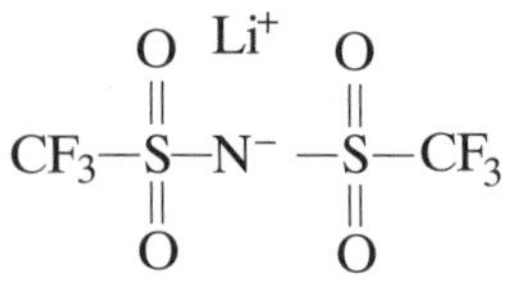

12.12

Organic Polymer Chain

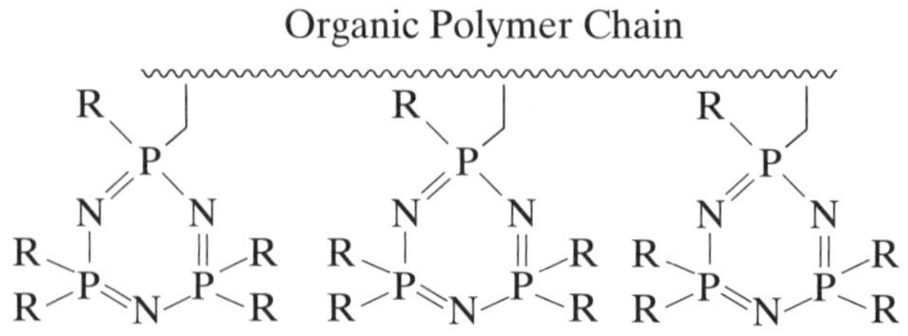

13

R=OCH$_2$CF$_3$ and/or OCH$_2$CH$_2$OCH$_2$CH$_2$OCH$_3$

Numerous analogs of MEEP have been synthesized, with longer side groups, branched side groups, crown ether side groups, and comb structures, and these provide some additional advantages for battery optimization, including, in some instances, fire resistance. The main disadvantage of the MEEP-type electrolytes is their higher cost in comparison to poly(ethylene oxide). Counterparts of polyphosphazene electrolytes have been developed with a polysiloxane backbone.

The longevity of a rechargeable battery depends on the number of charge–discharge cycles it can undergo before its performance deteriorates. At least 500–1000 cycles are needed before deterioration sets in. A drop in performance can often be traced to changes at the electrode–electrolyte interfaces. Failure of the cations to plate evenly on the lithium metal anode surface may lead to dendrite formation and eventually to internal short-circuiting. Chemical reactions between the solvent and the metallic lithium are also detrimental to battery performance. At the cathode, the expansion and contraction of the intercalation host particles as they receive and lose lithium ions eventually leads to a weakening of the contacts at the cathode-polymer interface. For these reasons the surface science of these systems is a subject of keen research interest. One of the main concerns about rechargeable batteries that contain metallic lithium anodes is the high reactivity of lithium metal in contact with water. This is a potential problem if the battery is damaged.

6. Lithium–Seawater Batteries

An even greater challenge involves the design and synthesis of electrolytes for lithium–seawater batteries. These are power sources designed to be stored for long periods of time and then activated by immersion in seawater for emergency use in marine environments or to power signaling devices. Lithium seawater batteries have the general structure shown in Figure 12.11.

The lithium–seawater battery is a *primary* cell (also known as a *lithium seawater fuel cell*), which consists of a metallic lithium anode surrounded by a polymer electrolyte membrane that is in contact with seawater. The cathode is located outside the polymer membrane and is immersed in the seawater. Lithium ions dissolve from

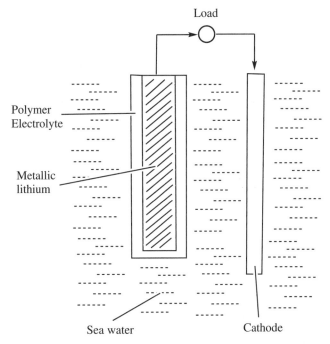

Figure 12.11. *The lithium seawater primary battery uses a metallic lithium anode surrounded by a lithium ion conductive membrane that prevents seawater from penetrating through to the lithium.*

the anode, cross the electrolyte, enter the seawater, and are discharged at a cathode. The polymer electrolyte membrane must possess two contradictory properties: (1) transport lithium ions but (2) prevent water from diffusing back to the anode, where it would react with the metallic lithium. Most lithium ion transport membranes are hydrophilic. Hence, designing a membrane that is both ion-conductive and hydrophobic is a serious challenge.

Although no entirely satisfactory membrane has been produced, some progress has been achieved through the use of polymers like structure **12.13**, which bear both hydrophobic and lithium ion–conductive side groups. In polymer **12.13** the backbone is a hydrophobic, flexible polynorbornane chain with cyclic phosphazene units attached as side groups. Each side group bears five functional units (R), which are lithium ion coordinative alkyl ether groups or hydrophobic trifluoroethoxy groups. The ratio between the two can be altered through simple changes to the synthesis procedure. This allows lithium ion conductivity and hydrophobicity to be balanced.

D. CAPACITORS AND SUPERCAPACITORS

Capacitors are energy storage devices based not on chemical transformations, but on a physical phenomenon. The structure of a traditional capacitor is illustrated in Figure 12.12a.

A traditional capacitor consists of a dielectric material such as poly(ethylene terephthalate) (Mylar) sandwiched between two metal plates or electrodes. A preferred dielectric material is a substance with a low electronic or ionic conductiv-

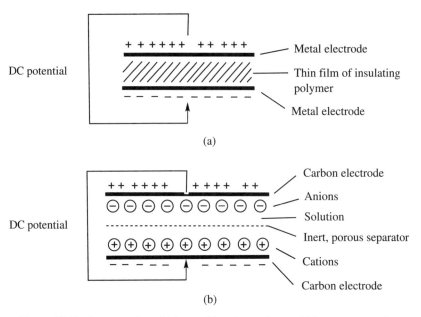

Figure 12.12. Cross section of (a) a traditional capacitor and (b) a supercapacitor.

ity—in other words, a good insulator. The application of a voltage across the two electrodes causes charge buildup on the plates, as indicated in Figure 12.12a. That charge is retained when the source of the voltage is disconnected. If the electrodes are now connected to a load, the stored power will be released. Because a capacitor contains no moving parts and no chemical reactions, the efficiency of power storage is potentially very high ($\geq 95\%$). Moreover, within reason, the higher the charging voltage, the more power will be stored. In practice, the amount of power that can be stored depends on the surface area and the ability of the dielectric material to resist internal charge leakage. The absence of chemical reactions means that fast charge and discharge times are possible (within milliseconds), and, unlike batteries, the charge–discharge cycle can be repeated an almost unlimited number of times.

A supercapacitor differs from a traditional capacitor in several significant ways:

1. A supercapacitor (also known as an *electric double-layer capacitor*) uses a liquid electrolyte in place of a thin polymer film. The electrolyte typically consists of a salt dissolved in an organic solvent.

2. The electrodes are high-surface-area carbon-based materials, and may be in the form of carbon micro- or nanofibers or carbon nanotubes.

3. Because the electrolyte is a liquid, separation between the two electrodes must be maintained by means of a porous separator; otherwise the device would be prone to internal short circuits. Power storage occurs when the externally applied voltage causes the dissolved anions and cations to congregate at the oppositely charged electrodes (Figure 12.12b).

Supercapacitors are charcterized by high power density. Moreover, they can release their energy very quickly. However, the voltage of each cell is limited to

~2.3 V to avoid electrolytic decomposition of the organic solvent. With these properties, it will be obvious why there is a growing interest in large, multilayer supercapacitors in hybrid gasoline–electric vehicles, where electrical energy is generated during braking and is stored in a supercapacitor, and where a short burst of power is needed to restart the engine. However, the energy density of a supercapacitor is below that of a typical battery, and the discharge voltage is directly related to the state of discharge of the device, Thus, the steady release of energy at a constant voltage over a long period of time requires special circuitry. However, for the temporary storage of energy, and the delivery of elecrical power over a relatively short period of time, supercapacitors have many advantages.

E. CHALLENGES FOR THE FUTURE

1. Challenges for Materials in Fuel Cells

Although fuel cell technology promises to solve many of the problems associated with both stationary and mobile electrical power generation, it is necessary to remember that this technology does not eliminate all the existing problems. If the fuel for a fuel cell is a hydrocarbon, the device will still produce carbon dioxide and, in this sense, will not provide an overwhelming environmental advantage over conventional combustion. Moreover, if hydrogen is the fuel of choice, energy must be expended to generate hydrogen from hydrocarbons, and carbon dioxide will still be produced. Hydrogen obtained through the electrolysis of water requires the expenditure of large amounts of electrical energy, and not enough of that energy can currently be produced directly from sunlight via solar cells (Chapters 10 and 14) to be commercially feasible. Nuclear power is perhaps the only source of electricity that could be used to generate hydrogen from water on a large scale, but this begs the question of why that electricity cannot be used directly for stationary purposes or stored in a battery for mobile applications.

There has been much debate about the relative merits of the three sources of fuel for fuel cells—pure hydrogen, hydrogen generated from hydrocarbons via an integrated catalytic converter, or methanol. Hydrogen is a cleaner fuel but is flammable or explosive if mishandled. Moreover, as mentioned above, the commercial production of hydrogen requires the expenditure of considerable energy either for the electrolysis of water or for reacting methane with water to form hydrogen and carbon monoxide or dioxide. Carbon dioxide is, of course, a climate-changing gas, and carbon monoxide is a catalyst poison. Thus, a hydrogen economy does not eliminate some of the drawbacks of current power plants or the internal-combustion engine. Methanol is an excellent fuel for portable devices like cell phones and laptop computers. Methanol could easily replace gasoline in the present energy distribution and storage infrastructure, but it generates carbon dioxide as a waste product, and its synthesis from oil requires the expenditure of energy. Methanol can be produced by the destructive distillation of wood, but this is not a viable source of sufficient quantities to replace gasoline.

From a materials standpoint, the proton conducting membrane is the heart of a low-temperature portable fuel cell. The membrane must be strong enough to withstand mechanical pressure, have enough proton-conducting acidic sites to allow rapid proton transport (10^{-2} S/cm or higher), be stable at elevated temperatures, and be able to resist the aggressive chemical reactions that take place near the cathode. These reactions generate highly reactive species such as peroxides and hydroxyl radicals. Inherent in this challenge is the requirement that the membrane should also resist dissolution in water, even though the higher the loading of acidic groups, the more hydrophilic and the more water-soluble it will become.

It will be clear from this chapter that much progress needs to be made in the design and synthesis of new materials for energy generation technology. Broad-scale fuel cell development for mobile applications is currently held up by the shortage of proton conductors with the thermal or electrochemical stability and high ionic conductivity to survive thousands of hours of operation, particularly at temperatures of 150 °C or higher. As in several areas of materials science, much emphasis in the future will probably be placed on composite materials such as ceramic ionic conductors combined with polymeric counterparts. It should also be noted that solid and gel ionic conductors are increasingly being investigated for uses in electrochromic devices and solar cells, topics that are described in Chapters 10 and 14.

2. Challenges in Battery Science and Technology

Development of a liquid-free rechargeable lithium battery is a major challenge given the low ionic conductivities of existing solid polymeric ionic conductors. Solvent-free polymers with lithium or magnesium ion conductivities of 10^{-3} S/cm or higher are needed to allow battery technology to advance. Because the organic plasticizers currently used to improve conductivity in polymer electrolytes are almost always flammable, a need exists for plasticizers that are nonflammable and nonvolatile and that, at the same time, enhance ionic transport. High-boiling-point, nonflammable liquids that dissolve lithium salts are alternatives to nonplasticized polymers. One of the main weaknesses of current commercial batteries that use intercalation anodes and cathodes is the low concentrations of cations that can be stored in these electrodes. This limits the storage capacity for a given weight of the device. Thus, a major challenge is to devise new intercalation electrode materials that have metal storage properties far higher than those of existing examples. An additional challenge is to design safe lithium batteries that use metallic lithium or other high-energy metals as the anode material rather than intercalation electrodes.

3. Challenges for Capacitors and Supercapacitors

Traditional capacitors have a long history, and it might be assumed that few improvements might be anticipated in the future. However, new materials with lower dielectric constants would reduce internal charge leakage and improve the power storage times of these devices. Also, methods to control the release of the electrical energy during discharge would increase the possibility that capacitors could be used more widely in place of batteries.

The amount of electrical power that can, in principle, be stored in a double-layer capacitor points to many possible uses in the future. "Load leveling," or the smoothing of the output from a power station, is an application where the short-term storage of electrical energy and its rapid release is an obvious advantage. A weakness of existing devices is the use of a liquid organic solvent, which might leak or generate excessive pressures at high temperatures or high voltages. Replacement of a liquid electrolyte by a solid polymer or low-solvent-gel electrolyte would help to solve this problem. Moreover, the storage capacity of supercapacitors depends on the surface area of the electrodes, which, in turn, controls the density of ions near the interface. Thus, the use of high-surface-area, electronically conductive materials such as carbon nanofibers, nanoparticles, or nanotubes (Chapter 17) is a topic with considerable promise.

F. SUGGESTIONS FOR FURTHER READING

1. Ratner, M. A.; Shriver, D. F., "Ion transport in solvent-free polymers," *Chem. Rev.* **88**:109–124 (1988).
2. Winter, M.; Brodd, R., "What are batteries, fuel cells, and supercapacitors," *Chem. Rev.* **104**:4245–4269 (2004).
3. Mauritz, K.; Moore, R., "State of understanding of Nafion," *Chem. Rev.* **104**:4535–4585 (2004).
4. Hickner, M. A.; Ghassemi, H.; Kim, Y. S.; Einsla, B. R.; McGrath, J. E., "Alternative polymer systems for proton exchange membranes," *Chem. Rev.* **104**:4587–4612 (2004).
5. Tarascon, J. M.; Armand, M., "Issues and challenges facing rechargeable lithium batteries," *Nature* **414**:359–367 (2001).
6. Bruce, P. G.; Vincent, C. A., "Polymer electrolytes," *J. Chem. Soc.—Faraday Trans.* **89**:3187–3203 (1993).
7. Armand, M., "Polymer solid electrolytes—an overview," *Solid State Ionics* **9–10**:745–754 (1983).
8. Scrosati, B., ed., *Second International Symposium on Polymer Electrolytes*, Elsevier: New York, NY, 1990.
9. Allcock, H. R.; O'Connor, S. J. M.; Olmeijer, D. L.; Napierala, M. E.; Cameron, C. G., "Polyphosphazenes bearing branched and linear oligoethyleneoxy side groups as solid solvents for ionic conduction, macromolecules," 23:7544–7552 (1996).
10. Rikukawa, M.; Sanui, K., "Proton-conducting polymer electrolyte membrances based on hydrocarbon polymers," *Progress in Polymer Science* **25**:1463–1502 (2000).
11. Li, Q. F.; He, R. H.; Jensen, J. O.; Bjerrum, N. J., "Approaches and recent development of polymer electrolyte membrances for fuel cells operating above 100°C," *Chemistry of Materials* **15**:4896–4915 (2003).
12. Carrette, L.; Friedrich, K. A.; Stimming, U., "Fuel cells: Principles, types, fuels, and applications," *Chem Phys. Chem.* **1**:162–193 (2000).
13. Kreuer, K. D., "On the development of proton conducting polymer membranes for hydrogen and methanol fuel cells," *J. Membrane Sci.* **185**:29–39 (2001).
14. Allcock, H. R.; Wood, R. M., "Design and synthesis of ion-conductive polyphosphazenes for fuel cell applications," *J. Polym. Sci., Part B. Polymer Physics* **44**:2358–2368 (2006).
15. Rolison, D. R.; Dunn, B., "Electrically conductive oxide aerogels: new materials in electrochemistry," *J. Materials Chem.* **11**:963–980 (2001).
16. Frackowiak, E.; Beguin, F., "Carbon materials for the electrochemical storage of energy in capacitors," *Carbon* **39**:937–950 (2001).
17. Sarangapani, S.; Tilak, B. V.; Chen, C. P., "Materials for electrochemical capacitors—theoretical and experimental constraints," *J. Electrochem. Soc.* **143**:3791–3799 (1996).
18. Arico, A. S.; Bruce, P.; Scrosati, B.; Tarascon, J. M.; Van Schalkwijk, W., "Nanostructured materials for advanced energy conversion and storage devices," *Nature Materials* **4**:366–377 (2005).

G. STUDY QUESTIONS

1. Discuss why a solid polymeric ionic conductor is less effective for ion transport than a liquid electrolyte. Be specific; give actual examples and numbers. For which applications would you choose one system rather than the other?

2. Discuss the requirements that would be needed to allow a solid oxide fuel cell to be employed for small-scale power generation—say, in a laptop computer. Specifically, what types of peripheral materials would be needed to make this device feasible?

3. Discuss how intercalation electrodes function as essential components in energy storage devices. What are the advantages and disadvantages for the use of these materials? Give at least four examples of materials that fall into this category and provide both formulas and solid-state structures.

4. Why is lithium a preferred element for use in rechargeable batteries? What other elements might be plausible replacements, and what would be their advantages and disadvantages?

5. The polyelectrolyte Nafion is a widely used material in prtable fuel cells. What is the formula of this material and its morphology? How might a replacement for Nafion be designed and synthesized? Be specific—explain in detail why you would change the chemistry of the materials structure.

6. The lithium seawater battery has proved to be a major scientific and technical challenge. Explain why this is so and discuss the properties needed for all the components for the battery to have a long lifetime.

7. Why are molten salts good ionic conductors whereas solid salts are not? Are there organic counterparts of molten inorganic salts? Discuss some of the challenges that exist to the use of these liquids in energy generation and energy storage.

8. Capacitors and supercapacitors are alternative devices for energy storage. What scaleup problems might be encountered for using these devices (a) for automotive power, and (b) for massive power storage or power plant power-smoothing? Which materials are strong candidates for widespread use in these devices? What are the hurdles to the use of capacitors to replace batteries?

9. The lifetime of some batteries can often be extended if they are not routinely completely discharged before being recharged. Survey the literature and discuss the materials issues that are involved. How do lithium batteries compare in this respect to nickel–cadmium, nickel–metal hydride, and lead–acid batteries?

10. Participate in a class discussion about the relative merits of the different fuel cell technologies, and consider their advantages or disadvantages compared to the combustion of coal, oil, natural gas, wind, or solar cells. Which of the fuel cell options provides the greatest insurance against a catastrophic decline in the availability of oil or natural gas, and why?

13

Membranes

A. BACKGROUND

The purpose of this chapter is not to give a comprehensive description of membrane science (which would be impossible in one chapter) but rather to provide a brief overview perspective and to illustrate the connections to other areas of materials science.

Membranes play a significant but often unseen role in many important devices and processes. These range from the polymer electrolyte membranes discussed in Chapter 12 to the use of membranes in biomedical devices, described in Chapter 16, and include many other applications. In this chapter we review some general aspects of membrane science that apply to their uses for the separation of gases, liquids, or solutes, and for the controlled release of biologically active molecules. In addition, we briefly consider the use of membranes as acoustic transducers used in telephones or sonar equipment.

Many industrial processes would not be economically feasible without the use of membranes for chemical or biochemical separations. The same applies to the biomedical sector, where the controlled release of pharmaceuticals is a key aspect of this technology. Many membranes are made from polymers, but some use ceramics, and a few make use of metals. There is also a close relationship between the chemistry of membranes and the science of surfaces (Chapter 15).

In general, a *membrane* is a thin film, plate, or wafer of material that separates two compartments of a device. A common feature of many membranes for separations and drug delivery is that they have special characteristics to optimize the transport of target molecules or ions across the material while, at the same time, retarding the transmission of other species. This selective transmission may be based on the molecular weight of the molecules being separated, including their size, shape, or compatibility or incompatibility with the membrane material. Such factors underlie the extraction of oxygen from air, the removal or organic pollutants from water, the separation of isotopes in gases, the removal of waste products from blood by hemodialysis, and the controlled delivery of drug molecules into the body. The use of membranes as sound transducers—to convert sound waves to electrical signals in microphones, or electrical signals to sound in speaker systems—is based on different principles that are considered at the end of this chapter.

Membranes can be divided into three classes—porous membranes, membranes that function by a chemical reaction between the membrane material and the target molecules, and membranes that are nonporous but transmit molecules by diffusion rather than chemical reactions.

B. POROUS MEMBRANES

1. Mechanism of Operation

Membranes that contains pores or tunnels that penetrate from one side of the material to the other are especially useful for gas separations and liquid filtrations. The membranes may be produced from inorganic oxides or from polymers. In ceramic porous membranes, the diffusion of gases is through rigid voids, tunnels, or galleries that penetrate the solid material. In such cases, diffusion may be controlled by the size of the pores, which restrict the gaseous flow of large molecules but allow smaller molecules to pass through. Alternatively, porous membranes may be used for liquid ultrafiltration. Pore sizes may range from nanometers to micrometers or larger, with the size in ceramic membranes controlled by synthesis techniques such as sol–gel methods, sintering of powders, removal of soluble template materials from within a composite solid, or electrochemical etching of an inorganic solid.

As an example, porous alumina membranes, with pore sizes in the micrometer–nanometer range, are produced by the acidic electrochemical oxidation of thin aluminum foils (Figure 13.1). Typically, pore diameters can range from 8 to 200 nm, with pore density in the range of 10^{11} pores/cm^2. For these structures to be used as membranes it is necessary to remove the unreacted aluminum foil that covers the closed ends of the tubular structure. Alumina porous membranes may also be used as templates for the deposition of metals in each pore. Dissolution of the alumina then leaves metal nanotubes or nanowires (see Chapter 17).

Porous polymer membranes are fabricated by (1) extracting one component of a polymer blend, (2) swelling the polymer in a solvent and then rapidly volatilizing

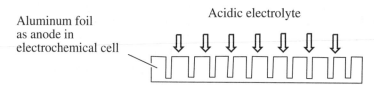

Figure 13.1. Porous aluminum oxide membranes can be produced by acidic electrochemical etching of the surface of aluminum foil.

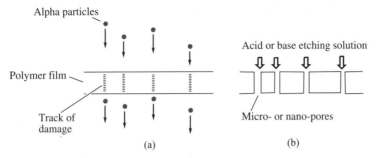

Figure 13.2. *Porous polymer membranes are produced commercially by exposure of a polymer film to α particles that penetrate the polymer, leaving behind a track of molecular damage. The damage track is more sensitive to etching by, for example, a strong base, which penetrates through the track and produces a cylindrical pore.*

the solvent, or (3) etching techniques. An example of process 3 involves exposure of a polymer membrane to α particles, which generate a path of damage through the material, a path that can be widened by etching with a reagent that penetrates preferentially along the damage track (Figure 13.2). Pore density then depends on the radiation exposure, and pore size is a function of the severity of etching. Such membranes provide a means for selecting the transmission of guest molecules on the basis of molecular size. Another approach is to produce a disk of stacked micro- or nanofibers, fill the space between them with a polymer, and then dissolve out the fibers. Alternatively, carbon nanotubes can be assembled in parallel arrays and used for nanoscale filtration. Tunnel clathrates and other nanoporous crystalline materials may also be adapted to produce porous membranes.

C. MEMBRANES THAT FUNCTION BY A CHEMICAL REACTION

Membranes of some metals, such as palladium, transmit hydrogen or deuterium at elevated temperatures (>300 °C) but reject other gases. This behavior is now ascribed to the facile dissolution of hydrogen in the metal through the formation of a transient hydride. The alternative explanation, that hydrogen can diffuse through the free volume in the metal lattice, is now considered to be less likely. Evidence for the participation of metal hydrides is provided by the dramatic change in crystal structure that accompanies hydrogen absorption, a change that is consistent with the formation of metal hydrides. Separation processes based on this phenomenon are so effective that very large-scale industrial separations of hydrogen from other gases are accomplished by this technique. For example, the reaction products formed by heating carbon monoxide, methane, or other hydrocarbons with steam at 900 °C are carbon monoxide, carbon dioxide, and hydrogen. The hydrogen can be isolated via palladium membranes. Interestingly, the alloying of about 20% silver with palladium yields a membrane that has an even higher permeability to hydrogen. The much heralded "hydrogen economy" could make extensive use of technology of this type of process.

D. NONPOROUS MEMBRANES THAT DO NOT REACT WITH PARTICIPATING MOLECULES

This is by far the largest class of membranes for gas or liquid separations and for the controlled release of small molecules. There are basically two different types of membranes that fall into this category—membranes fabricated from nonporous polymer films, and polymer matrices formed from a crosslinked polymer swollen by a solvent to form a gel. The distinction between these two types is sometimes unclear, but different mechanisms of operation are involved.

Polymer membranes are usually selective for the transmission of specific molecules because of three factors: (1) separation by differential solubility of the molecules (guests) at the surface of the membrane, known as "sorption"; (2) varied diffusion characteristics of different guest molecules or ions across the membrane; and (3) desorption of the molecules or ions at the other side of the membrane (Figure 13.3). In many gas or liquid separation systems, factors 1 and 2 have the most influence. For ion transport, all three factors are important.

Consider first the sorption factor at the leading face of a membrane. Gas or liquid molecules will interact with the surface of a membrane in different ways. Some will dissolve in the interface while others will not. In general, hydrophilic molecules will dissolve in a hydrophilic interface faster than will hydrophobic molecules, and vice versa. In practice, more is involved than a simple "like dissolves like" principle. Specific functional groups on one type of molecule may favor the attachment to and entry into a membrane surface. Some molecules may even react with the membrane material to give transient derivatives.

In addition, the ease of diffusion of specific molecules across a polymer membrane depends on several different factors. Two diffusion mechanisms are possible:

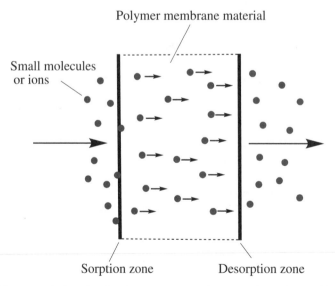

Figure 13.3. *The transmission of molecules across a membrane can be understood in terms of three processes—sorption, diffusion, and desorption—with separations accomplished if different molecules in a mixture respond differently to any of these three processes.*

one requiring diffusion through a solid polymer and the other involving diffusion through solvent molecules that are trapped within the membrane. In a solid polymer the guest molecules diffuse through the "free volume" between the polymer chains. Thus, the speed of diffusion depends on molecular size, the affinity of the guest for the host (strong interactions between the two may retard diffusion), and the freedom with which the polymer molecules can undergo conformational changes. Specifically, if the polymer is above its glass transition temperature, and the chains are able to undergo extensive molecular motions, the rate of guest diffusion will be faster than if the polymer is below the T_g. However, in practice, although the diffusion rate may be higher in a low-T_g polymer, the *selectivity* for transmission of different guests may be higher for a more rigid system that is below the T_g. Polymers above their T_g values may be less selective because the moving polymer chains allow both large and small molecules to diffuse through, whereas a rigid material may have relatively fixed and smaller "pore sizes."

In the case of membranes swollen by solvents, such as hydrogels or organogels Figure 13.4), the diffusion behavior may resemble the diffusion of molecules in solution, although the supporting polymer matrix can be an impediment to solute diffusion, especially if the degree of crosslinking is high or the membrane has a phase-separated structure. Thus, in gel membranes the relationship of the size of the diffusing species to the solvent-filled volume determines which molecules can penetrate the membrane and which cannot.

Desorption of molecules that have traversed the membrane will be affected by mass-action effects. For example, if no guest molecules are present in the space beyond the membrane, desorption will be easy. If, on the other hand, a high con-

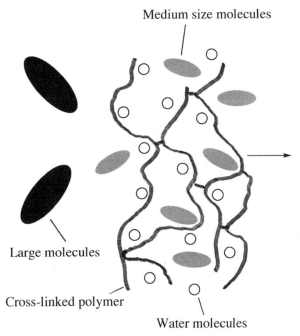

Figure 13.4. A hydrogel can function as a selection membrane if the free path through the crosslinked network allows some molecules to pass through, and others to be excluded.

centration of guest exists on the far side of the membrane, desorption will be difficult. Hence, removal of guest molecules as soon as they have crossed the membrane is a key requirement for facilitating the process. To this end, a common practice for liquid separations is to apply pressure to the liquid on the entrant side of the membrane and apply a vacuum on the emergent side. The vacuum ensures rapid removal of molecules by evaporation as they emerge from the membrane, a process known as *pervaporation*.

E. SPECIFIC EXAMPLES OF MATERIALS USED IN SOLID POLYMERIC MEMBRANES

1. Poly(dimethylsiloxane) Membranes for Oxygen and Carbon Dioxide Transmission

Membranes of silicone rubber have a high permeability for oxygen and a moderately high permeability for carbon dioxide. For this reason they have been investigated for uses in underwater breathing equipment. A classical experiment illustrates this principle. Here, a mouse within an underwater chamber is separated from the water by a silicone rubber membrane. Dissolved oxygen passes from the water through the membrane to keep the mouse alive, while carbon dioxide diffuses in the opposite direction. The high gas permeability of silicone rubber is ascribed to two factors: (1) the unusually high solubility of oxygen in the polymer, for reasons not fully understood; and (2) the extensive molecular motion that exists at room temperature in silicone rubber, as indicated by its very low T_g ($-130\,°C$), generates a high free volume for guest transport. Other low-T_g polymers, such as polyphosphazenes with fluoroalkoxy side groups, also have good oxygen permeability characteristics.

2. Desalination Membranes

The removal of salts from seawater is a major engineering challenge. Among the methods employed for this purpose is the use of membranes for reverse osmosis. Seawater is pumped under pressure through a series of membranes that allow water to pass through but retard transmission of the salts. Traditional desalination membranes are nonporous polymers, but newer membranes are micro- or nanoporous materials through which the high-pressure water travels faster than dissolved salts

3. Dialysis Membranes

Dialysis is a liquid-phase separation process that allows small molecules to pass through a nonporous membrane while preventing the passage of larger molecules. Dialysis is used in polymer research to purify macromolecules from small-molecule impurities, and it is a major technique in biochemistry for the purification of proteins and other large biomolecules. Hemodialysis is a process by which blood is purified of low-molecular-weight waste products by the use of membranes that allow small molecules to pass through while, at the same time, preventing the loss of biological macromolecules such as proteins or larger biostructures such as blood corpuscles. Ideally the material from which the membrane is constructed should have a high selectivity and a good biocompatibility with blood. This subject is relevant to the biomedical materials discussion in Chapter 16.

The traditional dialysis membrane material is cellulose triacetate (produced from cellulose and acetic anhydride), which can be obtained with different "pore sizes" to allow the passage of molecules with specific dimensions. The term "pore size" is a misnomer for most dialysis membranes because no pores are present. Instead, the term "molecular weight cutoff" is more accurate since it defines the size of molecules that will and will not traverse the membrane. Cellulose acetate films are plasticized with glycerol and water to overcome the brittleness of the films. Thus, the diffusion of small molecules through the membrane is probably through the water- and glycerol-filled free volume between the polymer chains. In this sense, these membranes come close to the gel membranes described below. Molecular weight cutoffs can vary over a wide range, with values of 10,000–40,000 Da typical for protein dialysis. The different cutoff characteristics are achieved through a variety of proprietary techniques. In principle, the amount of solvent-filled free volume can be controlled by the level of acetylation of the cellulose and the amount of crosslinking.

4. Membranes for Controlled Drug Delivery

This topic is also revisited in Chapter 16. *Controlled drug delivery* is a process that releases a limited amount of a drug either to the skin for transdermal delivery or internally for release into the muscle system or bloodstream. The operating principle is that a polymer membrane is chosen that has a moderate permeability to the drug in question so that the release rate into the body is constant over a certain time period. Thus, the objective is not to separate molecules but to control their diffusion.

Typical devices consist of a reservoir of a drug in an aqueous medium, which is separated from the exterior by a membrane. Different drug molecules have different permeabilities through various membrane materials. Thus, membranes are chosen to provide a predetermined delivery rate, which often depends on the molecular size and hydrophilicity of the drug and the character of the membrane material. A long-existing polymer, poly(vinyl acetate), has been widely used for this purpose.

F. GEL MEMBRANES

1. General Principles

A *gel* is a crosslinked polymer that is swollen by a solvent. Such materials can serve as reservoirs for the release of, for example, drug molecules, or they may be used as membranes for separations or for controlled drug delivery. The mechanism of action of a gel membrane usually depends on the diffusion of solutes through the liquid phase in the membrane. The polymer matrix may serve only to maintain the shape of the membrane. An exception was noted in Chapter 12, where the sulfonic acid groups attached to the polymer matrix in Nafion control the diffusion of protons through polymer electrolyte fuel cell membranes. The same may be true for the diffusion of lithium ions across a gel membrane in lithium polymer batteries. However, except for these special circumstances, the transmission of solute molecules through a gel membrane depends on the free path through the solvent component, and this, in turn, depends on the degree of cross-linking in the membrane.

2. The Special Case of Gel Membranes as On/Off Switching Systems

Numerous uses exist for membranes that can be switched from being permeable to being impermeable. An example is a device that automatically releases a drug molecule into the body in response to pH or temperature changes, or following the appearance of certain ions in the bloodstream. Many responsive membranes are solvent-swollen polymers that can expand or contract depending on an external stimulus. Some of the most interesting species in this category are hydrogels. In a water-swollen state a hydrogel membrane will allow the transmission of molecules through the water-filled volume. Some molecules may be excluded on the basis of the degree of crosslinking, with higher levels of crosslinking generating smaller free pathways for solute molecules to pass through.

However, if the gel collapses extruding the internal water, no solute molecules will pass through (Figure 13.5). There are four ways to bring about the collapse of a hydrogel: by changes in (1) temperature, (2) pH, and (3) cation charge, and by (4) the application of an electric potential. Hydrogels such as those formed from poly(ethylene oxide), poly(hydroxyethyl methacrylate) (HEMA), or the polyphosphazene MEEP (see Chapter 6, Table 6.1) have a lower critical solution temperature (LCST), which means that below a certain temperature they are fully expanded, but *above* that temperature they collapse and a membrane is no longer permeable. This behavior is a result of the balance of hydrophobic and hydrophilic components in the polymer and the change in hydrogen bonding to water as the temperature is raised.

Other hydrogels respond to pH changes. For example, a hydrogel formed from a water-soluble polymer with acidic side groups will be expanded at high pH but will collapse at low pH. The same gel will be expanded in solutions of monovalent cations, but will collapse through ionic crosslinking in the presence or di- or trivalent cations. Finally, if a gel derived from a polymer with acidic side groups contains cations that can be switched electrochemically from the monovalent to the di- or trivalent state, the passage of an electric current can bring about a reversible change in a membrane from permeable to impermeable. These changes also are illustrated in Figure 13.5. This same behavior can not only be used to control the passage of molecules through a membrane but also be employed to release molecules trapped in the hydrogel.

G. TESTING OF MEMBRANES

The evaluation of membranes is slightly different depending on whether a mixture to be separated is gaseous or liquid, and whether the membrane is designed for the controlled release of drug molecules.

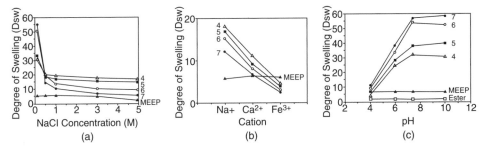

Figure 13.5. *Responsive hydrogel membranes derived from lightly crosslinked polyphosphazenes with both* —OCH₂CH₂OCH₂CH₂OCH₃ *and* OC₆H₄COOH *side groups. The different curves are for polymers with different ratios of the two side groups. For example, polymer 4 contains 30% of the carboxylate side groups while polymer 7 contains 94%. MEEP contains 100% of the alkyl ether side groups. The curve labeled "ester" is for a polymer where all the carboxylic acid groups have been esterified. (a) Contraction behavior of the hydorgels as the ionic strength of the aqueous medium is increased. (b) Contraction behavior as sodium ions in solution are replaced by calcium, and ferric ions. (c) Expansion of the membrane as the pH is increased. In the swollen state the membrane is permeable to small molecules. In the collapsed state, the pathways through the membrane are closed. These effects are all reversible. (From Allcock, H. R. and Ambrosio, A. M. A., Biomaterials, **17**:2295–2302 (1996).*

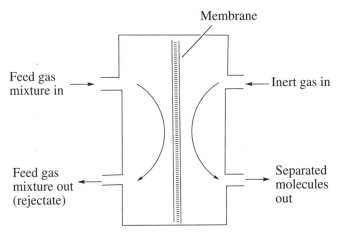

Figure 13.6. *Apparatus for the evaluation of gas separation membranes. The desorption step at the boundary of the membrane is accelerated by rapid removal of the emerging molecules in a stream of inert gas.*

1. Gas Separations

Measurements of gas permeability are often carried out with use of an apparatus shown schematically in Figure 13.6. A flowing mixture of gases is allowed to contact a membrane, and one type of molecule diffuses through the membrane faster than the others do and is captured by a flow of inert gas (the "purge"). The depleted gas flow on the entrant side of the membrane (the "rejectate") can now be recycled into the inlet stream. Two modes of operation are possible. In the first, the gas diffusing

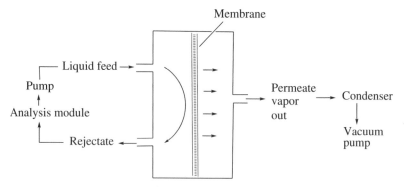

Figure 13.7. *Equipment for testing membranes for liquid phase separations. The desorption process at the downstream side of the membrane is assisted by volatilization of the molecules under vacuum.*

through the membrane is the purified species to be collected. In the second mode it is the impurity, and the rejectate then contains a higher concentration of the desired molecules.

2. Liquid Separations

A similar setup is used for liquid separations, as shown in Figure 13.7. Here, a liquid mixture is pumped past a membrane. The component that has the higher permeability through the membrane emerges and is removed by evaporation under vacuum. This accelerates the desorption step. That component is then condensed by cooling the collector vessel. Again, the depleted liquid that exits the device may be recycled to improve the overall efficiency of the separation.

3. Controlled Drug Release and Dialysis Membranes

Figure 13.8 illustrates the type of equipment used to evaluate the permeability of a membrane to a drug, a biomolecule, or any small molecule. The membrane is used as a separator between two halves of a cell, which is maintained at constant temperature by means of a temperature control jacket. Both compartments contain magnetic stirrer bars for efficient circulation.

Initially, an aqueous solution of a water-soluble dye is placed in one half of the cell, and pure water is placed in the other half. The rate of dye diffusion through the membrane is monitored by taking aliquots from the receiving side of the cell and measuring the concentration of dye by ultraviolet/visible spectroscopy. This establishes the general transmission behavior of the membrane to molecules with size and properties similar to those of the dye molecule. Similar experiments can then be carried out with the use of drug molecules in place of the dye. In this way, new membrane materials and new drugs can be matched to obtain appropriate release rates. The graph shown in Figure 13.8 illustrates the diffusion characteristics of the same dye molecules through two different membranes.

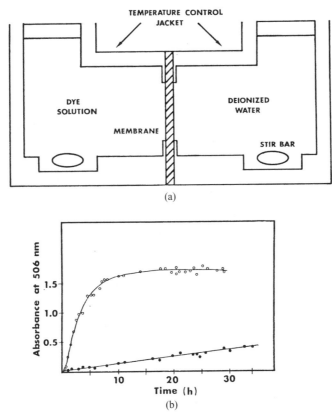

Figure 13.8. (a) Apparatus for evaluation of different membranes for small-molecule diffusion using a dye molecule as a model. In plot (b), the lower curve represents a cellulose acetate membrane; the upper curve, an experimental polyphosphazene membrane. Experiments of this type are useful for evaluating membranes for both drug delivery and dialysis applications.

H. SOUND TRANSDUCER MEMBRANES

These membranes offer a distinct contrast to those decribed above. They do not transmit ions or molecules. Their function is to either vibrate in response to an oscillating electric field or generate an oscillating electric current when they vibrate. The main requirement for membranes as sound (acoustic) transducers is that they be piezoelectric materials. These are polymers or ceramic-type materials that generate an electrical signal when stressed by the vibration of sound waves, or that themselves vibrate in response to oscillating electrical signals. Piezoelectric behavior arises because the solid material contains molecules, unit cells, or domains that have a dipole—in other words, the units are polarized, with a partial positive charge and a partial negative charge at opposite ends of the unit (Figure 13.9).

1. Principle of Operation

The coorientation of the dipoles may result from the normal crystal structure of the material, or the dipoles may need to be aligned by the application of a strong electric

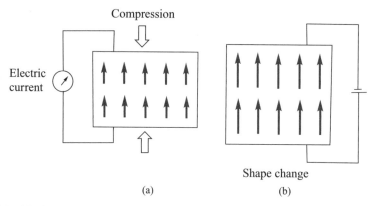

Figure 13.9. (a) The piezoelectric effect occurs when force is applied to a material that contains molecules or structural components of a solid that have an oriented dipole. Compression, and the resultant shape change, result in the generation of an electric current. (b) Conversely, application of an electric current causes a shape change (electrostriction). In this way sound waves impinging on a membrane are translated into an electric current, while the application of an oscillating electric current generates sound waves.

field to the molten material, followed by cooling below the melting or glass transition temperatures. When such a material is distorted by being twisted or mechanically deformed in some other way, a small electric field is generated. Conversely, when an electric field is applied, the material becomes mechanically deformed. If the applied electric field is alternating, the material will vibrate. This principle is utilized in speaker systems to convert electrical signals into sound waves, and in microphones to convert sound waves into electrical signals.

2. A Polymeric Example—Poly(vinylidene fluoride)

Although this polymer crystallizes in a conformation that aligns the C—F (carbon–fluorine) dipoles in one direction along the chain as shown in Figure 13.10, the dipole generated is insufficient to yield a high level of the piezoelectric effect. Instead, a film of the polymer must be stretched, a process that changes the conformation to an all-trans form (Figure 13.10). This orients the C—F dipoles in a direction at right angles (90°) to the axis of the skeleton. Even so, different polymer chains and different crystallites will be oriented randomly until the material is subjected to a strong electric field and the dipoles oriented at a temperature above the T_g. In this poled state piezoelectric coefficients of $\leq 30\,pM/V$ are possible. The polymer is easily fabricated into thin, flexible membranes, which makes it an almost ideal material for diaphragms in telephones and microphones. Thus, as a microphone diaphragm, the film vibrates in response to incoming sound waves and each microcrystalline domain will generate an electrical signal that matches the acoustic profile. This is amplified electronically for transmission. Conversely, an oscillating electric field applied to the membrane causes it to vibrate with the same frequency as the electrical signal and generate corresponding sound waves.

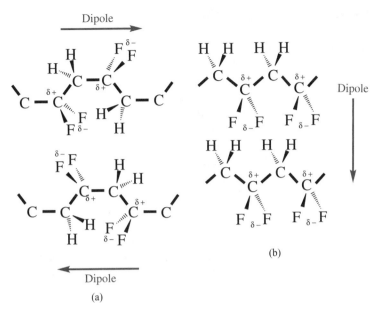

Figure 13.10. *The piezoelectric behavior or poly(vinylidene fluoride) depends on the polarity of the carbon–fluorine bonds. This generates a dipole at the molecular level for the cis–trans planar conformation (a), but the chain-packing arrangement in the solid state tends to cancel the dipole. However, when stretched into the all-trans conformation (b), the dipoles reinforce each other in the solid state and the material becomes piezoelectric.*

3. Ceramic-Type Piezoelectric Materials

Some inorganic solids such as quartz, barium titanate ($BaTiO_3$), and lead zirconyl titanate ($PbZrTiO_3$) have permanently aligned dipoles that are a consequence of the crystal structure. Others are fabricated into thin wafers after poling in the melt. Small quartz crystals have long been used as timing devices in watches and clocks because they vibrate with a characteristic frequency when subjected to an oscillating electric field. Cigarette lighters or propane torch igniters that generate a spark when struck with a spring-loaded hammer are other widely used piezoelectric devices. Sonar technology makes extensive use of piezoelectric sound generators and detectors. Ultrasonic transducers are used to detect defects in structural material. The profile of a surface being studied by AFM or STM microscopy (see Chapter 4) makes use of a piezoelectric crystal in the probe. The list of known inorganic piezoelectric materials is very long. For example, in addition to those mentioned above, piezoelectric behavior exists in potassium or lithium niobate, lithium tantalate, and calcium hydroxyapatite. This last material, which is the structural component of bone, is believed to stimulate bone growth through electric signals generated when a bone is stressed.

Lead zirconyl titanate is the benchmark piezoelectric material. It can generate piezoelectric coefficients of ≤300 pM/V. However, it is not a flexible material, and it is difficult to grow the large single crystals that are needed for some applications. Hence, the polymeric counterpart, poly(vinylidene fluoride), is preferred for some applications despite a piezoelectric coefficient that is 10 times less.

I. CHALLENGES FOR THE FUTURE

This chapter serves to compare several different aspects of membrane science and technology and acts as a conceptual bridge between the engineering, biochemical, and biomedical sectors. The study of membranes underlies large segments of materials science, and an awareness of this area is crucial not only for the topics described here but also for the fields of surface coatings, protective clothing, food storage and preservation, devices for the detection of explosives and biological hazards, and many other applications. In addition, there is a crucial need for increasingly sophisticated membranes to separate small molecules in the liquid and gaseous states, for air purification, for the purification of biological macromolecules, and to exclude oxygen and microorganisms from food products, as well as for a wide variety of sensors and detectors. These are major areas of technology that are increasingly dependent on the development of new materials for membrane fabrication.

A very strong argument for the growth of membrane science and technology is that it permits large or small-scale separations without the expenditure of large amounts of energy. In this sense it differs from current hydrocarbon refining and other distillation/fractionation processes, and this will undoubtedly be a driving force in the future.

Chemists, chemical engineers, and materials scientists can contribute to these developments through the design and synthesis of new ceramic membranes, new polymers and polymer composites, and new metallic membranes, as well as novel materials for sound detection and conversion of elecrical signals to sound. As emphasized elsewhere in this book, the synthesis work will require detailed attention to the characterization of the new materials, and collaborations between chemists, physicists, and engineers to optimize membrane performance.

J. SUGGESTIONS FOR FURTHER READING

1. Itoh, N.; Tomura, N.; Tsuji, T.; Hongo, M., Strengthened porous alumina membrane tube prepared by means of internal anodic oxidation, *Micropor. Mesopor. Mater.*, **20**:333–337 (1998).

2. Paul, D. R.; Yampol'skii, Y. P., *Polymeric Gas Separation Membranes*, John Wiley & Sons, New York, 1993.

3. Pinnau, I.; Freeman, B. D., *Advanced Materials for Membrane Separations (ACS Symposium Series) 876*, American Chemical Society, Washington D.C.

4. Kesting, R. E.; Fritzsche, A. K., *Polymeric Gas Separation Membranes*, Wiley-Interscience, 1993.

5. Kanellopoulos, N. K. (ed.), *Recent Advances in Gas Separation by Microporous Ceramic Membranes*, Elsevier, 2000.

6. Osada, Y., *Membrane Science and Technology*, CRC Press/Dekker, 1992.

7. Mulder, M.; Mulder, J., *Basic Principles of Membrane Technology*, Springer, 1996.

8. Sata, T., *Ion Exchange Membranes: Preparation, Characterization, Modification*, Royal Society of Chemistry, 2004.

9. Scott, K.; Hughes, R., *Industrial Membrane Separation Technology*, Springer 1996.

10. Nunes, S. P.; Peinemann, K.-V. eds., *Membrane Technology in the Chemical Industry*, Wiley VCH, 2001.

11. Svarovsky, L., *Solid-Liquid Separation* 4th ed., Butterwork Heinemann, 2000.

12. Freeman, B. D., "Basis of Permeability/Selectivity Tradeoff Relations in Polymeric Gas Separation Membranes," *Macromolecules*, **32**:375–380 (1999).

13. Wind, J. D.; Sirard, S. M.; Paul, D. R.; Green, P. F.; Johnson, K. P.; Koros, W. J., "CO_2-induced Plasticization of Polyimide Membranes: Part 2: Relaxation Dynamics of Diffusion, Sorption, and Swelling", *Macromolecules*, **36**(17):6442–6448 (2003).

14. Xu, W.; Paul, D. R.; Koros, W. J., "Carboxylic acid containing polyimides for pervaporation separations of toluene/iso-octane mixtures", *J. Membr. Sci.*, **219**(1–2):89–102 (2003).

15. Nagai, K.; Freeman, B. D.; Cannon, A.; Allcock, H. R., Gas permeability of poly[bis(trifluoroethoxy)phospazene] and its blends with adamantyl phosphazenes, *J. Membrane Sci.*, **172**:167–176 (2000).

16. Allcock, H. R.; Ambrosio, A. M. A., Synthesis and characterization of pH-sensitive poly(organophosphazene) hydrogels, *Biomaterials*, **17**:2295–2302 (1196).

K. STUDY QUESTIONS (for essays and class discussions)

1. Speculate as to why the electrochemical oxidation of aluminum foil leads to the formation of uniform tunnels. What other metals might behave in the same way?

2. The formation of porous membranes by α-track etching requires a choice of specific reagents for the etching process and selection of appropriate polymers. Choose four different polymer types and discuss what types of reagents might be used to etch the tunnels.

3. Glance ahead to Chapter 16 and the discussion of biomedical membranes. How does a thin polymer membrane control the release of drug molecules into the body? If the drug is hydrophilic or water-soluble, what type of membrane would you choose for this application? How would your thinking change if it were a hydrophobic drug?

4. Why is poly(vinylidene fluoride) used as a sound transducer membrane? What is it about the molecular or materials structure that makes it useful for this application? Under what circumstances would this material be used rather than a ceramic transducer, and vice versa?

5. In what sense can the behavior of a solid polymer electrolyte membrane of the type used in a lithium battery (Chapter 12) be understood in terms of adsorption, diffusion, and desorption of lithium ions? Explain the possible influence of these three factors on battery performance.

6. Assuming that a polymer electrolyte fuel cell membrane such as Nafion contains water absorbed into domains between hydrophobic regions, explain how protons might travel from one side of the membrane to the other.

7. A request has been issued by a government agency for proposals for new ways to separate methane from helium at natural gas wellheads. Write a proposal, explaining in detail how you propose to do this, what materials you would use, what the advantages of your method are, and why it is better than existing methods. Include a proposed budget to support three graduate students for 3 years and the costs of the chemicals and equipment that you would need to purchase.

8. What are the main practical disadvantages in the use of micro- or nanoporous filters to purify drinking water, and how would you correct them?

9. Consult an undergraduate-level physical chemistry text and describe some of the fundamental factors that might control the efficiency of separation of two gases such as hydrogen and deuterium or hydrogen and carbon monoxide during passage through nanoporous membranes. Is there a preferred size for the pores that penetrate the membrane?

10. What are some of the limitations to the use of polymer membranes for the large-scale separations of the products from an oil refinery, compared to distillation?

14

Optical and Photonic Materials

A. OVERVIEW

1. Passive versus Responsive Optical Materials

The optical properties of materials have been important almost from the beginnings of civilization. Windows, decorative glass, and jewelry are well-known examples. However, optical materials became the subject of intense interest when telescopes, binoculars, and eyeglasses came into general use, and they moved into mainstream technology with the burgeoning manufacture of camera lenses and prisms. Optical materials have also played a dominant role in the more recent changeover from electronic communications based on copper wiring to optical communications through optical fibers and photonic switches. Thus, materials with useful optical properties have now become a major area of research and technology based firmly in materials science in ways that are summarized in this chapter.

It is convenient to consider optical materials in two categories, which we will call "passive" and "responsive." The category of passive optical materials includes substances whose optical properties change relatively little when an electric field is applied or when the light intensity changes. However, the optical behavior of passive materials, like that of all substances, does vary with temperature and with the wavelength of light being transmitted. Responsive optical materials undergo a significant change in color, transparency, or refractive index under the influence of an electric field or a change in light intensity. This is superimposed on the changes due to variations in temperature or the wavelength of light.

Nearly all common optically useful materials are transparent—that is, they must transmit light rather than block or scatter it. Scattering results from a heterogeneous composition—for example, crystallites dispersed throughout an amorphous material or a mixture of any two or more materials that have different refractive indices. Materials that are transparent include single crystals (assuming the absence of imperfections), amorphous inorganic and polymeric glasses, and heterogeneous substances in which the domain size is smaller than the wavelength of light.

2. Importance of Refractive Index

The refractive index (RI) of a material (n) is defined by the speed of light in a vacuum divided by the speed through the transparent solid. It is also defined by the square root of the dielectric constant. Because the transmission of light in a vacuum is always faster than in a solid, the refractive index is a number larger than 1. However, refractive index values of zero or even negative values are becoming possible. The refractive indices of some typical optical materials and liquids are given in the Table 14.1.

In general, the refractive index depends on the density of electrons in the material. Solids that contain heavier elements have higher refractive indices than do those consisting of lighter elements. Thus, inorganic glasses that contain lead or other heavy elements, or polymeric glasses that contain chlorine, bromine, or iodine, have high n values. So too do organic materials with double or triple bonds.

TABLE 14.1. Refractive Indices

Compound	RI at ~600 nm and 20–25 °C
Liquids	
Water	1.33
Methanol	1.33
Octane	1.40
p-Dioxane	1.42
Ethylene glycol	1.43
Benzene	1.50
Toluene	1.50
Nitrobenzene	1.55
Aniline	1.59
Carbon disulfide	1.63
Inorganic Glasses	
Fluorite (CaF_2)	1.43
Crown glass (soda–lime glass)	1.52
Quartz	1.54
Flint glass (lead glass)	1.67
Polymeric Glasses	
Poly(methyl methacrylate)	1.49
Poly(ethylene terephthalate)	1.58
Polystyrene	1.59
Polycarbonate	1.59
Poly(*o*-chlorostyrene)	1.61
Various polyphosphazenes (depending on side groups and architecture)	1.38–1.76

Two important consequences of the refractive index are the phenomena of refraction and total internal reflection. These can be understood in terms of Snell's law, $n_1 \sin \theta_i = n_2 \sin \theta_{\text{refr}}$, and Figure 14.1. When a beam of light passes from a low RI medium into a high-RI solid two processes occur—reflection and refraction. Reflection may be minimized by the presence of an anti-refraction coating with a thickness similar to the wavelength of light (see discussion later). The refracted ray passes through the solid and emerges from the material such that the angle of incidence θ_i equals the final angle of refraction. If the two interfacial surfaces are not parallel (as in a prism or a lens), the *direction* of the emergent beam will differ from that of the entering beam, but θ_i will be the same (Figure 14.2).

High-RI glasses are of special importance in lens manufacture because the focal length of the lens can be shortened without the need to use surfaces with a high

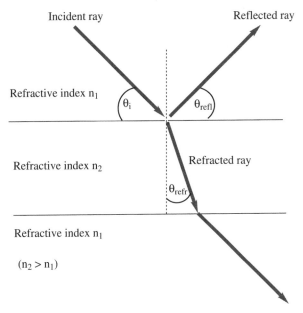

Figure 14.1. *The path of a light beam through a slab of transparent material depends on the refractive index of the surrounding (lower refractive index) medium (n_1) and that of the slab (n_2), where the angle of incidence (θ_i), the identical angle of reflection (θ_{refl}), and the angle of refraction (θ_{refr}) are related by Snell's law, $n_1 \sin \theta_i = n_2 \sin \theta_{refr}$.*

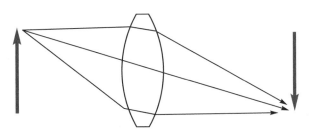

Figure 14.2. *Curved surfaces, as in a lens, follow the same refraction rules as flat slabs, but bring about focusing of an inverted image of an object.*

degree of curvature. Moreover, the correction of color dispersion becomes easier. However, the higher the curvature, the more acute are the manufacturing challenges.

Total internal reflection is a phenomenon that occurs when a beam of light traveling in a high-RI material impinges at a grazing angle on an interface with a lower-RI material, for example the interface between glass and air. This behavior is the basis for retention of a light beam within an optical fiber, as is discussed later in this chapter.

3. Optical Dispersion

Because the refractive index of a substance varies for the different wavelengths that constitute white light, the passage of a white light beam through a solid with non-parallel faces results in the separation of the beam into the different colors of the spectrum (Figure 14.3). Blue-violet light is refracted most, and red light, the least. For a single lens this means that there is no common point of focus for the different colors. Hence, an image focused by a single lens suffers from color fringing, which degrades the image. This is a major reason for the use of multilens assembles in camera and telescope systems.

Chromatic dispersion is defined in terms of the Abbe number v_d, as given by the equation

$$v_d = \frac{n_D - 1}{n_F - n_C}$$

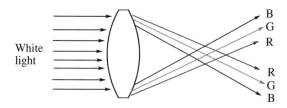

Spreading of the image due to chromatic dispersion to give color fringing

(a)

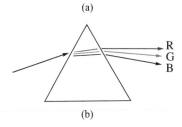

(b)

Figure 14.3. *The focus point of a lens (a) differs for the different wavelengths that make up white light due to the variation of refractive index with wavelength. This leads to color fringing of an image, which degrades overall sharpness. For a prism (b), the splitting of white light into the colors of the rainbow is a well-known phenomenon (see color insert).*

where n_D, n_F, and n_C are the refractive indices of the material for light at three wavelengths that are usually chosen from the *D, F,* and *C* Fraunhofer (Far-Field) emissions at 589.2, 486.1, and 656.3 nm, respectively. The *D* line corresponds to the main yellow emission from sodium. The *F* and *C* lines are the blue and red emissions that are part of the hydrogen emission spectrum. Thus, the Abbe number is a measure of the degree to which the different wavelengths of light are spread out as polychromatic light passes through a material. A higher Abbe number indicates a lower chromatic dispersion. Glasses with an Abbe number above 80 are considered to be low-dispersion materials.

For some optical applications, such as high-quality camera, binocular, or telescope lenses, it would be advantageous to have a high-RI lens that has a low chromatic dispersion. As shown in Figure 14.4, inorganic glasses are more likely to provide this combination of properties than are organic polymeric materials. Polymer glasses have refractive indices that typically range from 1.33 to 1.73, although specialized polymers, such as polyphosphazenes, sometimes exceed these values, but with high dispersions. Fluorinated polymers have low refractive indexes and low dispersions.

Low-dispersion inorganic glasses may be based on fluorite (CaF_2), which is both mechanically fragile and soluble in water, or on silicate glasses that contain barium or lanthanide cations. The addition of thorium to a silicate glass gives a high-RI material with low dispersion, but the radioactivity of thorium has limited the use of this material in more recent years. It is said that 99% of the approximately 2000 different optical glasses available for lens manufacture are used to correct for chromatic dispersion.

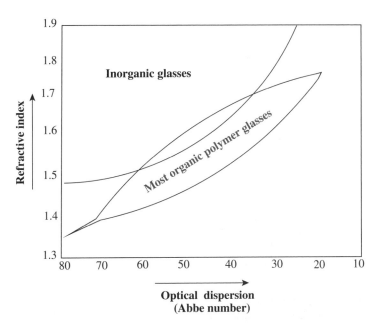

Figure 14.4. *Schematic plot of refractive index versus chromatic dispersion for a wide range of glasses. Note that for most glasses, chromatic dispersion rises with increasing refractive index. Inorganic glasses cover a wider span of optical properties than do polymer glasses. Fluorinated polymers tend to have low refractive indices and low chromatic dispersions.*

4. Optical Birefringence

A *birefringent* material is a transparent substance that has different refractive indices associated with the passage of light through different directions within the solid. When this situation exists, a beam of light entering the solid is split into two polarized beams—the "ordinary" (faster) beam n_o, and the extraordinary (slower) beam n_e. A well-known consequence is the formation of a double image when an object is viewed through a crystal of calcite ($CaCO_3$), quartz (SiO_2), sapphire (TiO_2), sodium nitrate ($NaNO_3$), or even ice. This is called "double refraction."

The phenomenon arises when the distribution of atoms and bonds, and hence the electron density, differs along different axes of the crystal. Because different axes have different refractive indices, one fraction of the light will proceed directly through the low-RI path through the crystal (n_o), while a second fraction will take a longer (slower) higher refractive index path before emerging (Figure 14.5).

The two beams will be polarized at right angles to each other (see later section in this chapter). The polarization effect can be seen easily if a film of Polaroid material is rotated between the eye and the crystal. First one component of the doubled image and then the other will be extinguished. Moreover, because of the different refractive indices and different path lengths, if the two emergent beams are combined, they will be subjected to constructive or destructive interference, depending on the thickness of the sample and the difference between the two refractive indices. Hence, for certain applications, like "quarter-wave plates" (discussed later), the precise thickness of the material becomes crucial for its utilization in optics. Thin plates of birefringent material can convert linear to circularized polarized light, or vice versa, as described in a later section.

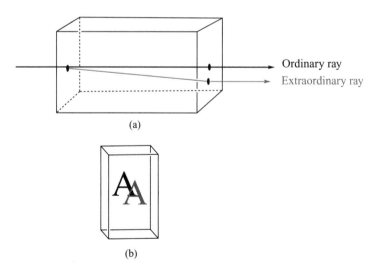

Ordinary ray
Extraordinary ray

(a)

(b)

Figure 14.5. *Optical birefringence occurs when the electron density in a solid differs for different spatial axes. This situation occurs most frequently for crystalline materials, but it may also occur with oriented liquid crystalline compounds or oriented noncrystalline polymers. A single beam entering the solid is split into two beams, one of which takes the most direct pathway (the "ordinary" faster, low-refractive-index path), and the other follows the slower, "extraordinary" high-refractive-index route (a). The result is a doubling of the image (b).*

Crystals are not the only materials that exhibit birefringence. Many polymers behave similarly, especially if they have been oriented by stretching. Regenerated cellulose films (cellophane), rigid-rod polymers, and liquid crystalline polymeric materials are examples. Small-molecule liquid crystalline phases often show the same behavior, again due to the different electron densities along different directional axes.

Birefringence also explains the brilliant colors that arise when crystals, many polymers, and liquid crystalline substances are rotated between crossed polarizers on a microscope stage. The different colors are a consequence of the constructive or destructive inteference effects for different wavelengths of light after passage through a birefringent material. On the other hand, totally amorphous glasses or elastomers give only a dark-field image when rotated between crossed polarizers.

B. PASSIVE OPTICAL MATERIALS

1. Materials and Devices for Passive Optical Applications

These represent by far the majority of optical materials used in technology. As discussed above, they range from transparent single crystals like quartz or mica to inorganic and organic amorphous polymeric glasses. The choice of a specific material depends, of course, on the demands of the application. However, in general, single crystals are employed where small size and limited fabrication options are acceptable. Inorganic glasses are best when thermal and radiation stability are primary requirements. Polymers are preferred for applications where ease of fabrication into unusual shapes, low cost, or lightness of weight are primary considerations. The disadvantages of single crystals revolve around size, shape, and fabrication problems. Inorganic glasses must be fabricated at elevated temperatures and can be engineered to precise shapes only by meticulous grinding and polishing. Problems often encountered with polymeric glasses are their relatively low T_g values and the consequent possibility of shape changes at elevated temperatures, and their tendency to generate colored decomposition products under intense ultraviolet irradiation. This last problem is especially acute for the adhesives used to cement lens or prism components together. The naturally occurring polymeric adhesives used in older lenses or laminates have also proved susceptible to microbial colonization and clouding.

2. General-Purpose Optical Materials

Windows in houses, automobiles, and aircraft represent one of the largest-volume uses for transparent materials. Silicate soda–lime glass (Chapter 7) is by far the most widely used window material for buildings and automobiles. It is produced in "annealed" and "tempered" modifications. The annealed material (cooled slowly) fragments into shards on impact and is thus dangerous for use in automobile or laboratory fume hood windows. Automobile windshields use laminated glass, with two glass layers cemented together with a thin film of poly(vinylbutyrate) adhesive, which limits the dispersion of glass fragments in an accident. Automobile side windows are tempered to place the two surfaces in tension (see Chapter7), which causes the glass to crumble into polygonal fragments on impact, with less potential for passenger injury. Polymer glasses are used in aircraft windows because of their impact resistance and lightweight characteristics. "Bulletproof" glass is made from a thick polymer such as polycarbonate, often laminated. Modern laboratory glassware is produced from borosilicate glass (Chapter 7) because of its superior thermal shock and chemical resistance.

3. Lenses and Prisms

Lens design, hitherto an art, is now a computerized science. As described above, single lenses are unsuitable for precision applications because of chromatic dispersion and the resultant color fringing of the image. It is also difficult to design and shape a single lens to eliminate optical aberrations, which are focus problems that are caused by the spherical profile of most normal lenses. For these reasons, good quality lenses for cameras, microscopes, binoculars, and telescopes are "compound" systems with several lenses clustered in the optical path or cemented together.

Chromatic abberation can be controlled by the use of two or three lenses in series. A doublet system consists of a convex lens with high refractive index (and thus a high chromatic dispersion) followed by a concave lens with a lower refractive index, as illustrated in Figure 14.6a. So-called apochromatic triplet lens systems provide an even better correction for all colors of the visible spectrum. Extra-low-dispersion glasses are used increasingly in high-performance lens systems.

Optical aberrations can be corrected in one of two ways: (1) by the use of complex, multilens assembles of the types shown in Figure 14.6b; or (2) by the incorporation of aspheric lenses. An aspheric lens has a surface that deviates from normal spherical geometry. By using aspheric elements, such as the one shown in Figure 14.6c, it is possible to bring about a high degree of optical correction with fewer lens elements than would normally be required. This reduces the weight of the overall lens assembly. A reason why this technique has only relatively recently become common is the difficulty of grinding inorganic glasses to aspherical profiles in a high-volume manufacturing environment, although complex profiles can be produced by diamond turning on a computer-controlled lathe. However, *polymer* glasses can be molded into aspherical lenses using a mold with the correct profile that has been produced by meticulous shaping. Thus, polymeric aspherical elements are now becoming increasingly common in high-performance lens systems.

An additional aspect of lens manufacture is the application of thin coatings of materials such as magnesium fluoride or zirconium oxide to the surfaces to minimize reflections and thereby increase light transmission to as much as 99.9%. Such coatings are applied by vapor deposition techniques to thicknesses that are close to the wavelengths of light.

Lenses in eyeglasses are now almost exclusively produced from polymers such as poly(methyl methacrylate) or polycarbonate because of their lightness of weight and safety considerations. Photochromic lenses, which change opacity with different light levels, are discussed in a later section of this chapter.

Many of the challenges in lens design also apply, to a lesser extent, to prism manufacture. Prisms are used in nearly all laboratory optical equipment. They are found in binoculars, telescopes, and single-lens-reflex cameras. For laboratory equip-

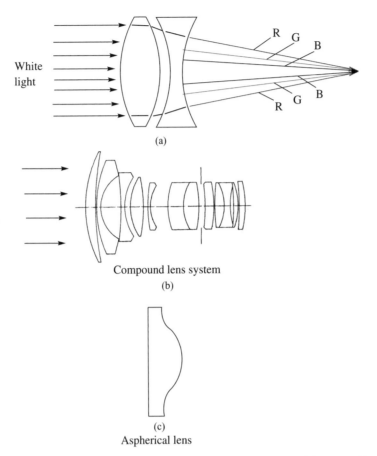

Figure 14.6. *(a) Correction of chromatic dispersion can be accomplished by a combination of a high-refractive-index, high-dispersion convex lens coupled with a low-refractive-index concave lens. (b) Complex lens assemblies used to correct both chromatic and optical aberations such as coma. (c) Aspheric lenses, often made from polymers, are employed increasingly to avoid the distortion problems associated with the use of spherical profile lenses, especially in wide-angle and zoom lenses used in photography. The aspherical profile allows correction for crucial "edge effects" in which those rays passing through the outer regions of a spherical profile lens are difficult to focus when wide apertures are employed (see color insert).*

ment, stability to intense visible and ultraviolet radiation usually mandates the use of inorganic glasses rather than polymers. The optical transmission window is often a critical factor in materials choice. For example, fused-silica prisms are required for the transmission of ultraviolet light. Both inorganic and polymeric glasses are used in binoculars and cameras, with the choice often depending on quality, cost, and the need to minimize weight.

4. Optical Waveguides

The subject of optical waveguides was introduced briefly in Chapter 7. Here, we emphasize that the prime requirements for long-distance optical signal transmission are freedom from color or microcrystallinity that would limit transparency in the near-infrared region, optical and chemical stability over periods of many years, and ease of fabrication into extremely long, thin, fibers. Fused silica is the material used almost exclusively, and is obtained synthetically rather than directly from sand.

The synthesis process is as follows. Silica sand is first converted to $SiCl_4$ by reaction with chlorine, and the $SiCl_4$ is purified by fractional distillation, before conversion back to SiO_2 by reaction with oxygen at high temperatures. The overall chemical sequence is shown in reaction 1.

$$SiO_2 + Cl_2 \longrightarrow SiCl_4 \xrightarrow{\text{distill}} SiCl_4 \xrightarrow{O_2} SiO_2 \qquad (1)$$

Pure   Ultrapure

$$GeCl_4 \xrightarrow{O_2} GeO_2$$

The fiber fabrication process begins with the extrusion of the ultrapure silica into a tube (a preform). Then, pure silica, doped with GeO_2, is vapor deposited on the inside of the fused-silica tube. This "preform" is then passed through a furnace, and the softened material is drawn out into thin fibers. The SiO_2 doped with GeO_2 collapses to a monolith during the hot drawing process to give a solid central core of germanium oxide–doped silica surrounded by the pure silica "cladding." After being drawn, each fiber has a core diameter of approximately 10 µm (Figure 14.7).

Because of the presence of the germanium, the central core has a higher refractive index than does the cladding (germanium contains more electrons than silicon does). Hence, total internal reflection of a light beam at the interface between the core and the cladding keeps the optical signal inside the core. The use of "synthesized" ultrapure germania–silica in this way allows the core to transmit more than 95% of the light over a distance of one kilometer, thus reducing the need for closely spaced amplifier devices for long-distance communication. Undersea cables, in particular, are critically dependent on simple, trouble-free amplifiers to help transmit signals over thousands of miles. The internal "amplifiers" consist of short regions of the fiber doped with erbium or yttrium. A light source outside the fiber causes a laser-like amplification process, which effectively amplifies the signals passing through. An advantage of this system is that, when different wavelengths are being

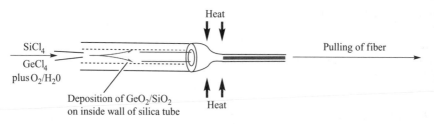

Figure 14.7. Fabrication of optical waveguides. A tube of pure silica is heated and drawn out into a fiber as a mixture of $SiCl_4$ and $GeCl_4$ together with water vapor or oxygen vapor, depositing a mixture of SiO_2 and GeO_2 on the inside of the tube. As fiber extension continues, a core of higher-refractive-index glass is formed surrounded by the lower-refractive-index silica.

transmitted down the same fiber, all are amplified simultaneously by passage through the device.

Polymers have been suggested as replacement materials for optical waveguides, especially where weight is a serious consideration. The lower transparency of most polymers limits their use to short-distance applications, such as in aircraft or automobiles. However, their incorporation into optical switching devices or optical integrated circuits is now well established. For these applications, slab or channel waveguides (Figure 14.8) are preferred configurations, and the ease of fabrication of polymers into thin sheets or as crosslinkable liquid monomers into channels is a significant advantage.

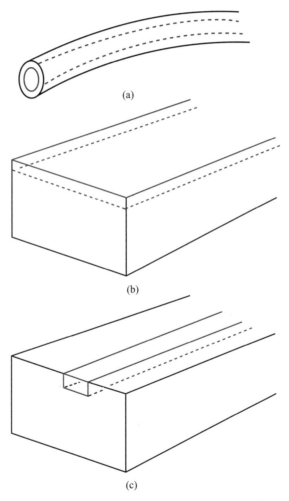

(a)

(b)

(c)

Figure 14.8. (a) Typical fiberoptic waveguide with a core of higher-refractive-index glass. (b) A slab waveguide often fabricated with a polymer slab as the high-refractive-index core. (c) A channel waveguide typically formed by microlithographic etching of a channel in a low-refractive-index material, followed by filling of the channel with the monomer of a high-refractive-index polymer and subsequent polymerization.

5. Waveguide Multiplex/Demultiplex Devices

Broadband digital information transmission over optical waveguides is practical only if more than one "information channel" is transmitted simultaneously over each fiber. This is accomplished by using as many as 50 different wavelengths of infrared light to carry digital signals down each waveguide. A typical situation uses eight or more different frequencies per fiber, with each wavelength separated by 1.6 nm. Each channel then carries 2.5 gigabits (Gb) of digital information per second, for a total of 20 Gb per eight-channel fiber. This is the equivalent to transmitting the words from 5000 average-length novels down each fiber every second.

A *multiplexer* is a device that combines the different wavelength signals from several sources into one fiber. This is accomplished by feeding 8, 16, 32, or more light beams from incoming fibers into a slab waveguide that has only one output.

Separation of the different carrier wavelengths with their individual signals at the receiving end is accomplished by a passive device known as a *wavelength-division demultiplexer*. The acronym DWDM denotes *dense wavelength-division multiplexing*, in which the maximum feasible number of different wavelengths are used. A passive demultiplexer is based on the principle that different optical path lengths within the device are used to separate the different incoming wavelengths and their signals. Figure 14.9 shows the general layout of such a device. A single fiber bearing the incoming signals on, say, eight different wavelengths leads into a slab waveguide. This divides ("lances") the beam into eight different, curved channels that require the beams to traverse different path lengths. At the output of each channel all the beams spread out to encompass the eight output waveguides, but interference between the beams channels a particular carrier wavelength into a specific output channel. This arrangement separates the wavelengths and directs their signals uniquely into one of the eight receiver circuits.

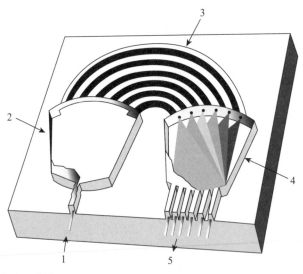

Figure 14.9. Optical multiplexers are passive devices for channeling several different wavelength signals from a number of optical fibers into one optical fiber. The demultiplexer (shown here) uses eight channel waveguides with different path lengths to separate the incoming signals into the different wavelength signal components.

Assembly of a device of this type is accomplished by standard photolithographic methods to fabricate curved channel waveguides, which are then filled with polymers that have the appropriate refractive indices. The system can be expanded so that each of the eight outputs from one demultiplexer can be routed into a separate eight-channel splitter. One advantage of this passive layout is that no electrical connections are required. However, because the refractive indices of the various components vary with temperature, the devices must be maintained at a constant temperature to work effectively.

6. Optical Filters

An *optical filter* is a transparent disk or plate that removes specific wavelengths from a polychromatic light beam such as white light. The principle is shown in Figure 14.10.

Thus, a yellow filter blocks blue light, a cyan (blue-green) filter will remove red light, and a magenta (bluish-red) filter will remove green light. This illustrates the complementarity of the primary colors (blue, green, and red) to the secondary colors (yellow, magenta, and cyan). Filters of a primary color will block transmission of its secondary counterparts. Filters that allow visible light to pass through, but block ultraviolet light, are widely used in photography, as are colored filters that emphasize specific tones in monochrome and color photography. Laboratory optics experiments often employ specialized filters, for example in fluorescence microscopy or Raman spectroscopy. Optical microscopy also makes use of filters to render colored objects more visible. Flat-panel television and computer screens employ blue, green, and red filters in front of different pixels, so that any color can be generated by combinations of the three primary colors.

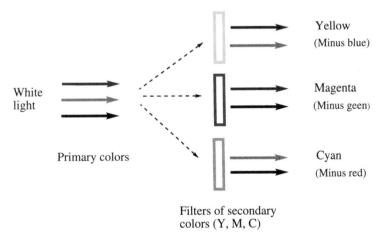

Figure 14.10. Absorption filters remove certain wavelengths of light from polychromatic light. Contrast this with a nonlinear optical material that changes one wavelength to another. Thus, filters of the primary colors, red, green, and blue, remove all wavelegths except those colors. Filters of the secondary colors, yellow, magenta, and cyan, allow two of the primary colors to be transmitted. Filters of the primary colors that cover individual pixels are used in flat-panel liquid crystal displays (see later) and in the sensors in digital cameras (see color insert).

Filters may be made from inorganic glasses or from polymers; the former have the advantage of color stability and abrasion resistance, while the latter provide ease of fabrication and better impact resistance. Some optical filters are hybrid structures with a polymeric filter material sandwiched between two glass sheets. High-quality filters need to have absolutely parallel surfaces because they are usually employed in conjunction with high-quality lens or prism systems. All-glass filters are colored by the addition of color-generating elements (usually transition elements) to the molten glass. Polymer filters use organic dyes to achieve the same effect. The choice of a chromophore is a challenge for filter design because sharp spectral cutoff characteristics are often desired, but can only rarely be obtained.

It is important to distinguish between an optical filter, which *removes* specific wavelengths from the incident light, and materials that *change* the wavelength of an incident beam. These last substances are considered in Section C, below.

7. Optical Polarizing Filters

Normal light, such as direct sunlight, is unpolarized. This means that the electric vector plane-wave structure of light is reproduced all around the 360° beam axis (see Figure 14.11).

A polarizing filter screens out all plane waves except those that lie in one orientation. A polarizer is like a nanoscale grating or venetian blind—a series of narrow slits that allow only one mode of polarization to pass through.

Polarizing filters have many uses:

1. They reduce the intensity of transmitted light, typically by about 25–30%, which is why they are used in sunglasses. They also reduce glare which is often polarized.

2. Polarizers are employed in imaging processes to eliminate reflections. Reflected light from metals, vegetation, glass, or water is often polarized. Hence, if a polarizing filter is rotated in front of a camera lens to the point that it is at right angles to the reflected plane of polarization, the reflection will disappear from the image. This intensifies colors and allows objects that are underwater or behind glass to become clearly visible.

3. The juxtaposition of two polarizing filters serves as an optical switch or attenuator. The first polarizer allows only one plane of polarization to be transmitted. If the second polarizer is at right angles to the first, no light will pass through (Figure 14.11b). Only if one of the polarizers is rotated so that the two polarization axes are parallel will light be able to penetrate. In practice, because each polarizer is not 100% effective, rotation of one component of a polarizer pair can be used as a mechanically simple way to vary the *intensity* of a light beam.

This phenomenon is widely used in microscopy to highlight birefringent crystalline or liquid crystalline materials, many of which rotate the plane of a polarized light beam. Light polarized by a first filter beneath the sample passes through the material of interest and is then "analyzed" by passage through a second, crossed polarizer. Normally light would not penetrate through the second polarizer; however, if the

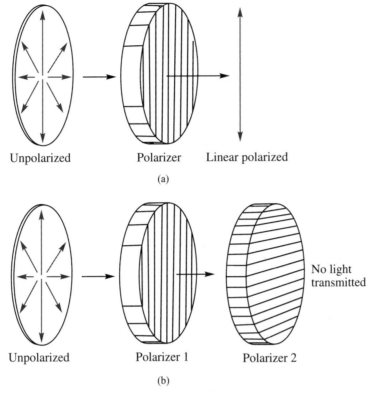

Unpolarized   Polarizer   Linear polarized
(a)

Unpolarized   Polarizer 1   Polarizer 2
(b)

Figure 14.11. *(a) A polarizer is a sheet of material with linear molecules or nanocrystals oriented in one direction. These serve as a filter to remove all components of unpolarized (i.e., normal) light except those light waves that lie in the plane of the polarizer "grating." (b) Two crossed polarizers prevent the passage of light except when they are separated by a material that can rotate the plane of polarization. Such chiral materials include many organic small molecules, spiral helix polymers, and a number of inorganic crystals. Chiral liquid crystalline organic molecules are used for this purpose in flat-panel screens in television and computer screens.*

sample rotates the plane of polarization, it will stand out as a brilliant image against a dark background.

The ability of a liquid crystalline material to rotate the plane of polarized light is employed in flat-panel displays, where the orientation of liquid crystalline molecules either blocks light or allows light to pass through a polarizing screen. A similar effect is employed to reveal areas of stress in transparent structural glass or polymers.

How are polarizing filters made? Typically they are composed of long, rigid molecules or needle-like nanocrystals, such as long organic molecules, inorganic halide, or iodine crystals, embedded in a polymer film. The molecules or crystals are oriented parallel to each other by stretching the film or by softening it and applying a strong electric field. This generates the "slits" needed to isolate one plane of polarization. The effectiveness of the alignment process determines the degree of orientation and hence the sharpness of optical cutoff as the polarizer is rotated.

"Circular " polarizing filters are now recommended for use with the optics of many modern cameras to avoid problems with exposure and automatic focusing

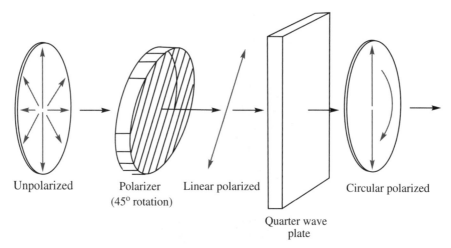

Figure 14.12. *Circular polarizing filters are formed from a linear polarizer backed by a quarter-wave plate fabricated from a birefringent material such as quartz that is engineered to a precise thickness.*

features. A circular polarizer consists of a linear polarizer backed with a "quarter-wave plate" derived from a birefringent material. The principle is shown in Figure 14.12.

Light waves are composed of electric and magnetic vectors that are offset $90°$ (a quarter of a wavelength) to each other. They can be visualized as moving through space in the form of oscillating right- and left-handed helices. Plane (linear) polarization occurs when the left- and right-handed electric vectors are combined in a way that cancels out their magnetic vectors. Thus, the "circular" nature of the light is lost on passage through a linear polarizer.

The birefringent quarter-plate component is a material such as crystalline quartz or calcite that has two different indices of refraction. Thus, plane-polarized light entering the plate will be divided into two components, one traveling along the lower-RI path and the second following the (slower) higher-RI route. The plate material is fabricated to a carefully adjusted thickness that causes the image from the higher-RI path to be retarded by one-quarter of a wavelength relative to the low-RI component when rotated $45°$ to the primary polarization plane. This reconverts the plane-polarized beam to circular polarization (Figure 14.12). Thus, the recircularized light no longer behaves abnormally when passing through the half-silvered mirrors in optical devices.

C. RESPONSIVE OPTICAL MATERIALS

1. General Observations

Both passive and responsive optical materials are widely used in technology. However, responsive materials such as liquid crystals, photochromic materials, and nonlinear optical substances, combined with passive materials, are prominent in some of the most rapidly expanding areas of photonic science and technology.

Responsive materials change their optical properties under the influence of an electric or magnetic field or a change in light intensity. As such, they lie at the heart of devices such as optical telephone switches, optical transistors and microprocessors, liquid crystalline flat-panel screens in computers and television sets, variable-density sunglasses and windows, and materials that double or triple the frequency of laser light.

2. Liquid Crystalline Materials

This subject was introduced in Chapter 5. Ordinary crystals melt cleanly and go from the solid state to an isotropic liquid over a very small temperature range (1 or 2 degrees). Liquid crystalline (LC) substances occupy a semiordered state of matter in a temperature range between the solid and liquid phases. The liquid crystalline temperature range extends from a solid/liquid crystalline transition (T_{lc}) to the temperature at which all order is lost [T_m or T_c (clearing temperature)], where the material finally becomes a clear isotropic liquid. These changes can often be detected by DSC experiments (Chapter 4). Additional transitions may occur within this temperature range as one liquid crystalline type is converted to another.

Liquid crystalline behavior is associated with molecules or parts of molecules that possess disk-like or rod-like shapes. These shapes favor the assembly of semiordered arrangements above the temperature at which the ordered three-dimensional structure of a crystal breaks down. Thus, liquid crystalline molecules have a tendency to stack in the quasiliquid state, thereby generating viscosity, opalescence, and several other useful properties. The liquid crystalline state is known as a *mesophase*, and molecules that favor formation of this state are called *mesogens*. Several examples of mesogens are shown in Figure 5.5 (in Chapter 5).

Mesogens can assemble in a number of different ways. First, it is necessary to distinguish between *lyotropic* systems, which are formed from mesogens in various solvents, and *thermotropic*, systems in which no solvent is present. We will focus on thermotropic systems. Figure 5.6 illustrates a number of different assembly modes that may be detected in the liquid crystalline state. Note that the figure shows possibilities for small molecules, but in most cases these arrangements are also possible for rigid-rod macromolecules or for flexible polymer chains that bear mesogens in the side groups. In this last situation, the polymer chains occupy space in random conformations and the mesogens assemble independently irrespective of the behavior of the main chain. For this to happen, it helps to have each mesogen connected to the main chain through a flexible spacer group to decouple the orientation of the mesogen from the motions of the skeleton.

One of the most useful properties of liquid crystalline materials is that the component mesogens can be *aligned* by exposure to an electric or magnetic field. They may also undergo spontaneous alignment in contact with a surface that has parallel ridges and valleys. Alignment in an electric or magnetic field is facilitated if the mesogen is a polar or polarizable unit. In other words, if the electrons in a mesogen are concentrated toward one end of the molecule either spontaneously or after application of an electric field, the molecule will align itself with the field (Figure 14.13).

This behavior is facilitated if one end of the molecule bears an electron-withdrawing group such as NO_2, F, Cl, or Br, while the other end has an electron-

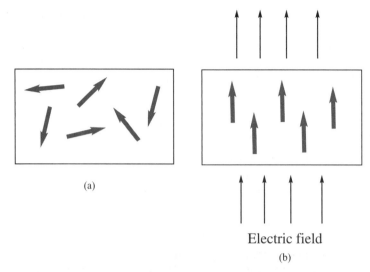

(a)

Electric field

(b)

Figure 14.13. *Illustration of (a) the random orientation of mesogenic molecules in the absence of an electric field and (b) alignment of the same molecules in the presence of an electric field. In a liquid crystalline flat-panel display the electric field is provided by the electrode that underlies each pixel.*

supplying group such as OR or NR_2. The opportunity for induced polarization is amplified if the mesogen has delocalized electrons such as those in aryl rings or linear polyene structures.

The tilted phases formed by chiral mesogens (chiral nematic, chiral smectic C) are permanently polarized and are described as *ferroelectric*. They are important because they respond more quickly to the electrical switching of an electric field than do nematic systems. In liquid crystralline display devices the mesogens are in contact with a surface that has been grooved by abrasion in one direction. This orients the mesogens. Application of an electric field then reorients the mesogens to change the optical transmission characteristics.

Another principle used in liquid crystalline displays is to sandwich the mesogens between two polarizing filters oriented 90° to each other. If the low-energy state of the mesogen is twisted (i.e., chiral), it will twist the plane of polarized light from the first polarizer through 90° so that the light beam passes though the second polarizer. When the electric field is applied, the mesogens become aligned, lose their chiral packing, and the light is unable to pass through the second polarizer. Each pixel on an LC screen is controlled by a separate electrode to switch the light transmission on or off. Different pixels are covered by a red, blue, or green filter so that the overall effect is a full-color image (Figure 14.14).

Side-chain liquid crystalline polymers (Chapter 6, Figure 6.6) have some potential advantages over small-molecule mesogens for uses in specialized display screens or temperature indicators. One aspect is the observation that some small-molecule mesogens may not become liquid crystalline *until* they are linked as side groups to a polymer chain. This is called the "polymer effect" and it reflects the influence of the main chain in anchoring the motions of the mesogen. If the mesogen is linked directly to the main chain, liquid crystallinity may not occur. But the insertion of a flexible spacer group, such as an oligoethyleneoxy unit, gives the mesogen sufficient

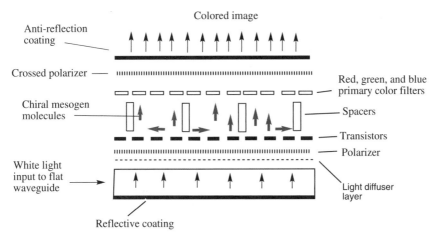

Figure 14.14. *Simplified cross section of a flat-panel liquid crystalline display showing the chiral mesogens that respond to the electric fields generated by the pixel transistors to cut off or allow passage of light through the red, green, or blue filters. The image and its colors depend on the pattern of transistors activated at any given instant (see color insert).*

freedom that liquid crystal behavior becomes possible. The polymer effect may also raise the temperature and increase the temperature range over which the system is liquid crystalline. This can be a considerable advantage for display devices that must operate at elevated temperatures.

3. Photochromic Materials

A *photochromic* material changes color or darkens reversibly when exposed to strong light. The best-known examples are eyeglasses that darken outdoors but bleach when the wearer moves indoors. Other uses include windows that darken or lighten in response to the intensity of daylight, holography devices, and rewritable compact disks (CDs) or digital videodisks (DVDs).

Many different types of materials manifest photochromic behavior. However, they can be divided into two categories—those based on inorganic materials and those that involve the photochromisn of organic small molecules. These latter systems utilize organic polymers as the supporting matrix.

A classical system that is typical of the inorganic category involves borosilicate glass, which contains dispersed minute crystals of a silver halide. These crystals are 50–150 Å in size, and are generated by annealing the glass/silver halide solid solution at specific temperatures. The glass is virtually colorless under indoor illumination but darkens when exposed to ultraviolet light outdoors. These materials were used in the earliest photochromic eyeglasses.

The mechanism of color change has been debated extensively. One of the most widely accepted explanations is that ultraviolet light disrupts the colorless silver halide crystals into opaque silver metal particles and halogen gas. The gas is trapped within the microenvironment of each crystal and recombines with the silver to reform silver halide once the ultraviolet irradiation ceases. Copper halides are apparently used along with the silver halide to optimize the photo response. The color of the darkened glass depends on the size and shape of the silver particles. The color change is infinitely reversible. Hence, inorganic photochromic lenses do not deteriorate with repeated use. However, glass lenses are heavier than their plastic counterparts and are less impact-resistant. Numerous other inorganic photochromic systems are known. For example, phosphine-doped tungsten and molybdenum oxide films show reversible photochromism, as do laminates of nickel dihydroxide and titanium dioxide.

However, most of the photochromic systems used in modern sunglasses are organic compounds dissolved in a polymer glass such as polycarbonate. These are lighter in weight than their glass counterparts, are more impact-resistant, and can be easier to fabricate, but they are more prone to damage by abrasion. Of the many organic molecules that undergo reversible photochromism, the most widely used are based on the spiropyran/merocyanine transformation shown in reaction 2.

(2)

Ultraviolet light induces cleavage of a bond in the spiropyran with simultaneous formation of a double bond that is conjugated with the rest of the molecule. This changes the spectral absorption from the ultraviolet region to the visible. At low light levels the process is reversed and the color bleaches. The color of the darkened form depends on the substituents linked to the aryl ring of the spiropyran. For example, use of a naphthyl unit rather than a phenyl ring in the compound moves the absorption spectrum of the merocyanine form to longer wavelengths in the visible. Attempts are made to balance the spectra of two or more photochromic molecules to yield a broad-spectrum gray or brown color.

A problem with sunglasses based on a molecular rearrangement is the relatively slow response to both darkening and bleaching. This is due to the very high local viscosity within a glassy polymer matrix, which slows the rearrangement and its reversion. Basically, each photochromic molecule occupies a restricted cage within the solid polymer, and a rearrangement that increases the molecular volume or changes the shape of the chromophore will slow the rearrangement process. Attempts have been made to overcome this problem by linkage of the photochromic molecules to flexible oligomers that are dispersed throughout the glass matrix as particles with sizes less than the wavelength of light. This size restriction is needed to prevent scattering of the light. Another idea is to trap the photochromic molecules inside micelles that are dispersed throughout the polymer. An alternative approach is to laminate a flexible (low-T_g) polymer that contains the photochromic molecules between two layers or a rigid glass such as a polycarbonate.

Another problem with sunglasses or other devices based on photochromic organic molecules is that the response tends to weaken after a certain number of darkening-lightening cycles. This may be due to oxidation or to side reactions. Useful lifetimes of 2–3 years are quoted in the literature, but many users have detected no falloff in effectiveness after considerably longer periods of time.

The same considerations apply to the photochromic materials used to record digital information on CDs or DVDs. These are fabricated by vapor deposition of the photochromic molecules on to the surface of a preridged polycarbonate disk. The compounds used may be reversibly photochromic [for CD-RW (read–write) or DVD-RW disks] or irreversibly photochromic (for CD-R or DVD-R disks). An important consideration is the absorption wavelength for writing data on the disk (blue laser light) and the wavelength for reading the data (longer-wavelength visible or infrared).

A photochromic system that utilizes a different principle is based on the change in absorption spectrum that occurs when electrons in highly delocalized, fused polyaromatic molecules are promoted to an excited triplet state following absorption of ultraviolet light. Conversion from the colorless ground state to the colored triplet state is instantaneous. The excited electrons then return to the ground state slowly, emitting visible light as they do so (phosphorescence). Because the light intensities needed to actuate the color change are high, and the bleaching step is slow, these materials have more utility as flash protection lenses than as ordinary sunglasses.

4. Nonlinear Optical Materials

a. The Phenomenon. *Nonlinear optics* (NLO) deals with the ways in which the electromagnetic field of light interacts with the electromagnetic field generated by the electrons in a molecule or material. NLO phenomena become most obvious when light is generated by a powerful pulsed laser, and the polarization, frequency, phase, and even the path of the light beam are changed by passage through a solid.

A host of different and useful phenomena are induced by NLO materials. These include a *doubling or tripling of the frequency* of a laser beam. This is called *second-* or *third-harmonic generation* (SHG or THG). A second phenomenon is the *Pockels effect*, in which the application of a direct current electric field to the NLO material changes its refractive index. This means that the direction of a light beam passing through the material can be controlled by the strength of an electric field. Thus, an incoming light beam from an optical fiber may be switched into one of several outgoing waveguides. The effect is also used to switch a beam on or off to convert a digital electronic signal into an optical counterpart (see discussion later). Other NLO effects include *difference frequency generation* (DFG), in which an emerging light beam has a frequency halfway between those of two incoming beams. *Amplification* of an optical signal is also possible by interaction with a second beam. *Optical rectification* is also possible. Another consequence of NLO activity is the *Kerr effect*, in which the refractive index of a material varies with light intensity.

b. Origins of NLO Behavior. A *polarizable material* is one in which the electrons can be displaced through a molecule or the bulk material rather than being confined to individual atoms or bonds. Practically important NLO behavior becomes possible if the electron movement is mainly along one axis rather than another. This can be favored by the native crystal structure that exists for certain inorganic solids, or by processing a polymeric solid in a way that causes individual molecules to point in the same direction. The displacable electrons may interact with and resonate with the oscillating electric component of incoming electromagnetic radiation. The resultant electron oscillations in the solid will be either at the fundamental frequency of the incoming light or at a higher-frequency harmonic (Figure 14.15).

These interactions by an initial photon of incoming radiation will alter the overall electron polarization, and the subsequent quanta of incoming light will encounter an electron distribution slightly different from that experienced by the initial encounter. The individual *molecular* consequences of this are described by the terms in equation 1 (below). The consequences for the bulk material are represented by equation 2, which gives the bulk polarization P, where the χ^i terms are the susceptibility coefficients and P is the permanent dipole moment:

$$p_i = \Sigma \alpha_{ij} E_i + \Sigma \beta_{ijk} E_j E_k + \Sigma \gamma_{ijkl} E_j E_k E_l + \cdots \tag{1}$$

$$P_i = \Sigma \chi^1_{ij} E_i + \Sigma \chi^2_{ijk} E_j E_k + \Sigma \chi^3_{ijkl} E_j E_k E_l + \cdots \tag{2}$$

In these equations P is the polarization, which depends on the applied electric field E (or the oscillating electric field of light). The χ^1 term represents the behavior of a "linear" (i.e., conventional) material as it interacts with a photon of light. It is related to the normal refractive index, and is a function of the number of electrons in the material. The second-order term (χ^2) represents the effect of combining the resonance of two photons within the same optical "reactor" (i.e., a crystalline solid or an assembly of organic molecules). The third-order term (χ^3) represents of the interaction of three photons. Thus, χ^1 refers to light passing through a material without any change of frequency, χ^2 represents a doubling of the frequency and χ^3, a tripling. The intensity (amplitude, χ, of the emergent beam) decreases in the order $\chi^1 > \chi^2 > \chi^3$, which is why strong laser beams are needed to detect the χ^3 and higher-level emissions.

From the viewpoint of molecular or materials design, it should be noted that χ^1 and χ^3 arise from all solids irrespective of their internal symmetry. *However, χ^2 behavior is possible only if the internal structure of the solid is non-centrosymmetric.* Most molecular solids have a centrosymmetric arrangement of atoms or molecules. Thus, relatively few materials give rise to second-harmonic generation χ^2. Some of these are naturally occurring or synthesized inorganic crystals. Others, especially organic, organometallic, or polymeric species, must have the internal structure aligned by a process of "poling" in order to generate an asymmetric arrangement.

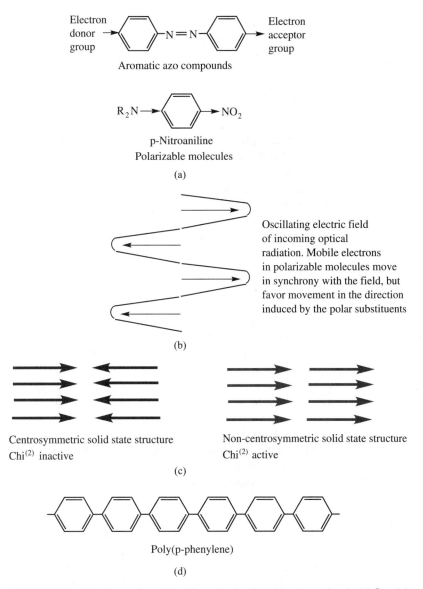

Electron donor group → Aromatic azo compounds ← Electron acceptor group

R_2N → → NO_2

p-Nitroaniline
Polarizable molecules

(a)

Oscillating electric field of incoming optical radiation. Mobile electrons in polarizable molecules move in synchrony with the field, but favor movement in the direction induced by the polar substituents

(b)

Centrosymmetric solid state structure
Chi$^{(2)}$ inactive

Non-centrosymmetric solid state structure
Chi$^{(2)}$ active

(c)

Poly(p-phenylene)

(d)

Figure 14.15. *(a) Two typical organic polarizable molecules that show second-order NLO activity when oriented along the same axis in the solid state, together with a representation of the oscillating electric field vector of incoming light (b) that causes the delocalized electrons to oscillate back and forth within each molecule in synchrony with the field. Beause electron movement in the direction induced by the polar substituents is favored, there will be a delay before the electrons relax. Thus, the next photon will encounter a different electronic state. This will allow two photons to combine in the system and double the frequency of the light that is subsequently emitted. However, χ^2 behavior is found only for solids that have a noncentrosymmetric structure. The molecule shown in (d) also has a delocalized electron system, but is not χ^2-active. However, it is χ^3-active because of the extensive opportunities for electron delocalization.*

Polymeric second harmonic materials offer some advantages compared to inorganic crystals, not the least of which is ease of fabrication into thin films or channel waveguides. These materials tend to be either unsaturated organic small molecules dissolved in a transparent polymer glass, or polymer molecules with polarizable side groups linked to the polymer backbone. Other polymers have unsaturated organic units in the backbone. Examples are shown in Figure 14.15.

c. Inorganic NLO Crystals. A major advantage of inorganic crystals as NLO materials for χ^2 activity is that the required internal asymmetry is already present in the crystal structure. Thus, they do not have to be poled, and the NLO behavior is stable over long periods of time. They are also transparent over a wide spectral range. Typical inorganic χ^2 materials are lithium niobate, $LiNbO_3$, potassium dihydrogen phosphate, KH_2PO_4, and β-barium borate (β-BaB_2O_4). However, limits exist to the size of crystals that can be grown and to their fabrication into the intricate shapes needed for optical switches and other devices.

d. Organic NLO Materials. Although the traditional materials for NLO investigation have been inorganic crystals, considerable emphasis more recently there has been on the use of electron-delocalized polarizable molecules incorporated into solids. Two of the simplest of these are *p*-nitroaniline and aromatic azo compounds (Figure 14.15), which are both polar and polarizable. Crystals of these compounds are centrosymmetric and therefore χ^2-inactive. However, when dissolved in a molten polymer such as polystyrene and poled (see discussion below) to orient the small molecules, and then cooled below the T_g, the material becomes NLO-active.

Polymers with covalently linked NLO side groups have been studied in some detail. χ^3 activity can arise from polymers that have delocalized electrons in the backbone, such as poly-*p*-phenylene or polyenes. Orientation by poling is not needed, but the χ^3 NLO effect is quite weak.

e. Poling. Corona poling is a process in which polar or polarizable organic or organometallic molecules or side groups become oriented along the same axis in a strong electric field. This normally occurs only when the compound is in the molten or quasiliquid state above its glass transition temperature. The experimental setup is shown schematically in Figure 14.16. For polymer films above the T_g, roughly 10 kV are needed to achieve effective orientation. Because the oriented molecules will become randomized again once the electric field is turned off, it is necessary to "freeze" the material to below the T_g before the field is interrupted.

The poling process can be applied to polarizable small molecules dissolved in an inert polymer matrix, such as polystyrene, or to polymers that bear polar or polarizable side groups. As with their liquid crystalline counterparts, orientation is facilitated if flexible spacer groups disconnect the orientation of the active side groups from the thermal motions of the main chain.

A problem with most poled solids is that, even after being cooled below the T_g, some thermal molecular motion is still possible. Thus the NLO activity generated by poling is slowly lost over time. Photochemical crosslinking of the poled side groups before the field is turned off retards this process.

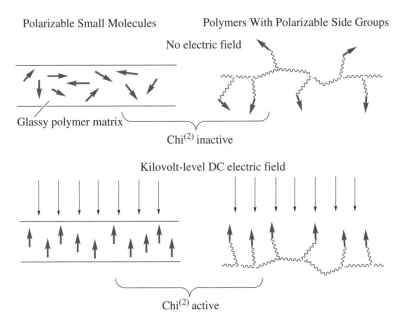

Figure 14.16. *Poling of polar, polarizable small molecules dissolved in an inert polymer matrix above its glass transition temperature allows the NLO molecules to become aligned. Rapid cooling to below T_g then fixes the guest molecules in the oriented position and allows second-order NLO effects to occur. However, sufficient molecular motion exists at room temperature that the guest molecules soon become randomly oriented again, and the NLO properties are lost. If the NLO active units are linked via flexible spacers to a polymer chain, a similar effect can be achieved by poling. The linkage of the NLO unit to a polymer may slow the speed of thermal randomization, especially if the polymer is crosslinked during the poling process. However, special techniques such as layer-by-layer assembly on a surface may be needed to completely stabilize the NLO response.*

f. Orientation by Self-Assembly. Second-order NLO films can also be prepared from polar molecules by self-orientation at a surface. The formation of a noncentrosymmetric coating is accomplished through layer-by-layer assembly of the molecules on a surface by Langmuir–Blodgett techniques. Either crosslinking of the molecules or the packing stabilization on a surface then prevents or retards randomization.

g. Devices

(1) Second- and Higher-Order Harmonic Generation. Certain limits exist to the light frequencies that can be generated by lasers. Thus, it is often necessary to double the frequency of laser light in order to access colors between those of existing lasers, or to produce short-wave (blue) radiation that is difficult to generate directly. The following widely used SHG materials are employed to halve the classical laser wavelengths given in parentheses: LiO_3 (806, 1319 nm), $KNbO_3$ (860, 980, 1319 nm), KH_2PO_4 (1064, 1319 nm), β-BaB_2O_4 (1064, 1319 nm), and $LiNbO_3$ (1319 nm). For example, β-barium borate is used for second-, fourth-, and fifth-harmonic generation from neodymium YAG lasers; second-, third-, and fourth-harmonic generation, from titanium sapphire lasers; and frequency doubling, tripling, and mixing of dye lasers. Inorganic crystals such as these are employed in preference to organic polymeric

species because they are generally more stable to intense laser light and because the size and shape restrictions associated with crystals do not impede mounting at the front of a laser.

Note that, although direct third-harmonic generation to triple the frequency of light is possible, in practice, frequency tripling is usually accomplished by combining frequency-doubled light with the fundamental emission frequency of the laser to generate a summed frequency equivalent to a χ^3 ouput.

(2) Electrooptical Switching. The ability of an electric field to change the refractive index of an NLO material allows the direction of an optical beam to be controlled as it passes through the material. This is illustrated in Figure 14.17a,b.

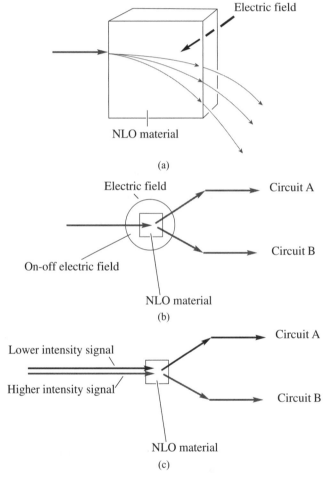

Figure 14.17. *Three devices that depend on nonlinear optical behavior. (a) Because the refractive index of an NLO material depends on the strength of an applied electric field, the path of a light beam can be altered by changes to the strength of the field. This allows an incoming signal from one optical fiber to be routed to any of several output circuits. (b) An optical signal may be switched from one circuit to another by turning the electric field on or off. (c) For NLO materials in which the refractive index depends on the intensity of the light beam, switching from one circuit to another is possible by varying the intensity of the optical signal.*

One use of this phenomenon is the rapid switching of a light beam into different outgoing waveguides by variations to the potential of the applied electric field. The utility of this for optical telephone switching is obvious.

(3) Intensity-Triggered Switch. Certain NLO materials change their refractive indices depending on the intensity of a light beam. This provides a means for separating signals of different intensity into separate circuits (Figure 14.17c)

(4) Mach–Zehnder Interferometer. This is a device that converts an electrical on/off (ie. digital) signal into an optical counterpart. One important application is in the conversion of electrical signals from a telephone or computer into optical signals for long-distance transmission. The principle behind the device is shown in Figures 14.18.

An optical beam from a waveguide or laser enters the device from the left, before being split into two beams, A and B. Beam A continues unaffected through the top circuit. Beam B is directed through the NLO material where an electric potential is applied to slow the beam by raising the refractive index of the material. The potential is chosen such that the phase of the light beam is slowed by $180°$. This means that, at the right-hand junction point, the phases of the two beams exactly cancel each other out and no light gets through. Thus, by switching the electric potential off or on, the light beam that leaves the device will now bear the identical digital information. Thus, an electrical digital signal is converted to an optical digital signal. The speed of the response is limited only by the speed of the electrical on/off signal. The same device can also be used to control the intensity of a light beam by retarding the active beam by less than $180°$ to generate a sinusoidal output.

A Mach–Zehnder interferometer is typically constructed from channel waveguides etched lithographically into a resist material and filled with a non-NLO polymer (passive circuit) and an NLO material (switching circuit). Metal electrodes are deposited on the NLO section by sputtering or vapor deposition.

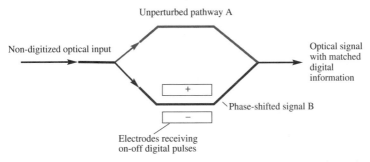

Figure 14.18. *Schematic diagram of a Mach–Zehnder interferometer. The incoming light beam is split into two different circuits, one of which passes through an electric field, which increases the refractive index of the NLO material and slows the beam. The field strength is such that the phase of this beam is retarded by exactly 180° relative to the unperturbed beam, thus extinguishing the signal. An on off digital signal in the electric field is thus transferred to the emerging optical beam.*

5. Electrochromic Devices

An *electrochromic* material is one in which the color or transparency can be changed reversibly by the application of an electric field. In a sense, liquid crystalline devices are electrochromic, but the term is normally used for materials that apply other principles.

The main uses for electrochromism are in windows to control the amount of light entering a room via an electric rheostat, and for rear-view mirrors and sunroofs in automobiles. Electrochromic eyeglasses are an alternative to photochromic glasses since the user can control the transparency at will. Electrochromic pixels in flat-panel display devices have been considered for development, but they are not yet as far advanced as liquid crystalline devices.

The simplest electrochromic system consists of two transparent conductive electrodes [typically indium–tin oxide (ITO) deposited on glass] separated by an organic liquid that contains a dissolved organic compound such as viologen (reaction 3). Application of an electric current causes reduction of the colorless viologen dication to a violet-colored radical cation at the cathode. However, greater color intensity can be achieved if a second electrochromic molecule, such as an anion derived from a thiazine or a phenylenediamine, is also present that forms a colored species when *oxidized* at the anode.

$$CH_3-\overset{+}{N}\underset{X^-}{\diagdown}\!\!\!\!\!\!\diagup\!\!\!\!\!\!\overset{+}{N}\underset{X^-}{\diagup}\!\!-CH_3 \quad \xrightarrow{\;e^-\;} \quad CH_3-N\diagdown\!\!\!\!\!\!\diagup\!\!\!\!\!\!\overset{+}{N}\underset{X^-}{\diagup}\!\!-CH_3 \quad (3)$$

Colorless   Blue-violet

When the electric current is interrupted, the two compounds mix and neutralize each other by electron transfer, and the color bleaches. A system like this is ideal for automobile rear-view mirrors, but not for architectural windows, where the electric current would be needed continuously to maintain the darkening effect.

Most of the other electrochromic designs use solid inorganic electrochromic materials such as tungsten oxide (WO_3) or nickel hydroxide [$Ni(OH)_2$] as reducible or oxidizable layers. These may be deposited on transparent or reflective electrodes by vapor deposition or by sol–gel processes. Anodic or cathodic systems are accessible, but the strongest optical density changes are obtained when both electrodes bear electrochromic layers. The central layer between the photochromic layers is a material that is a good ionic conductor but a poor electronic conductor such as a lithium salt dissolved in a liquid or in a polymer (see Chapter 12). A device configuration is shown in Figure 14.19 (see also reaction 4).

$$WO_3 + x(Li^+)\,e^- \quad \longrightarrow \quad Li_x W^{VI}_{1-x} W^V_x O_3 \qquad (4)$$

Colorless   Blue

When electrons are injected into the primary electrochromic electrode, the process shown in reaction 4 is believed to occur. Reversal of the electric potential causes bleaching of this layer. The secondary electrode exists for two purposes: (1) it serves to balance the charge during the darkening part of the cycle and to accept electrons during the bleaching reaction; and (2) if it is a material that becomes colored when

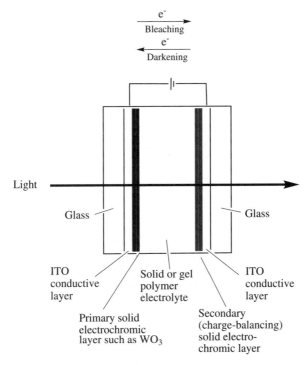

Figure 14.19. *A transmission electrochromic device suitable for an electrochromic window.*

oxidized but colorless when reduced, it will intensify the color change brought about at the primary electrode. Iridium hydroxide behaves in this way. The advantage of this type of system is that electric current is needed only for switching back and forth between colored and transparent states, but not for maintaining either state. Hence, this arrangement is ideal for windows in buildings because it offers serious electrical energy savings.

Alternative configurations include devices with an electrochromic film on one electrode only and an electrolyte such as lithium iodide that serves as an electron sink. This system bleaches when the current is turned off.

In addition to metal oxides and hydroxides, organic polymeric semiconductor materials such as polythiophene, polypyrrole, and polyaniline also undergo electrochromic changes when reduced.

6. Thermochromism

Many chemical compounds and materials change color reversibly when heated. Thermotropoic liquid crystals behave in this way, and this is the basis of creams that, when spread on the skin, give different colors to indicate different temperature regions.

Thermochromism differs from the other responsive systems described in this chapter in the sense that heat rather than light or electricity generates a reversible color change. Nearly all colored materials are thermochromic to some degree, but some striking examples are mercury(II) iodide (red to yellow at 126 °C) and zinc oxide and lead oxide (reversibly white to yellow on heating).

Examples of thermochromism are also to be found among conductive polymers, mercury complexes, cobalt coordination complexes, vanadium oxide, and certain organic dyes. Many different mechanisms give rise to thermochromism. They range from the thermal promotion of electrons to new energy levels in metal coordination complexes to reversible molecular rearrangements in molecules such as spiropyrans (see Section C.3). Uses suggested for thermochromism range from solid-state thermometers, heat shields, and heat responsive windows to heat-activated digital recording disks.

7. Light-Emitting Materials

Light-emitting solids are key components of many different optical devices. They include light-emitting diodes, electroluminescent materials, and other electron-to-photon conversion devices. These materials are discussed in the context of semiconductors, and details will be found in Chapter 10.

D. CHALLENGES FOR THE FUTURE

It will be clear from the information presented in this chapter that the field of optical materials is extremely broad and is at a rapidly evolving stage. Although most of the developments until relatively recently were in the field of totally inorganic materials, future developments will involve polymers either on their own or in combination with inorganic materials. These changes are currently being driven by the needs of the telecommunications sector. Of particular interest are new materials with high refractive indices and low chromatic dispersion. Also needed are materials other than silica that are transparent to a wide spectrum of radiation from the 157-nm region in the ultraviolet, through the visible, and into the near infrared. The bringing together of semiconductor technology and optical materials science is a field of growing importance. Finally, a need exists for a wider range of materials that show nonlinear optical, electrochromic, or thermochromic properties. These are all areas where collaborations between chemists, physicists, and engineers are essential for meaningful progress to be made.

E. FINAL COMMENTS

This chapter contains introductory information on many different aspects of chemistry, materials, optics, and device design. It illustrates the vital importance of optical materials in modern society and the possibilities for future developments in this area. It also illustrates the ways in which modern science and engineering draw on a diverse portfolio of different molecules, materials, and principles to achieve a practical result. In this chapter we have neglected much of the physics that underlies these topics, and for further information on these aspects, the reader is referred to more specialized literature.

F. SUGGESTIONS FOR FURTHER READING

1. Hecht, E., *Optics* 4th ed., Pearson Education, Inc., 2002.

2. Williams, C. S.; Becklund, O. A., "*Optics—A short Course for Engineers and Scientists*", Wiley-Interscience, Hoboken, NY, 1972.

3. Broer et al, Optical Waveguide Materials, MRS Vol 144 EMS Library TA401.3.M39.

4. Shibaev V. P. et al, "*Liquid Crystalline and Mesomorphic Polymers*", Springer, 1994.

5. Shelby, R. A.; Smilth, D. R.; Schultz, S., "Negative refractive index", *Science* **292**:77 (2001).

6. Van Krevelen, D. W., "*Properties of Polymers; Their Correlation with Chemical Structure; Their Numerical Estimation and Prediction from Additive Group Contrbutions*", 3rd Ed.; Elsevier, New York, 1990. (Tables of data for the calculation of refractive indices).

7. Hornak, L. A., Ed., "*Polymers for Lightwave and Integrated Optics: Technology and Applications*", Marcel Dekker, New York, 1992.

8. Crano, J. C., "*Organic Photochromic and Thermochromic Compounds*", Plenum Press: New York, 1999.

9. Bosshard, C.; Sutter, K.; Prêtre, P.; Hulliger, J.; Flörsheimer, M.; Kaatz, P.; Günter, P., "*Organic Nonlinear Optical Materials*", Gordon and Sreach, Switzerland, 1995.

10. Joannopoulos, J. D.; Villeneuve, P. R.; Fan, S., "Photonic crystals: putting a new twist on light", *Nature* **386**:143–150 (1997).

11. Moerner, W. E.; Grunnet-Jepsen, A.; Thompson, C. L., "Responsive optical materials". *Ann. Rev. Mater. Sci.* **27**:585 (1997).

12. Monk, P. M. S.; Mortimer, R. J., *Rosseinsky, "Electrochromism—Fundamentals and Applications*", VCH Publishers, Weinheim, 1995.

13. Mortimer, R. J., "Organic Electronchromic Materials", *Electrochimica Acta* **44**:2971–2981 (1999).

14. Mortimer, R. J., Electrochromic Materials, *Chem. Soc. Reviews* **26**:147–156 (1997).

15. Avendano, E.; berggren, L.; Niklasson, G A.; Grenqvist, C G.; Azens, A., "Electrochromic Materials and Devices: Brief Survey and New Data on Optical Absorption in Tungsten Oxide and Nickel Oxide Films", *Thin Solid Films* **496**:30–36 (2006).

G. STUDY QUESTIONS

1. Design a new material to be used as an optical waveguide. The target material should be at least as clear as silica, but easier to fabricate, and also preferably lighter in weight. Justify the logic that underlies your design.

2. As a class project, design a future generation digital single-lens reflex camera specifying the materials you would use for the color senor, the mirror, lenses, lens barrel, prism, and body. Explain why the materials chosen are preferred. Describe the weaknesses of each material for the above components listed above. How would your new design move the technology forward?

3. Explain how a very thin coating of one material on a glass reduces the amount of light reflected. Why is this important?

4. Review the differences between optical filters and nonlinear optical materials. Under what circumstances would you use one rather than the other to change the color of a light beam?

5. The lens of the eye changes its focal length by altering its shape. Design a nonliving lens that does the same thing. Pay special attention to the materials you would use and how the shape change would be effected. For what applications might it be useful? What problems mighty you anticipate in the functioning of such as device?

6. Under what different circumstances would it be appropriate to switch an optical beam from one circuit to another using (a) a simple slab of an NLO material with two electrodes; (b) a Mach–Zehnder interferometer, or (c) a switch that uses the refractive index change of a material as the temperature is changed? What are the advantages and disadvantages of each method?

7. Why does a color change occur when a spirogyran is exposed to ultraviolet light? Speculate on why the photochromic response of eyeglasses that contain this compound might weaken with time.

8. Discuss in some detail how a linear optical polarizer is fabricated and how it works. Compare it with the structure and function of a circular polarizer.

9. The production of silica-based optical waveguides (fibers) requires meticulous attention to materials quality and processing techniques. Describe these requirements and discuss the reasons for them.

10. The sensors in digital cameras and flat-panel TV and computer screens use red, green, and blue filters to achieve full color effects, but digital printers and the producers of book and magazine pages use yellow, magenta, cyan (and sometimes black) inks to produce a color picture on paper. Explain this anomaly.

11. Why are inorganic glasses used for long-distance optical waveguides, but polymer glasses are not? However, polymer glasses are increasingly used in optical switches and lens systems. Explain the reasons for these differences.

12. You have been assigned the task of designing a new material for optical switching applications for use in optical integrated micro- or nanocircuits. The material must retain its properties for at least 20 years. Review the design principles involved and explain specifically why you have targeted the material of you choice.

15

Surface Science of Materials

A. PERSPECTIVE

Surface science, like the field of membranes discussed in Chapter 13, is a subject that is important to many different areas of materials science. These areas include biomedical materials, semiconductors, adhesion, optics, and membrane science, to name only a few. Virtually all the applications of polymers and metals depend in some way on their surface behavior. Aspects of surface science have been introduced in other chapters, and the objective here is to consider the subject in general terms. Thus, in this short chapter we bring together a number of these different aspects and summarize some of the underlying principles.

In order to understand the behavior of surfaces, it is necessary to know something about the chemistry and physics of a given interface. Materials that have surfaces that are similar in chemical composition to the bulk material, may generate surface properties that can be understood from a knowledge of the internal structure. For example, the crystal structure of a metal may be the key to its behavior as a heterogeneous catalyst. However, for most materials the surface composition differs from that of the bulk material, and the chemical nature of the interface is a key to the properties. For example, a metal oxide or hydroxide layer at the surface of a metal will control the sensitivity to further oxidation, the extent of continued corrosion, the adhesion of surface coatings, and the wear of bearings and other sliding surfaces. For ceramics, the presence of surface hydroxyl groups will allow particles to coalesce when heated and will be responsible for the separation behavior of compounds passing through a chromatography column. A wider variety of surface groups are present on polymers, and these are discussed later in this chapter.

Introduction to Materials Chemistry, by Harry R. Allcock
Copyright © 2008 by John Wiley & Sons, Inc.

Surface properties that underlie many applications include the hydrophobicity or hydrophilicity of the material, as well as its acidity, basicity, and coordination power, all of which are connected with the presence of interfacial groups such as hydroxyl, amino, oxide, nitride, alkyl or aryl, acidic units such as COOH or SO_3H, and the character of the interface. By "character" we mean the thickness of a surface layer, the degree to which it adheres to the underlying material, and its porosity and permanence.

The important surface characteristics of a material are only rarely the consequence of a single layer of atoms at the interface with the environment. More often they depend on structural features that lie buried below the surface layer. Thus, it is important to know the depth of the interfacial layer, whether expressed in the range of Angstrom units, nanometers, or micrometers. This information may be obtained by optical microscopy of sections, by XPS, or by TIR-IR methods.

B. SUMMARY OF CHARACTERIZATION METHODS

Surface science is heavily dependent on analytical characterization methods. Some of these techniques are described in Chapter 4. For example, the surface of a metal can be explored by techniques that range from X-ray photoelectron spectroscopy (XPS) to scanning electron microscopy (SEM) (see Chapter 4). Polymer surfaces may be investigated by the same methods plus total internal reflectance infrared spectroscopy (TIR-IR). Surfaces of ceramics, which often bear Si—OH or Al—OH groups, and the surfaces of polymers can be examined by reaction chemistry, XPS, or SEM methods. Surface planarity, which is an important aspect of both integrated circuit semiconductor technology and nanotechnology, can be addressed by the use of ellipsometry. Of particular importance are optical microscopy, scanning electron microscopy (SCM), X-ray photoelectron spectroscopy (XPS), scanning tunneling microscopy (STM) and atomic force microscopy (AFM), and contact angle investigations. The application of theory to the study of surfaces has also been productive. Optical microscopy is applicable to surfaces of all materials. XPS, SEM, STM, and AFM methods are particularly useful for metal and semiconductor surfaces, powders, and micro- and nanofibers of ceramics and polymers.

C. SURFACES OF METALS

1. Important Aspects

What are the surface characteristics that play such a significant role in metal science and technology? The sensitivity of a metal surface to oxidation, corrosion, or adhesion to coatings plays a dominant role in applications such as automotive, marine, or architectural engineering. The resistance of a metal surface to corrosion is a key property in electrical switches, dentistry, and catalysis. Questions that need to be answered for metals are the chemical nature of any surface layer (the presence of oxides, nitrides, hydroxides, etc.) and the depth and stability of the interfacial layer. Is the surface layer electrically conductive or insulating? Does it coordinate to molecules of the types found in surface coatings? If a second metal is deposited on the surface of the first by vapor deposition, sputtering, electrodeposition, or dipping the first metal into a bath of the molten second metal, is the crystal structure of the second metal influenced by the crystal structure at the surface of the first? What is the strength of adhesion between the two?

2. Etching of Metal Surfaces

The process of removing metal from a surface by immersion of the metal in an acid bath has for many years been the basis of both artistic techniques (intaglio printing) and metal surface cleaning. In addition, valuable information about the surface of a metal can be obtained by etching. By this we mean treatment with a reagent, such as an acid, that will dissolve the outer regions of the surface. Ideally, the attack on the surface will be selective so that grain boundaries or other morphological features will be revealed. Metal surfaces can also be studied by reactive-ion etching, in which ions derived from a plasma remove both surface oxides and the metal beneath.

An especially valuable technique for the etching of metal surfaces is oxidative electrochemical etching, which involves using a metal as an electrode in an electrochemical cell. As described in Chapter 13, aluminum can be oxidatively etched in acidic solution to form aluminum oxide nanotubes. The reaction begins at the surface and extends deep into the metal. A similar process using titanium metal foil generates a layer of semiconductive columnar titanium dioxide nanotubes at the surface.

3. Heterogeneous Catalysis by Metals

Many chemical processes are catalyzed by metal surfaces. Classical example reactions include hydrogenations of alkenes to give alkanes or reduction of nitro componds to amines. Metals that are commonly used in heterogeneous catalysis include finely divided nickel, iron, and platinum. Because of the economic impact of heterogeneous catalysis by metals, considerable research effort is devoted to understanding the mechanisms of these processes. A common feature of nearly all these catalyses is the coordination of reactant molecules to the metal surface, followed by reactions that are mediated by the surface features. Thus, the surface of a metal particle often contains ledges and steps that represent the emergence of crystal structure features at the interface (Figure 15.1). These steps may present unsatisfied valences or unused coordination sites to the external environment such that unsaturated molecules or hydrogen will be held on the surface in orientations that are favorable for reactions. Grain boundaries and dislocations that emerge at the surface may also participate in these processes. Clearly, the smaller the metal particle, the greater is the ratio of surface to volume and hence the more effective is the catalysis for a given weight of metal. Catalyst "poisoning" can occur when impurity molecules that have a strong affinity for the coordination sites are preferentially adsorbed and block the catalytic reaction.

4. Metal Surfaces and Vapor Deposition, Sputtering, or Solution Reactions

The process of chemical vapor deposition has been mentioned in earlier chapters, particularly in connection with semiconductor fabrication and optical materials. In

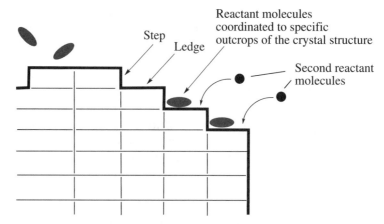

Figure 15.1. *Heterogeneous catalysis at metal surfaces often involves coordination of reactant molecules to specific sets of metal atoms that emerge at the surface and present unused coordination sites to reactant molecules.*

this process volatile organometallic compounds are decomposed at elevated temperatures in vacuum, and the metal atoms are allowed to impinge on a surface. This process allows very thin films of a metal to be deposited to form wires, coatings, mirrors, and a wide range of other features. "Sputtering" is a related process in which a metal is vaporized at high temperatures and allowed to deposit on a surface. The preparation of gold coatings for SCM imaging uses this method. An advantage of the sputtering method is that contamination from nonmetallic species in the vacuum chamber is avoided. Finally, some thin metallic surfaces can be deposited on other materials by chemical processing. The best known of these is the deposition of silver mirrors by the exposure of glass surfaces to an aqueous solution of a silver salt that contains a reducing agent such as formaldehyde or glucose. Finally, use of a metal electrode in an electrochemical cell allows a second metal to be electrodeposited on the first. A well-known example is chromium plating on steel or nickel, and of silver plating on nickel.

A characteristic of deposited metal surfaces is their high reflectance, which results from the uniform thickness of the metal. However, an important additional characteristic is that most of these methods can be used to coat three-dimensional objects such as complex surfaces being prepared for SEM analysis.

D. CERAMIC SURFACES

1. Oxide Ceramic Surfaces

The surface chemistry of oxide ceramics is usually dominated by the presence of Si—OH or Al—OH groups. Both of these are hydrophilic, acidic units, which can be neutralized by exposure to aqueous base. The presence of surface hydroxyl groups is responsible for the ability of powders of many oxide ceramics to undergo sintering by condensation crosslinking when they are compressed at moderate temperatures. As discussed in Chapter 7, the same surface groups determine the behavior of clay particles in media with different pH values. The surface structure is also important for the functioning of alumina powder to separate compounds dissolved in organic media in chromatography procedures, because of the differential interaction of organic molecules with acidic surfaces.

2. Chemical Modification of Glass Surfaces

The same acidic hydroxyl groups that dominate the surface chemistry of many ceramic particles also play a role in the behavior of glass surfaces. For example, many chemical reactions carried out in glass equipment suffer from side reactions that arise from hydrolysis-type processes at the glass–solvent interface. This can be avoided by prior treatment of the glass with molecules such as trimethylchlorosilane to give hydrophobic, pH-neutral surfaces. This process is shown in reaction 1. A dramatic illustration of the effectiveness of this procedure is to compare the pH of water in contact with fiberglass. Material with untreated surfaces will slowly dissolve in water to give a basic solution. The same material surface treated with trimethylchlorosilane will not. Related chemical reactions are employed to covalently bind bioorganic molecules to glass, processes that are important in micro-separations technology, and for the adhesion of surface coatings to glass.

$$\left|-\text{OH} \xrightarrow[-\text{HCl}]{\text{Me}_3\text{SiCl}} \right|-\text{SiMe}_3 \qquad (1)$$

3. Nonoxide Ceramic Fiber Surfaces

As discussed in Chapter 7, several of the nonoxide ceramics produced by pyrolysis of inorganic polymers are sensitive to reaction with atmospheric oxygen to form silicon oxide materials. This problem is particularly acute for thin fibers or whiskers of silicon carbide or silicon nitride. In the case of silicon carbide fibers, this surface oxidation process is utilized to protect the interior of the fibers from further reaction with the atmosphere. For this type of surface reaction to be useful, the oxide coating must be strongly adherent to the core of the fiber.

4. Ceramic Decomposition by Pollutants

A final example of the importance of surface chemistry in ceramic technology is the ability of a masonry surface to withstand the effects of atmospheric pollutants. Architecturally important buildings throughout the world are crumbling because of attack by acidic atmospheric pollutants on the stone. The two most sensitive masonry materials are sandstone and limestone. Sandstone consists of silica particles cemented together by calcium carbonate. Limestone is mainly calcium carbonate. Sulfurous and sulfuric acids are major atmospheric pollutants, and these dissolve the calcium carbonate and initiate the materials breakdown. Attempts to retard these processes include coating the ceramic with transparent hydrophobic polymers.

E. POLYMER SURFACES

1. General Aspects of Polymer Surfaces

The ability of a polymer surface to resist the ingress of water, organic solvents, bacteria, gas molecules, and the overall hydrophobicity or hydrophilicity are crucial properties in automotive or marine engineering, biomedicine, electrical engineering, membrane design, and microelectronics.

The surface of a polymer depends on the chemical composition of the component macromolecules, the physical state of the surface (glass, elastomer, etc.), and the method used for processing and fabrication. For example, different polymer surfaces may contain a wide variety of chemical species that range from hydrophobic alkyl, aryl, and fluorinated organic units, to hydrophilic hydroxyl, amino, or carboxylate units, and they may also bear the products of atmospheric oxidation such as carbonyl, hydroperoxide, or epoxy units. However, the property-determining groups that actually populate a surface often depend on the method of fabrication, as discussed below. An understanding of the factors involved in polymer surface behavior is an essential requirement for designing new polymers in a rational way.

2. Unusual Characteristics of Polymer Surfaces

Because polymer molecules are in constant motion, especially above the glass transition temperature, the surface of a polymer may change over time in response to the environment with which it is in contact. For example, dry air is considered to be hydrophobic, and hydrophobic components of the polymer molecules may migrate to a polymer/air surface. The converse happens if the surface is in contact with water. It is this feature of "surface turnover" that distinguishes polymers from other materials.

One of the main methods for studying surface turnover is to measure the contact angle of water droplets or a hydrophobic liquid on the surface. The technique is described in Chapter 4. Thus, a material that is experiencing surface turnover may undergo a change in contact angle while a water drop is sitting on the surface. Note that there is a rough correlation between the contact angles and the hydrophilicity or hydrophobicity of the polymer side groups or skeletal units. Thus, chemical intuition is a good starting point for understanding polymer surface behavior.

Note also that the process of surface turnover has both positive and negative aspects for technology. A positive aspect is that it provides a method to fine-tune a surface by bringing it into contact with a hydrophilic or hydrophobic medium. The drawback is that a process such as adhesion may be compromised by a change in surface character after a bond has been formed between two different materials.

3. Chemical Modification of Polymer Surfaces

The surface of a polymer clearly depends on the molecular stucture of the component macromolecules, and the most obvious way to change the surface character is to use a different polymer derived from a different monomer. However, it often happens that an ideal target material needs to have two different sets of properties—one for the materials interior and another for the surface. For example,

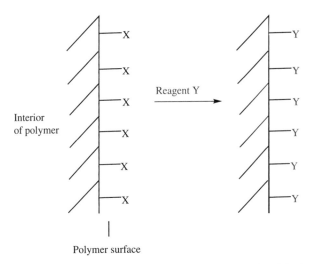

Figure 15.2. *Illustration of the process of surface modification by wet chemistry in which a reagent in solution displaces polymer side groups at the interface, thus changing the surface character.*

a biomedical material may require a hydrophobic interior to discourage water absorption or bacterial colonization, but a hydrophilic surface for interaction with specific proteins or cells. It is usually unwise to depend on surface turnover to generate these composite properties. Moreover, the application of surface coatings is a relatively short-term solution. However, chemical reactions that alter the covalently bound groups at the surface, *while leaving the interior unchanged*, may provide the require combination of properties. The process is illustrated schematically in Figure 15.2.

Most polymer surfaces can be modified by wet chemistry, but some polymers are more useful in this respect than others. Classical polymers such as polyolefins with methyl side groups can be surface oxidized to introduce carboxylic acid or hydroxyl surface units. Polyethylene is surface-fluorinated to produce a hydrocarbon-resistant fluorinated surface. Polyacrylates may be surface hydrolyzed to generate carboxylic acid groups. However, polymers such as poly(tetrafluoroethylene) are more difficult to modify in this way, perhaps because of the unreactivity of carbon–fluorine bonds and the high hydrophobicity. Polyphosphazenes have proved to be particularly suitable for surface reaction chemistry mainly because the side groups can be displaced by other units fairly easily. For example, the surface trifluoroethoxy side groups in the polymer $[NP(OCH_2CF_3)_2]_n$ (Chapter 6) can be replaced by treatment with other alkoxide units or hydroxl groups. Related polymers with methylphenoxy side units are readily surface-sulfonated to introduce acidic functionality. Similar processes yield functional sites for the linkage of biologically active molecules.

4. Polymer Surfaces in Offset Lithography Printing

Chapter 10 drew a comparison between the process of photolithography used in semiconductor fabrication and the technology of rotary offset printing. Modern printing plates depend almost exclusively on polymer surface chemistry, and on the ability of a photochemical reaction to produce an image consisting of hydrophilic and hydrophobic domains distributed over a surface. The printing process then involves the spreading of a hydrophobic ink over the rotating plate and its temporary retention by the hydrophobic regions of the plate and its repulsion by the hydrophilic area, prior to transfer to the paper. Similar processes can be used in soft contact printing for the printing of micro- and nanoscale features on a surface. An example would be the mass-production printing of organic semiconductor circuits using a rotary press.

5. Plasma Modification of Polymer Surfaces

A *plasma* is a hot gas of reactive ions and radicals generated by a radiofrequency electric field. When a plasma impinges on a polymer surface, the reactive species in it can bring about a wide range of changes. These effects can include melting, erosion, or the reaction of surface units with species in the plasma gas. Two types of plasmas are commonly used for surface modification processes—vacuum plasmas and atmospheric pressure "environmental" plasmas. Of these, the latter are often the most useful for ease of polymer surface alteration, although the surface chemistry may be more complex. Figure 15.3 shows the essential features of an atmospheric plasma device.

An example from the author's laboratory involves surface modification of films of $[NP(OCH_2CF_3)_2]_n$. Different gases in an environmental plasma have been used to modify the surface of this polymer. Changes in the surface chemistry are evident from the following water contact angles: untreated surface, 103°; oxygen plasma, 5°; nitrogen plasma, 40°; methane plasma, 68°; and tetrafluoromethane/hydrogen plasma, 151°. These values indicate the respective introduction of carboxylic acid or hydroxyl groups, amino, graphitic, or perfluoro units into the surface structure of a hydrophobic polymer film.

6. Influence of Polymer Fabrication Method

Spin casting, melt casting, spray coating, dipping, and fiber or nanofiber fabrication all generate different surface properties. Often these changes are not due to chemical modifications but reflect differences of crystallinity, surface roughness, or in some

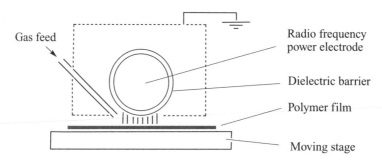

Figure 15.3. *Cross section of an environmental plasma generator. The plasma generated from the inflowing gas reacts with the surface of a polymer to replace surface groups by new species.*

cases, the result of surface turnover to bring specific polymer component units to the interface. This is a common result with block copolymers.

7. Micro- and Nanofiber Surfaces

Because fibers are often employed in composite materials and in biomedicine, as well as in textiles, the surface properties are important. The properties of thin fibers, with their large surface area:volume ratios, are often more noticeably affected by surface characteristics than are slabs, blocks, or even films of polymers. The following examples illustrate this fact:

1. If a fiber has been produced by extrusion of a polymer solution into a nonsolvent, the outer regions of the fiber will coagulate first and will then become roughened by contraction as the core solidifies at a later stage. Thus, fibers produced by this method will have a rough surface that may or may not be smoothed out by subsequent heating and stretching. A rough surface increases the adhesion between a fiber and a matrix material in a composite. The same polymer produced by a melt-extrusion technique may have a smooth surface, which might be more appropriate for use in a textile.

2. Fibers are particularly responsive to modification by surface chemistry or environmental plasma techniques. Surface reactions take place rapidly, and must be controlled carefully to prevent the reaction from penetrating into the core.

3. Nanofibers produced from hydrophobic polymers are more hydrophobic than are films or microfibers. They can have water contact angles that are between $100°$ and $160°$, which is the region of superhydrophobicity. The image shown in Figure 15.4 illustrates the shape of a water droplet on a nanofiber mat of the polymer $[NP(OCH_2CF_3)_2]_n$. The water contact angle of a droplet on this mat is $50–60°$ higher than on a flat fim of the same polymer. Superhydrophobic

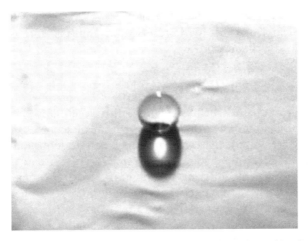

Figure 15.4. *A water droplet on a superhydrophobic mat of nanofibers formed by electrostatic spinning of a fluorinated polyphosphazene. The contact angle is ~160°, which demonstrates that the actual contact between the droplet and the polymer is minimal.*

nanofiber mats are so water-repellent that droplets of water almost bounce off the surface. The superhydrophobicity is ascribed to the air trapped between the fibers. The water droplet has only minimal contact with the polymer surface and is essentially suspended in air.

8. Role of Block Copolymers at Surfaces

It follows from the previous comments that block copolymers with blocks that have different hydrophobic or hydrophilic character will not only phase-separate in the bulk of the material but may also reorganize in the surface region to expose one block or the other to the environment. This process can be assisted by annealing the polymer at different temperatures in contact with a particular environment. Another alternative is that the cylindrical or layer morphology (Chapter 9) may emerge on specific faces of the solid, in which case an overall amphiphilic character may be generated.

F. SURFACES OF SEMICONDUCTORS

1. Oxidation of Semiconductor Silicon

The oxidation of silicon wafers to produce a surface layer of adherent silicon dioxide is a key step in integrated electronic circuit manufacture (Chapter 10). However, this same process changes a hydrophobic surface to one that is hydrophilic. Thus, photolithographic imaging, followed by selective etching away of the silica, enables the formation of hydrophobic/hydrophilic images on a wafer surface. In principle, this technique could be used for nanoscale contact printing.

It should be noted that organic semiconductors generate a new set of problems that involve surface chemistry. This is because most organic semiconductors are sensitive to oxidation, a process that converts them to insulators. Thus, a technological challenge is to identify electrically insulating surface coatings that prevent oxygen transmission. Considerable research has been devoted to the subject of oxygen-impermeable polymer films for food preservation, and this work is clearly relevant to the semiconductor challenge.

2. High-Surface-Area Semiconductors

Transparent semiconductors such as titanium dioxide are key components of dye-based solar cells (Chapter 14). An essential feature of such cells is the need for a very high surface area on which the dye molecules can be deposited and that interact efficiently with the electrolyte solution. The high surface area can be achieved in one of three ways: (1) by sintering of titanium dioxide nanoparticles to form a porous electrode; (2) by vapor deposition techniques that favor the growth of finger-like, micro- or nanoscale columnar TiO_2 features from a surface; or (3) by formation of a nanoscale columnar pattern on the surface by oxidative electrochemical etching of metallic titanium foil (Figure 15.5). These different surface features provide different modes of interaction between the surface and, for example, dye molecules or a liquid electrolyte.

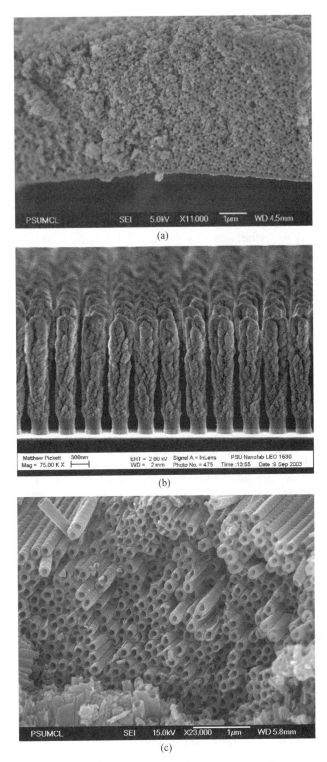

Figure 15.5. *Three different nanoscale semiconductor surface structures. (a) Titanium dioxide nanospheres fabricated into a semiconductor electrode surface sponge for solar cell applications (courtesy of T. Mallouk, The Pennsylvania State University). (b) Fingers of vapor deposited titanium dioxide (courtesy of M. Horn, The Pennsylvania State University). (c) Nanotubes of titanium dioxide produced by electrochemical etching of titanium foil (courtesy of C. Grimes, The Pennsylvania State University).*

G. ASSEMBLY OF MOLECULES ON SURFACES

1. Langmuir–Blodgett Techniques

A primary method for the preparation of self-assembled monolayers is to allow amphiphilic small molecules to organize themselves on the surface of a liquid. This procedure makes use of a Langmuir–Blodgett trough, which is a miniature water-filled swimming-pool-type device—typically about 30×60 cm in area. One or two drops of a solution of the small-molecule compound in a volatile organic solvent are placed on the water surface. The solution then spreads across the surface as the solvent evaporates, with the hydrophilic end of each molecule oriented toward the water and the hydrophobic end pointed upward toward the air. The surface film can be compressed by movement of a barrier as the surface pressure is measured by a device known as a *Wilhelmy balance*.

At a certain point the pressure rises as the molecules form an increasingly close packed monolayer. Example molecules that respond to this technique are linear aliphatic carboxylic acids. The monolayer formed in this way may then be transferred to a solid surface by dipping a flat material (e.g., a glass microscope slide) into the liquid–air interface or by withdrawing the slide from beneath the liquid into the air. If the solid interface is hydrophilic, the hydrophilic ends of the molecules will attach to that surface. If the surface is hydrophobic, then the hydrophobic end will adhere to the surface and the hydrophilic units will point outward. Successive dipping and withdrawal of the slide will generate multiple layers in which the hydrophobic and hydrophilic components are adjacent to each other to form bilayer structures (see Figure 15.6).

Amphiphilic *high polymer* molecules can be transferred to a surface in a similar way by Langmuir–Blodgett techniques, but the structures formed are more complex than those generated by small molecules. Polymer molecules may form coiled or entangled structures floating on the water surface, often with the hydrophilic components statistically oriented toward the aqueous surface. However, only when the polymer chains are short is it possible to construct ordered features that can be transferred to a solid surface. Nevertheless, the ability to produce thin layers of high polymer in this way is extremely useful for multiple layer-by-layer assembly of nanometer-thick layers of different polymers that may be useful in controlled drug delivery, nonlinear optical, or other devices.

2. Self-Assembly on Gold Surfaces

A surface reaction that has grown in importance with the development of nanoscience and nanotechnology (Chapter 17) involves the coordination of thiols to a gold surface. Gold is one of the few metals that does not react with the atmosphere, and thus it presents an unadulterated layer of metal atoms to the environment. Sulfur is an element that has a strong ability to coordinate to gold atoms. Thus, exposure of a gold surface to a solution of a thiol in an organic solvent results in a coherent coverage of the surface by a *self-assembled monolayer* of the thiol molecules oriented as shown in Figure 15.7.

The alkyl groups are typically oriented at an angle of $30°$ to the gold surface. The orientation of the aliphatic chains resembles the packing of linear hydrocarbon

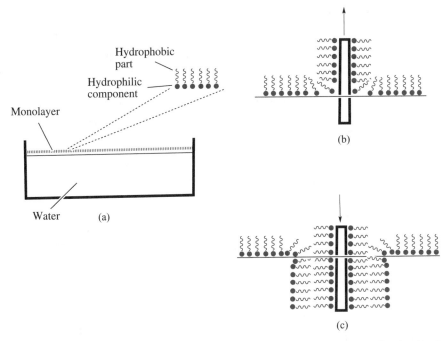

Figure 15.6. *(a) A Langmuir–Blodgett trough with a monolayer of amphiphilic molecules forming a monolayer on the water surface. (b) Withdrawal of a glass slide with a hydrophilic surface causes the monolayer to be deposited on the slide with the hydrophilic component of each molecule at the glass surface and the hydrophobic units exposed to the air. (b) Subsequent immersion of the monolayer slide back into the trough generates a bilayer coating with the hydrophilic side of the layer now exposed to the outside. Multiple bilayer coatings can be formed by repeating this process.*

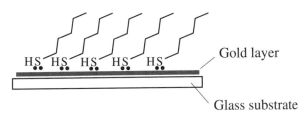

Figure 15.7. *Schematic illustration of alkylthiol molecules forming a self-assembled monolayer on a gold surface. The packing pattern is driven by strong coordination between the sulfur atoms and the gold, and by the side-by-side packing of the trans-planar alkyl groups.*

molecules in the solid state—that is, the methylene chains assume a trans-planar conformation and pack together efficiently to minimize the overall energy. Clearly, the thickness of the thiol layer is related to the length of the alkyl chains. However, if the surface coverage is less than the maximum required for close packing, the chains may point in a variety of different directions.

Gold–thiol surfaces can be used as a patterning process in which a resist protects parts of the surface from coordination. Terminal functional groups on the thiol may then provide a mechanism for linkage to bioactive molecules.

3. Layer-by-Layer Assembly

A useful method for forming nanoscale alternating layers of two polymers on a surface is the process of layer-by-layer assembly. One technique uses two different polymers, a macromolecule that bears negative charges and another with positive charges. The substrate (e.g., a glass slide or another polymer) is initially coated with the first polymer (say, the cationic macromolecule). This surface is then dipped into a solution of the anionic polymer, which forms a monolayer on top of the first. Immersion of this construct in a solution of the cationic polymer causes deposition of a monolayer of this polymer, and so on. In this way, multiple layers can be assembled. Numerous uses for these multilayer structures have been proposed, including drug release matrices and thin-film electrolytes for batteries and fuel cells.

A second alternative is to build a surface structure by covalently binding successive layers of polymers to build a multilayer sandwich. This method has been used successfully to construct stable nonlinear optical layers from organic molecules.

4. Surface Patterning by AFM

Another method for patterning small molecules on a surface is to actually move them into position using an AFM tip. Nanoscale constructs spelling out names of investigators and laboratories have been produced in this way. This technique has also been employed to line up metal nanospheres to form nanowires, and to assemble nanomachines such as model vehicles with spherical molecules that function as wheels.

H. ADHESION AND SURFACE CHEMISTRY

1. General Characteristics of Adhesion

Adhesion is a materials process that is crucial to virtually all areas of technology. Two examples that illustrate the breadth of this subject are the fabrication of aircraft using adhesives instead of rivets, and the development of adhesives instead of surgical sutures and clips to connect living tissues in surgery. Adhesion has been mentioned throughout this book, and the brief discussion here is intended to summarize a few of the main points.

Adhesion between two surfaces is a complex process that depends on at least two phenomena: (1) covalent or coordination bonding between the two surfaces and (2) a process of physical interlocking of three-dimensional features on the two interfaces. It is not always clear which of these two processes predominates for a given adhesive. However, for simplicity these two mechanisms are considered separately.

2. Chemical Bonding as a Source of Adhesion

As discussed at the beginning of this book, covalent bonds are among the strongest forces that maintain the integrity of molecules. Similarly, covalent bonding between two surfaces can be the basis of the strongest type of adhesion. For this to be possible, both surfaces need to have functional groups at the interface. Polymer surfaces with interfacial hydroxyl or amino groups are appropriate for covalent bonding to adhesives with epoxy functional units. Crosslinking of the adhesive is usually required to generate the strongest binding between the two components. Metal surfaces, if they are oxide-free, can form strong coordination bond to polymers that bear amino, thio, or other units with lone pairs of electrons. Ceramic surfaces with interfacial hydroxyl units will form strong bonds to polymers with Si—Cl or acid chloride groups.

3. Physical Bonding of Surfaces

Porous materials or highly sculptured surfaces can form a tight physical bond to an adhesive. It seems reasonable to suppose that van der Waals forces play a major role in this type of adhesion. The success of this method depends on the ability of a liquid adhesive to penetrate into the surface pores or to conform to the features of a rough surface. Adhesives that consist of polymers dissolved in a volatile organic solvent function in this way, as do aqueous phase bonding agents used for wood and similar porous materials. In such cases the strength of the bond depends almost entirely on the strength of the bulk adhesive and its ability to maximize ita area of contact with a surface. If the solvent in an adhesive penetrates into the surface of a polymer, it may be possible for the polymer molecules in the adhesive to become entangled with those at the polymer surface. This would enhance the physical adhesion between the two components. One common form of adhesive uses an elastomer to conform to the features on a surface and to form a physical bond through a large area of contact. Two separate solids, both coated with elastomer, are then brought together. The elastomer molecules interpenetrate, and crosslinking of this polymer then strengthens the union.

I. RELATIONSHIP TO OTHER MATERIALS TOPICS

1. Soft Contact Printing

Soft contact printing is a technique in which a stamp is used to print an image on a surface (Figure 15.8). The normal procedure for fabrication of the stamp is to prepare a negative mold by photolithography on silicon. A polysiloxane film is then cast on the silicon surface and subsequently crosslinked. The silicone is then peeled from the mold to give a raised relief image. The raised sections may then be inked and the inked image transferred to a substrate by light pressure. The difference from standard macroscopic printing is that the raised relief image on the stamp may be miniaturized down to the micro- or even nanoscale. This opens the possibility of low-cost mass production of integrated electronic circuits (hydrophobic conductive polymers), nanoscale etching masks, and other advanced applications. Because the traditional material used for a soft contact stamp is silicone rubbber—poly(dimethylsiloxane)—the features on the raised relief image are highly hydrophobic. Hence, the most obvious application is for the transfer of hydrophobic "inks" or metal nanoparticles to a receiver substrate. The use of this technique to print features consisting of solutions of organic polymeric semiconductors is well

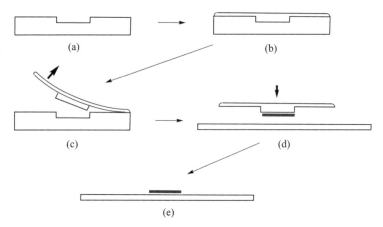

Figure 15.8. *Soft contact printing or soft lithography requires the formation of a stamp of an elastomeric polymer produced on a patterned, raised relief surface. Microlithography techniques of the type used in semiconductor fabrication permit the preparation of molds with micro- or even nanoscale features that can be transferred to the stamp and used for repetitive printing of electronic circuits.*

advanced. So, too, is the use of soft lithography to pattern surfaces for biological screening devices. In principle, almost any elastomer could be used for soft contact printing, with the material chosen to ensure hydrophobic or hydrophilic surface characteristics. Surface chemistry, as described above, can be employed to fine-tune the image on a surface for special applications such as the printing of enzymes. In one case, microscale ridges of a hydrogel were soft-contact-printed and seeded with nerve cells for use in detection devices for toxic compounds.

2. Biomedical Materials Surfaces

As mentioned in Chapter 16, the surface of a biomaterial has a profound effect on blood and soft tissue compatibility. Specifically, the response of blood to a materials surface determines the propensity of that material to induce thrombus (clot) formation. Blood is an unstable, complex fluid that contains proteins, platelets, red blood cells, and dissolved salts and lipids. It is stable as long as it is moving through the circulatory system through blood vessels lined with endothelial cells.

Soft tissues and bone are less sensitive to a biomaterials surface, yet even here the materials surface matters. Surfaces that are rough or are composed of materials that erode to give toxic products cause tissue death and often become isolated by the body within a capsule, while giant cells attempt to attack the foreign object. The result will be tissue irritation and rejection of the implanted device. The least damaging interactions appear to be connected with highly hydrophobic surfaces such as those provided by fluorocarbon polymers or silicone rubber or with inert metal surfaces such as titanium or gold.

A major concern in biomaterials research is the resistance of an implanted material to colonization by bacteria or fungi. Hydrophilic materials seem to be particularly susceptible to this problem, although the deposition of proteins, polysaccharides, or general cell debris on a hydrophobic surface may also be a prelude to microbial colonization. To this end, polymers are under development that have antimicrobial molecules grafted to the surface, although the efficacy of this technique has yet to be proved. In many biomaterials applications, the best prospects are for solids or elastomers that have one type of material constituting the interior and another, quite different, material constituting the surface. This can be accomplished either by surface reaction techniques or by the use of block copolymers that phase-separate to put one block at the surface.

J. SUGGESTIONS FOR FURTHER READING

1. Nuzzo, R. G.; Allara, D. L., "Adsorption of bifunctional organic disulfides on gold surfaces," *J. Am. Chem. Soc.* **105**:4481–4483 (1983).

2. Frey, S.; Shaporenko, A.; Zharnikov, M.; Harder, P.; Allara D. L., "Self-assembled monolayers of nitrile-functionalized alkanethiols on gold and silver substrates," *J. Phys Chem. B* **107**:7716–7725 (2003).

3. Bain, C. D.; Troughton, E. B.; Tao, Y.-T.; Evall, J.; Whitesides, G. M.; Nuzzo, R. G., "Formation of monolayer films by the spontaneous assembly of organic thiols from solution onto gold," *J. Am. Chem. Soc.* **111**:321–335 (1989).

4. Kumar, A.; Biebuyck, H. A.; Whitesides, G. M., "Patterning self-assembled monolayers: Applications in materials science," *Langmuir*, **10**:1498–1511 (1994).

5. Xia, Y. N.; Whitesides, G. M., "Soft lithography," *Angew. Chem. Int Ed.* **37**:550–575 (1998).

6. Nishimura, S.; Abrams, N.; Lewis, B. A.; Halaoui, L. I.; Mallouk, T. E.; Benkstein, K. D.; Van de Largematt, J.; Frank, A. J., "Standing wave enhancement of red absorbance and photocurrent in dye-sensitized titanium dioxide photoelectrodes coupled to photonic crystals," *J. Am. Chem. Soc.* **125**:6306–6310 (2003).

7. Allcock, H. R.; Steely, L. B.; Kim, S. H.; Kim, J. H.; Kang, B.-K., "Plasma surface functionalization of poly[bis(trifluoroethoxy)phosphazene] films and nanofibers," *Langmuir* **25**(15):8103–8107 (2008).

8. Barrett, E. W.; Phelps, M. V. B.; Silva, R. J.; Gaumond, R. P.; Allcock, H. R., "Patterning poly(organophosphazenes) for selective cell adhesion applications," *Biomacromolecules* **6**:1689–1698 (2005).

9. Lee, K. W.; McCarthy, T. J., "Synthesis of a polymer surface containing covalently attached triethoxysilane functionality: adhesion to glass," *Macromolecules* **21**:3353–3356 (1988).

10. Petty, M. C., *Langmuir-Blodgett Films: An Introduction*, Cambridge University Press, 1996.

11. Ullk, J. M.; Mera, A. E.; Fox, R. B.; Wynne, K. J., "Hydrosilation-cured polydimethylsiloxane networks: Intrinsic contact angles via dynamic contact angle analysis," *Macromolecules* **36**:3689–3694 (2003).

12. Allcock, H. R.; Fitzpatrick, R. J., "Functionalization of the surface of poly[bis(trifluoroethoxy)phosphazene] by reactions with alkoxide nucleophiles," *Chem. Mater.* **3**:450–454 (1991).

13. Marks, T. J.; Ratner, M. A., "Design, synthesis, and properties of molecule-based assemblies with large second-order optical nonlinearities," *Angew. Chem. Int Ed.* **34**:155–202 (1995).

14. Jiang, X.; Ortiz, C.; Hammond, P., "Exploring the rules for selective deposition: Interactions of model polyamines on acid and oligoethylene oxide surfaces," *Langmuir* **18**:1131–1143 (2002).

15. Zheng, H.; Rubner, M.; Hammond, P., "Particle assembly on patterned plus/minus polyelectrolyte surfaces via polymer on polymer stamping," *Langmuir* **18**:4505–4510 (2002).

16. Kim, S. H.; "Fabrication of superhydrophobic surfaces," *J. Adhesion Sci. Technol.* **22**:235–250 (2008).

K. STUDY QUESTIONS

1. The catalysis of organic reactions at the surface of metals such as platinum or palladium is one of the most important processes in chemical research and technology. Why are these metals used in preference to other catalysts? What is the role of finely divided platinum in fuel cells? Are there other metals that can be used under different conditions? Discuss these issues in a short essay.

2. Describe the characterization techniques which would be needed for the study of the surfaces of metals, glass, organic polymers, and finely divided ceramic spheres.

3. How might a surface produced by melting a material followed by cooling differ from one prepared by chemical vapor deposition of the same substance? Explain the reasons for any differences.

4. A water droplet placed on a certain polymer surface forms beads with a contact angle of 100°. Some time later, when the same experiment is carried out, the contact angle is only 50°. What has happened to change the surface and what might you do to reverse the change? How could you modify this material to retain the high contact angle over a long period of time?

5. Most oxide ceramics, including glass, have a hydrophilic surface. Why is this the case? What would your recommend in order to make the surface permanently hydrophobic? Suggest reasons for wanting to make glass hydrophobic.

6. After consulting the literature, explain why "layer-by-layer" deposition of polymers gives surfaces that differ from those produced simply by solution spin casting of films.

7. Why is gold so widely used for the formation of self-assembled monolayers of organic molecules? Which types of organic molecules are especially suited for this process, and why? Suggest possible applications.

8. Design an adhesive that will strongly bond wood to aluminum. Explain the fundamental principles on which your design depends. Suggest actual materials for the adhesive and explain the reasons for your choice. What factors might cause this adhesive to fail?

9. Explain the difference between soft contact printing and traditional letter-press printing. What are the advantages and disadvantages of each process. Could you accomplish similar results using an ink-jet printer to print electronic circuits from soluble organic semiconductors?

10. Damaged regions in most mammalian tissues are hydrophilic and are difficult to repair with conventional adhesives. Yet one polymer—poly(cyanoacrylate) (also known as "wonder glue" or "elephant glue")—will cause tissues to stick together so strongly that the surrounding tissue will tear before the adhesive bond breaks. Speculate about how this adhesive may work. Is there a reason why this adhesives is not used instead of surgical sutures in medical practice?

16

Biomedical Materials

A. SPECIAL REQUIREMENTS FOR BIOMEDICAL MATERIALS

Materials have been used in medical practice since prehistoric times. Today, materials science plays a crucial role in many aspects of biomedical science and technology. Modern biomedical materials fall into two main categories: (1) those selected for their resistance to breakdown in a biological environment and (2) those chosen for their ability to decompose to harmless products when implanted on the body. The field of biostable materials can be further subdivided into solids for hard tissue replacement and those for soft tissue applications. The latter group includes elastomers for cardiovascular applications, hydrogels for optical prostheses, and membranes for controlled drug delivery. Bioerodible (hydrolytically unstable) materials are of interest as tissue regeneration matrices, surgical sutures and screws, and controlled drug delivery devices. These options are summarized in Figure 16.1. Within all these categories are traditional materials that were optimized for some other use, and a few newer substances that have been designed specifically for biomedical applications.

What are the special property requirements for medical materials? The first is the need to avoid any detrimental response by the human body to a "foreign" substance—whether the material is implanted in the body or used in an external device ("extracorporeal"). Satisfying this requirement is not a trivial challenge, as will be seen in later sections of this chapter. The detrimental response may be as basic as rejection of an object because of its shape, roughness, or lack of flexibility, or the response may be based on more complex factors such as toxicity, protein–surface interactions, or colonization by microorganisms. In addition to this primary requirement for biological compatibility, a material must be optimized for a different set of physical properties for each individual medical application. For example, the

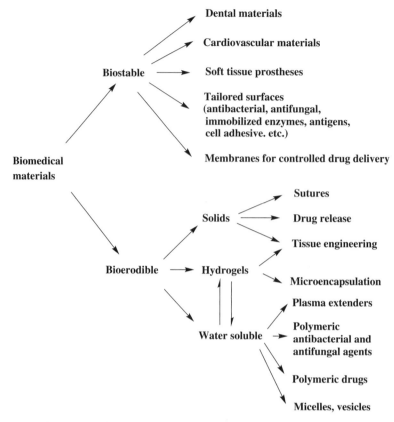

Figure 16.1. Chart of different biomedical properties and applications.

property combinations needed for a bone reinforcement material are quite different from those required for artificial blood vessels or hemodialysis membranes.

However, the biggest challenge in this field involves the decision between using long-existing materials that are available in large quantities versus newer materials specifically designed to possess a significantly improved set of properties. At present, long-existing materials borrowed from other technologies far outnumber those that have been designed and synthesized primarily for biomedical uses.

The issue of materials stability in the aqueous environment of the human body is an important factor for many applications. Clearly, cardiovascular, optical, and renal devices should ideally last without decomposition for the life of the patient. However, a growing number of medical materials are used *because* they decompose slowly in water to give harmless products. Absorbable surgical sutures, temporary frameworks for bone, skin, or liver regeneration ("tissue engineering"), and materials for the controlled release of drug molecules or vaccines are examples.

The following section reviews examples of some widely used biomaterials. These are separated into those that are biostable and those that are bioerodible. Later sections deal with the ways in which different materials are used in specific medical applications, followed by some brief observations on fabrication and testing, and comments on some unsolved problems in this field.

Many materials types and their properties that are important in biomedicine are discussed in other chapters. For example, membranes for biomedical applications are described in Chapter 13. Polymers are discussed in Chapter 6. Surfaces that are optimized for biomedical applications are covered in Chapter 15. Orthopedic composites are mentioned in Chapter 9. The purpose of this chapter is to provide an overview of the biomedical materials field, and the reader is encouraged to integrate the material covered elsewhere in this book with the information provided here.

B. TRADITIONAL BIOMEDICAL MATERIALS

Metals, ceramics, and polymers are used as biomedical materials. Of these, polymers are being increasingly employed in medical devices, although metals and ceramics have important specialized applications.

1. Metals

The requirement of corrosion resistance limits the number of metals that can be employed in internal medicine. The commonest examples of metals in use are gold, silver, and silver–mercury alloys ("silver amalgam"), titanium, and stainless steel.

Gold is widely used in tooth restoration because of its inertness, malleability, and ease of fabrication by melting. Modern tooth restorations often use gold bonded to a ceramic outer coating to mimic the appearance of natural tooth enamel. The high cost of gold is, of course, a limiting factor.

Silver is less expensive than gold, but more prone to tarnishing and corrosion. However, when alloyed with mercury, it is one of the most widely used materials for dental fillings. When first mixed, the alloy is moldable at room temperature as it is forced into the tooth cavity, where it sets to a tough antibacterial solid that can last as an effective restoration for 50 or more years. Fears about the long-term release of mercury into the body seem to be unfounded.

Stainless steel has for many years been the material of choice for use as pins and spikes to reinforce bone fractures. It has high strength and is corrosion-resistant. Stainless steel is also heavy, and this is a definite drawback. Some concern exists about the possible long-term release of transition metals such as chromium or nickel into the body, but the beneficial properties are considered to outweigh any potential disadvantage.

Titanium is a metal that overcomes nearly all of these drawbacks. It is lightweight and corrosion-resistant. However, as discussed in Chapter 8, titanium is more difficult to fabricate than stainless steel and is more expensive. Both stainless steel and titanium are employed as the ball joint in artificial hips.

One often-overlooked advantage to the use of metals in medicine is that they are opaque to X rays. Thus, their location and condition can be monitored by conventional X-ray methods. Metals are also easy to sterilize by heat, chemicals, or gamma radiation. A problem with the use of metals in orthopedics is that they tend to trigger warning devices at airports and some public buildings.

2. Ceramics

The main uses for oxide ceramics in medicine are in replacements or fillings for teeth, in bone repair materials, or in artificial hip joints.

Calcium hydroxyapatite (HAP) [$Ca_{10}(PO_4)_6(OH)_2$] is the mineral component of both bone and tooth enamel (Chapter 7). Indeed, various composites of HAP with polymers have wide use in dental restoration and bone repair cements. HAP-based ceramics can, under special conditions, be bioerodible, but the rate of erosion is very slow. Inorganic composites such as zinc oxide and phosphoric acid are used as tooth fillings. Porcelain tooth restoration and denture ceramics are well known. Glass–ionomer tooth restoration cements are based on silicate–acrylic acid composites.

Nonoxide ceramics such as glassy carbon or silicon carbide have been proposed for both dental and orthopaedic applications, but their use is still limited. Glassy carbon is used in artificial heart valves (see Section C.1).

3. Polymers

The flexibility, elasticity, and ease of fabrication of most polymers make them an obvious choice for many biomedical applications. In physical properties they can be chosen to mimic the characteristics of many different types of living tissues, from soft tissues to bone. In addition, their surfaces can be modified to enhance biocompatibility. Moreover, a few polymers degrade hydrolytically to relatively harmless products.

However, some disadvantages also exist. For example, two of the main lines of attack by the body on foreign objects are hydrolysis and oxidation reactions, and many otherwise biostable polymers are susceptible to oxidative breakdown. Moreover, polymers tend to be more prone to colonization by bacteria or fungi than do metals or ceramics. Another disadvantage of many polymers is that they are too fragile for sterilization by heat. Thus, exposure to γ rays or chemicals is the preferred sterilization method. A major problem for some polymers is that they contain plasticizers and/or monomer or catalyst residues, and leaching of these into the body can give rise to a variety of serious physiological problems. Many of the early evaluations of polymers as biomaterials neglected to take this factor into account.

Nearly all the different classes of polymers have been investigated as biomaterials. The descriptions presented in the following paragraphs are intended for purposes of illustration, and are by no means comprehensive. Seemingly biostable polymers are considered first, followed by bioerodible materials. Their structures are shown in Table 16.1.

a. Biostable Polymers

(1) Polyethylene. This hydrophobic polymer (structure **16.1** in Table 16.1) is used in the socket of artificial hip joints. Its biomedical utility is based on the fact that it is impervious to biological fluids, can be fabricated easily by melt or machining techniques, and appears to be bioinert, at least over the short term. However, surprisingly, given its chemical structure, evidence exists that its molecular weight and strength begin to decrease within months after implantation, possibly by oxidative breakdown.

TABLE 16.1. Examples of Polymers Used in Biomedical Materials Research and Clinical Practice

Structure Number	Name	Formula
16.1	Polyethylene	$-(CH_2-CH_2)-_n$
16.2	Poly(tetrafluoroethylene)	$-(CF_2-CF_2)-_n$
16.3	Poly(methyl methacrylate)	$-[CH_2-C(Me)(COOMe)]-_n$
16.4	Polyesters	$-[O-CH_2-CH_2-O(O)C-\langle\bigcirc\rangle-C(O)]-_n$
16.5	Polyamides	$-[NH-[O(CH_2)_6NH-(O)C-\langle\bigcirc\rangle-C(O)]-_n$
16.6	Poly(dimethylsiloxane)	$\left[\begin{array}{c} Me \\ \| \\ O-Si \\ \| \\ Me \end{array}\right]_n$
16.7	Cellulose acetate	Acetylated glucose polymers
16.8	Hydrophobic polyphosphazenes	$\left[\begin{array}{c} R \\ \| \\ N=P \\ \| \\ R \end{array}\right]_n$ $R = -OCH_2CF_3, -OC_6H_5$, etc.
16.9	Polyurethanes	$-[RNHC(O)O-(CH_2)_xOC(O)NH]-_n$ often copolymerized with silicones
16.10	Poly(ethylene glycol)	$-(CH_2-CH_2O)-_n$
16.11	Poly(hydroxyethyl methacrylate)	$-(CH_2-C(Me)C(O)OCH_2CHOH)-_n$
16.12	Poly(vinylpyrolidone)	$\left[\begin{array}{c} -CH_2-CH- \\ \text{(N-pyrrolidinone ring)} \end{array}\right]_n$
16.13	Hydrophilic polyphosphazenes	$\left[\begin{array}{c} R \\ \| \\ N=P \\ \| \\ R \end{array}\right]_n$ $R = -OCH_2CH_2OCH_2CH_2OCH_3$, (a) $-O-\langle\bigcirc\rangle-COONa$, etc. (b)
16.14	Poly(acrylic acid)	$-(CH_2-CH(C(O)OH))-_n$
16.15	Collagen	$\{[NHCH_2C(O)]_x-[NHCH(R)C(O)]_y\}_n$
16.16	Alginates	$(C_6H_8O_6)_n$ mannose copolymers
16.17	Poly(lactic–glycolic acid)	$-\{(OCH_2-CHCO)_x-(OCH(Me)CO)_y\}-_n$
16.18	Polyanhydrides	$-[O-R'-C(O)OC(O)-R'']-_n$
16.19	Polycaprolactone	$-[O(CH_2)_5C(O)]_n-$
16.20	Poly(trimethylene carbonate)	$-[O(CH_2)_3OC(O)]_n-$
16.21	Polyphosphazenes with amino acid ester or sugar side groups	$\left[\begin{array}{c} R \\ \| \\ N=P \\ \| \\ R \end{array}\right]_n$ $R = -'NHCHR'COOEt$, glucosyl, etc.

(2) Poly(tetrafluoroethylene) (PTFE, Teflon, Gore-Tex). This polymer (structure **16.2** in Table 16.1) is widely used as an inert biomedical material. It is one of the most hydrophobic polymers known (contact angle to water ~100°), and it is thus resistant to penetration by water or microorganisms. Monolithic Teflon is a relatively inflexible, machinable version of the polymer. Gore-Tex is a flexible, highly porous modification used in arterial replacement tubes. Velours (felt-like) surfaces with PTFE nanofibers are of some interest as platforms for cell spreading. PTFE is one of the few materials that appear to be stable enough in a biological environment to last the lifetime of a patient. The main drawback of PTFE is the difficulty of fabrication since it does not melt below 327°C and it is soluble in various solvents only at temperatures close to its melting point. Thus, sinter fabrication is one of the few alternatives for device preparation.

(3) Poly(methyl methacrylate) (PMMA). PMMA (structure **16.3** in Table 16.1) was once used in a variety of biomedical devices, but is now employed mainly in hip replacement surgery to cement the metal ball joint spike into the end of the femur, a process that is accomplished by in situ free-radical polymerization of the monomer. Some questions exist about the toxicity of the monomer released into the body during this procedure.

(4) Polyesters. The principal polyester used in biomedicine is poly(ethylene terephthalate) (structure **16.4** in Table 16.1), also known as Dacron©, Terylene©, or Mylar©. It is a flexible, high-melting-point (258°C) polymer that has good chemical resistance, and is melt-spun into strong fibers or fabricated into clear films. As a woven cloth or a velour, it is a favored material for use in tissue reinforcement.

(5) Polyamides. Although nylons are produced on a vast scale for textiles, their initial use as biomaterials has been largely discontinued due to their susceptibility to hydrolytic breakdown, often to give toxic products. However, synthetic polypeptides produced by the polymerization of amino acids are useful as bioerodible polymers.

(6) Poly(dimethylsiloxane) (Silicone Rubber) (PDMS). This polymer (structure **16.6** in Table 16.1) has a long history of successful use as a hydrophobic elastomeric biomaterial. It has been utilized in prosthetic heart valves, catheters, subcutaneous devices for the slow release of steroids, and numerous other applications. PDMS tends to absorb lipids from blood, which causes it to soften over time. The legal issues that surrounded silicone breast implants failed to implicate *high-polymeric* silicones as toxic species. This polymer is one of the most useful elastomeric biomaterials.

(7) Cellulose Acetate. This is a long-existing polymer (structure **16.7** in Table 16.1) produced by acetylation of cellulose by acetic anhydride, and it is used as a film-forming material in general technology. Its main use in biomedicine is as a semipermeable membrane in hemodialysis equipment. This is a relatively short-term use, and so issues of long-term hydrolytic and oxidative stability do not arise.

(8) Hydrophobic Polyphosphazenes. Several of these polymers (structure **16.8** in Table 16.1), especially those with fluoroalkoxy or aryloxy side groups, have generated considerable interest as potential hydrophobic alternatives to silicone rubber or Teflon-type fluorocarbon materials. Fluorinated phosphazene polymers do not absorb lipids, and their structures can be tuned to give either elastomers or film-formers. Unlike Teflon, they are soluble in many organic solvents and can be fabricated easily. Their contact angles to water are higher than that of PDMS and are similar to that of Teflon. As nanofibers, they are even more hydrophobic (~160°). Blood and tissue compatibility tests suggest that they are good candidates for cardiovascular applications, catheters, pacemaker coatings, and related devices. These polymers are examples of newer candidates that show some advantages over traditional materials.

(9) Polyurethanes. Although some polyurethanes (structure **16.9** in Table 16.1) are unstable to water and oxidation over periods of years, others (especially those with poly(dimethylsiloxane) and alkyl ether block structures) are believed to be biostable. However, their main attribute is elasticity combined with flex strength. Polyurethane films can be flexed millions of times before materials fatigue sets in. Hence, they are used as membranes in blood pumps, and have FDA approval for this application.

(10) Poly(ethylene glycol) (PEG). This polymer (structure **16.10** in Table 16.1) is a water-soluble, biostable, macromolecule that has limited FDA approval for use in human biomedicine. It is crosslinked to give hydrogels, is used as a viscosity-enhancing agent, and can be attached to surfaces to make them hydrophilic. The safety of this polymer depends on the molecular weight. High polymers cannot be excreted through the kidneys, and they accumulate in the spleen and liver. Low-molecular-weight linear polymers (MW below 1000–2000) are excreted.

(11) Poly(hydroxyethyl methacrylate) (HEMA). This (structure **16.11** in Table 16.1) is another water-soluble polymer with a long history of use in experimental biomaterials research, especially as a material for conversion to hydrogels. Its hydrogels have been developed for uses in contact lenses and intraocular lenses, and, like PEG, it produces hydrophilic surfaces when grafted onto other materials.

(12) Poly(vinyl pyrolidone) (PVP). PVP (structure **16.12** in Table 16.1) is a water-soluble, hydrolytically stable macromolecule that was used in Germany as a blood plasma substitute during World War II. Because it in nonbioerodible it accumulates in the spleen and liver, but with seemingly benign results. Its main uses today are in semipermeable membranes and in cosmetic products.

(13) Water-Soluble Polyphosphazenes. Phosphazene polymers with alkyl ether side groups or with alkoxy or aryloxy side units that bear hydroxyl or carboxylic acid groups are soluble in water and are nonbiodegradable (structures **16.13** in Table 16.1). The most heavily investigated are the MEEP-type systems (structure **16.13a** in Table 16.1) with methoxyethoxyethoxy or similar side chains. Like PEO and HEMA, they can be crosslinked to form hydrogels that absorb water below a critical solution temperature (T_c) but extrude the water above this temperature. Devices

are being developed that use this phenomenon to deliver drugs through thermally or pH-responsive membranes.

(14) Poly(acrylic acid). This is a nonbiodegradable polymer (structure **16.14** in Table 16.1) that is soluble only in basic aqueous media. Its main biomedical use is in dental restoration resins in combination with silicates. In solution it is a polyelectrolyte and is capable of triggering a strong antigenic (i.e., rejection) response by the body's immunological system.

b. Bioerodible Materials

(1) Collagen. The idea that materials derived from living things might be effective biomaterials has an intuitive appeal. Heart valves from pigs (porcine valves) have been implanted routinely and successfully in humans. However, there are limits to the use of biologically derived materials. The human body has a sophisticated surveillance and rejection system that prevents widespread use of donor human or animal tissues. Moreover, the transmission of bacterial or viral infections is an ever-present possibility.

Collagen is a protein that is present in a wide gamut of mammalian tissues ranging from skin and internal organs to bone. The protein is a copolymer that consists mainly of 80–85% glycine residues. It can be processed by hydrolysis to break crosslinks and reduce the molecular weight, and can be regenerated in different shapes and recrosslinked if necessary. Collagen has been employed for many years in the form of "catgut" in surgical suture fibers for the closing of wounds. Its main disadvantage is that it does not erode quickly in the body. Hence, the sutures must be removed by a second operation. Hydrolyzed collagen is "gelatin" used in food products. Solutions of low-molecular-weight collagen are used as adhesives for repairing perforations in the eardrum.

(2) Alginates. These are water-soluble, bioerodible, mucopolysaccharides extracted from seaweed and kelp. Alginates (structure **16.16** in Table 16.1) are used as thickening agents in the food and healthcare industries, and they have a long history of safe biocompatibility. Because alginates can be crosslinked ionically by calcium ions, they have been studied for conversion into microspheres for controlled drug delivery.

(3) Poly(lactic-glycolic acid) (PLGA). Of all the available bioerodible medical materials, PLGA (structure **16.17** in Table 16.1) has received by far the most attention. This polymer was developed commercially in the mid-1960s initially as a bioerodible surgical suture material. There are several versions of PLGA, with different ratios of the two monomer residues and various molecular weights. Synthesis is via the ring-opening polymerization of the "glycolide" cyclic dimers of glycolic and lactic acids. Poly(glycolic acid) hydrolyzes faster than does poly(lactic acid), presumably because it lacks the protective methyl group at the α position. Thus, different co-monomer ratios yield polymers with different rates of hydrolysis. The ultimate hydrolysis products are glycolic and lactic acids. Both are species normally present in the body, but both are acidic. Thus, tissue healing around a suture or implant may be inhibited by the low pH. PLGA is now used in experimental drug delivery

devices, bone cements, microspheres, tissue engineering scaffolds, and many other applications. It has FDA approval for certain uses.

In addition to the monomer ratios in the polymer, the speed of erosion depends on the surface area exposed to the aqueous medium (smaller objects erode faster), the degree of polymer crystallinity (crystallinity slows hydrolysis), and the shape of the object. For example, because the surface area:volume ratio of a sphere increases as a sphere contracts in size, spherical devices may not release a drug in a "zero-order" (i.e., constant-release-rate) manner. Note that for drug delivery applications and for uses where strength must be retained even as hydrolysis proceeds, the erosion should take place at the surface and not within the bulk material. Bulk erosion could lead to the catastrophic release of a drug or premature failure of a suture. However, for tissue engineering applications, bulk erosion may be needed to promote uniform cell replication and colonization.

The main defects of PLGA are the limited range of hydrolysis rates and the acidity of the hydrolysis products (reaction 1). The acidity factor is more of a problem for tissue engineering than for drug delivery or surgical devices.

$$\{[-OCH_2C(O)]_x - [OCH(CH_3)C(O)]_y\}_n$$

$$\xrightarrow{H_2O} HOCH_2C(O)OH + HOCH(CH_3)C(O)OH \qquad (1)$$

(4) Polyanhydrides. Until relatively recently, polyanhydrides (structure **16.18** in Table 16.1) were an orphan of the synthetic polymer world because of their hydrolytic sensitivity. They are formed by condensation reactions of dicarboxylic acids, and they hydrolyze back to those acids in contact with water. However, the sensitivity to water depends on whether the polymer backbone contains aliphatic or aromatic residues; the latter have a slowing effect on the hydrolysis rate. Intermediate hydrolysis rates are possible by using both aliphatic and aromatic residues in the same polymer. Polyanhydrides have been used for the localized delivery of a powerful antitumor agent in the treatment of an aggressive brain cancer.

(5) Polycaprolactone and Poly(trimethylene carbonate). These are two polymers that are receiving increased attention as bioerodible materials. Polycaprolactone (structure **16.19** in Table 16.1) hydrolyzes at a slower rate than does PLGA to give $HO(CH_2)_5C(O)OH$. The elastomeric polycarbonate (structure **16.20** in Table 16.1) is also used to slow the rate of hydrolysis or other bioerodible polymers, particularly in block copolymer structures.

(6) Bioerodible Polyphosphazenes. The types of side groups connected to a polyphosphazene chain govern the stability of the polymer to aqueous media and thus its capacity for bioerosion (see structure **16.21** in Table 16.1). Side groups such as amino acid esters, glucosyl or glyceryl units, and imidazole groups sensitize the polymer to hydrolysis. Those polymers with ethyl glycinate or ethyl alanate side groups have been studied in the greatest detail. The hydrolysis products are the amino acid, ethanol, phosphate, and ammonia; the latter two products provide a pH-buffered medium. The glycine derivative hydrolyzes faster than does its alanine counterpart, which in turn is more sensitive to hydrolysis than the phenylalanine

derivative. Thus, the ratio of the two or three different side groups in a polymer controls the overall erosion rate. Other amino acid ester side groups allow property tuning over an even wider range. These polymers have been investigated for controlled drug delivery and especially for bone regeneration applications (see Section C.5 and C.6).

C. MATERIALS FOR SPECIFIC MEDICAL APPLICATIONS

This section provides a summary of how and why different materials are utilized in specific medical devices. The purpose is to give examples that illustrate major areas of medicine where traditional materials are used and where new materials are needed. The emphasis is on cardiovascular materials, sutures and other surgical devices, drug and vaccine delivery, orthopedic applications, and tissue engineering.

1. Cardiovascular Materials

a. General Features. A variety of materials are used in the following cardiovascular devices—artificial heart pumps and valves, replacement blood vessels, stents, pacemakers, heart–lung machines, and renal dialysis equipment. In all these devices (except pacemakers), materials come into intimate contact with circulating blood. Blood is a complex fluid that contains proteins, salts, red and white corpuscles, and antibodies dissolved or suspended in water. It is an unstable fluid in the sense that it is capable of coagulating in response to trauma or irritation or the presence of an "unnatural" material. Coagulation, or thrombus formation, follows a chain reaction of molecular responses often preceded by the deposition of proteins on the surface.

Although it was once widely believed that hydrophilic surfaces provide the best protection against the coagulation process, it has been found in practice that hydrophobic surfaces such as those found in poly(tetrafluoroethylene), silicones, fluorinated polyphosphazenes, or polyesters provide the best protection. Explanations range from the role of the surface in attracting anticoagulent proteins like albumin, to the ability of the host's endothelial cells to colonize a hydrophobic surface and insulate blood from the polymer. Control of surface properties is discussed in Chapter 15.

b. Prosthetic Heart Valves. These are widely used in cardiovascular surgery. They fall into three categories—porcine valves (from pigs), caged ball valves, and leaflet valves. The leaflet valve group is divided into those with a single flap or those with two flaps in a trapdoor configuration. A caged ball valve and a leaflet valve are shown in Figure 16.2.

Porcine valves (mainly collagen) are obtained from the animal and are then stabilized (crosslinked) with formaldehyde before being sutured into the heart tissue. These valves give rise to the fewest complications, but have shorter lifetimes than do mechanical valves. Caged ball valves consist of a silicone rubber sphere retained in a stainless-steel cage, with the device sutured to the heart tissue through a Dacron© cloth surround. An example was shown in an earlier chapter (Figure 1.3). The ball moves back and forth in the cage in response to fluid flow. Ridging of the

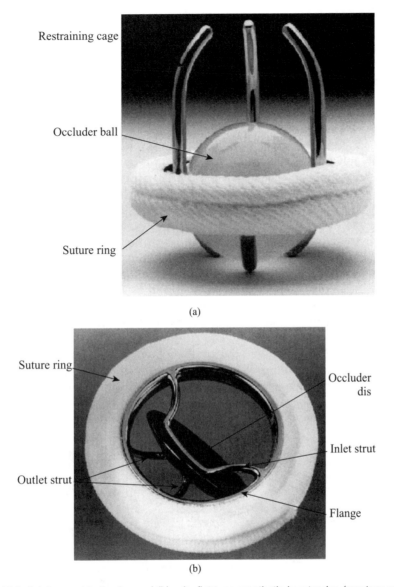

Figure 16.2. *(a) A caged ball valve and (b) a leaflet-type prosthetic heart valve (courtesy of Baxter Edwards CVS and Medtronic Heart Valve division).*

silicone sphere due to repeated impact with the cage, perhaps exacerbated by the absorption of lipids, is a long-term problem. These valves have the highest incidence of failures due to blood clotting (17% within 5 years, 44% within 15 years). Bileaflet valves made from pyrolytic carbon have the lowest rate of thromboembolism (13% after 5 years, 35% after 15 years), but are more prone to mechanical failure (by leaflet fracture) than are porcine or caged ball valves. Single-leaflet valves fall in the middle of these ranges of reliability. Porcine valves are as quiet as human valves, caged ball valves are somewhat more noisy, and leaflet valves are the noisiest.

c. Artificial Heart Pumps. Although the prospect of permanently implanted replacements for a damaged heart have been explored for many years, this technology is still at an early stage. Progress is limited by both materials problems and by the methods available to power such devices in ways that allow freedom of untethered travel by the patient. Thus, most of the progress in this field has been made in the development of pumps that can take over the function of a damaged heart for a limited period of time in preparation for a heart transplant or other surgical procedures. Blood pumps are produced in several different designs (Figure 16.3). Both one- and two-ventricle designs have been developed. Earlier models used an elastomeric polyurethane membrane enclosed within a hollow titanium alloy or rigid polyurethane sphere. The high flexural strength of the polyurethane membrane was a leading factor in the choice of this polymer, which must survive 60–80 flexures per minute for as long as the device is in use. Pulses of compressed air move the membrane and pump the blood through valves into the arterial system. A major challenge is to deliver pulses of compressed air from an external air pump via a flexible, reinforced tube through the patient's thorax to the blood pump. More recent models are driven by an electric motor with power supplied by induction though the skin. This eliminates the need for an air line, gives the patient more mobility, and reduces the incidence of infections, some of which have been responsible for failure of the implanted pumps and death of the patients through toxic shock.

In addition to flexible or rigid polyurethanes and titanium alloys, the materials used in artificial heart pumps include the polyester Dacron©. Connections of the pump to the patient's cardiovascular system are via Dacron cloth cuffs, sometimes embellished with a velour of the same polymer to assist colonization by endothelial tissue. Epoxy components are also used. Apparently a major problem with all pump

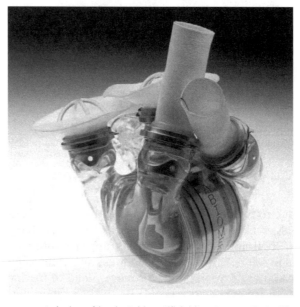

Figure 16.3. *A more recent design of implantable artificial heart pump that utilizes an electric drive, with the power supplied by transduction through the skin (courtesy of Abiomed Corp.).*

designs is thrombus (clot) formation, especially at locations where the flowing blood passes from a smooth surface to a rough one at the connections to the living circulatory system. Differences between the compliance (i.e., flexibility) of the pump outlets and the living blood vessels are part of the problem.

It will be apparent that the design of new materials for heart valves and pumps is a major challenge that has not yet been overcome. Reliance by researchers and manufacturers on long-existing materials rather than a focus on the development of specifically designed new materials is part of the problem.

d. Replacement Blood Vessels and Arterial Reinforcement Materials. Some of the same problems that have been mentioned for valves and blood pumps also apply to tubes used to replace damaged blood vessels. However, tubes made from the fluorocarbon polymer Gore-Tex are reported to be more effective than those from other polymers. The porosity of this material favors the deposition of proteins that inhibit blood clotting, and these, in turn, provide a base layer for colonization by the patient's own endothelial cells. As with the other devices mentioned above, discontinuities in the fluid flow at the points where the graft is sutured to the living blood vessels can favor thrombus formation due to the different compliance characteristics. Replacement of sections of wider blood vessels has been more successful than replacement of narrower vessels due to problems with suturing and the greater danger of blockage with the narrower tubes.

Blood vessels with sections that have ballooned from excessive blood pressure or damage to the vessel wall (aneurisms) are corrected using a reinforcement fabric of Dacron wrapped around the outside of the expanded section. The polyester fabric does not come into direct contact with the flowing blood, and provides a long-term solution for many patients.

e. Stents. A *stent* is a small metal cage that is inserted into a narrowed section of an artery to keep it open after a blockage has been removed by angioplasty (expansion of a balloon within the artery). The cage is inserted in a collapsed configuration, and is then expanded by means of a balloon device or via shape memory. Metals used for the cage include corrosion-resistant cobalt–chromium or nickel–titanium alloys, tantalum, or 316L stainless steel, occasionally gold-plated to improve X-ray visibility. However, metal stents have a tendency to promote restinosis (scar tissue formation) and thrombosis in one out of three patients. Attempts to reduce this problem have involved coating the metal with a thin polymer layer that slowly releases an immunosuppressive, antiscarring drug. In a clinical trial this resulted in a reduction of stent-related blockages from 21% to 8.6%. However, future stents may be fabricated from bioerodible, shape-memory, spring-like polymers that could cause fewer long-term problems, especially if combined with the drug release option.

f. Renal Dialysis and Blood Oxygenation. These processes are a routine part of medical practice, using machinery that has been developed over many years. Yet, even here improvements are possible as new materials become available. *Hemodialysis* is the process by which urea and other water-soluble waste products are removed from the blood. The process is described from a membrane perspective in Chapter 13. Traditional equipment uses a cellulose acetate membrane to separate the blood from an aqueous salt solution. The membrane must be permeable to small molecules but not to proteins or blood corpuscles. The greatest challenge is in miniaturization of these devices to produce portable or even implantable units. Miniaturized devices often consist of a bundle of hollow fibers or narrow tubes in which blood and the salt solution flow countercurrent to each other. Although heparin can be administered to protect against blood clotting, an anticoagulent, semipermeable membrane with a more rapid permeability and selectivity than cellulose acetate would be a significant improvement.

Blood oxygenation during surgical procedures is accomplished with the use of a heart–lung machine that provides a pumped bypass around the heart and lungs coupled with a device for blood oxygenation. The device is quite complex. Pumping is achieved by several roller pumps and a centrifugal pump. The roller pumps squeeze blood through plasticized poly(vinyl chloride) or silicone rubber tubes. Oxygenation is accomplished either by bubbling the gas through the blood or by use of an oxygen-permeable membrane. Filters are installed in the system to remove gas bubbles. Thus, the number of blood/materials contact zones, and the opportunities for clot formation, are quite large, and anticoagulants must be administered. Polymers with hydrophobic surfaces, perhaps supplemented with surface-bound anticoagulants, could bring about significant improvements.

g. Pacemaker Materials. Pacemakers to control the heart rhythm via an electric pulse are implanted below the collarbone. More than 2 million people have been fitted with these devices. Although pacemakers do not come in direct contact with blood, they do interact with sensitive tissues. The outer casing of many pacemakers is made of titanium (lightweight and corrosion-resistant). The electric leads that connect the device to the heart muscle must be coated with a water-repellent, biocompatible, electrically insulating polymer.

2. Surgical Sutures, Clips, and Staples

The widespread use of poly(glycolic/lactic acid) (PLGA) as absorbable surgical sutures has been mentioned earlier. The same polymer is used for surgical clips and staples used to close wounds and assist with surgical procedures. The PLGA clips replace stainless-steel counterparts that were the traditional materials.

3. Orthopedic Materials

The repair of broken bones accounts for a large percentage of surgical procedures carried out worldwide. Simple fractures are treated by immobilization of the bone to allow natural healing. More complex fractures require reinforcement by permanently implanted titanium or stainless-steel spikes. The most advanced research in this field involves bone regeneration by tissue engineering, a topic that is discussed later in this chapter.

Hip replacement materials were mentioned earlier. The main materials currently in use are titanium, polyethylene, and poly(methyl methacrylate). A problem with metal ball joints is that the spike-type connection to the femur tends to loosen over time because of stress response differences between the bone and the metal. Hence tissue engineering approaches may be preferred in the future.

4. Optical Materials in Medicine

Optical materials for eyeglasses and contact lenses were discussed in Chapter 14. Implantable intraocular lenses are derived from flexible polymers or hydrogels that can be rolled into a small enough profile to be inserted into the lens cavity via a syringe-style device, to then unwind to give the final lens configuration. Thus, shape-memory characteristics are important. New materials are needed that will inhibit protein deposition and bacterial colonization.

5. Controlled Drug and Vaccine Delivery

The traditional administration of drugs by oral delivery or intramuscular injection leads to widely oscillating concentrations of the drug in the body (Figure 16.4).

Methods to smooth out the concentration swings are made using several different approaches. These include diffusion of the drug across a membrane, release of the drug by erosion of a hydrolytically unstable polymer, and the use of vesicles and micelles (Figure 16.5).

a. Membranes. Membranes have been used for many years in hemodialysis equipment and controlled drug delivery, and this topic is addressed in Chapter 13. The wider aspects of controlled drug release are summarized in Figure 16.5a–c. In practice, one of the first examples of membrane control of drug release was the Ocusert® device for the slow release of pilocarpine—a drug that lowers the internal eye (ocular) pressure for patients with glaucoma. The device consists of two membranes of a poly(vinyl acetate) copolymer separated by a thin gasket, and an aqueous solution of pilocarpine and an alginic acid viscosity enhancer. The device, about the size of the nail on the little finger, is placed under the lower eyelid, and the slow diffusion of the drug through the membrane provides a continuous supply to the eye over a 1-week period. Transdermal release of nitroglycerine through a similar

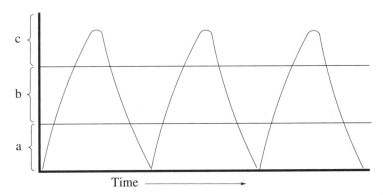

Figure 16.4. *Variation of the concentration of an injected or ingested drug in the body as a function of time. The sawtooth profile reflects an initial concentration at the target site that is ineffective (a), but rises through the effective zone (b), and then peaks as it reaches toxic levels (c). The concentration then declines through the effective concentration range into the ineffective region. This must then be followed by another injection. An ideal drug release profile would maintain the drug concentration within the effective level throughout the therapy.*

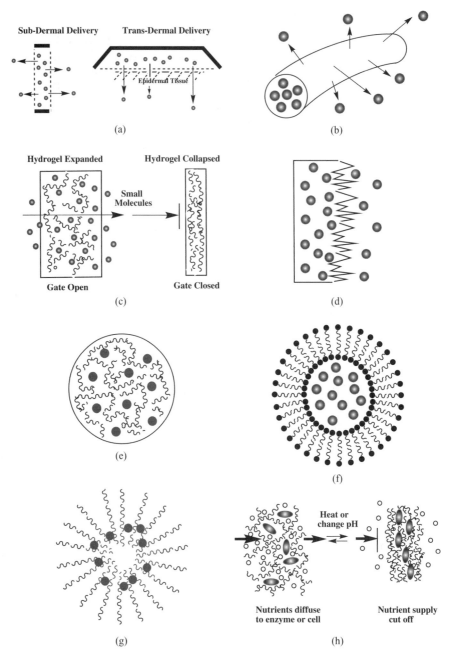

Figure 16.5. *Different designs for controlled drug delivery devices. (a) A membrane is employed to control the release of the drug. (b) A polymer tube or cylinder serves as a diffusion membrane for controlled drug release. (c) A responsive membrane that opens or closes as the pH, temperature, or ion strength is changed allows control of the drug release rate. (d) Hydrolytic erosion of a polymer provides a mechanism for drug release. (e) Microspheres are used to immobilize drugs and vaccines for protection through the stomach or for inhalation deep into the lungs. (f) Bilayer vesicles can serve as carriers for hydrophilic drugs for delivery to sites that are distant from the point of injection. (g) Micelles allow hydrophobic drugs to be solubilized and delivered to targeted sites via the bloodstream. (h) Diagram showing how a responsive hydrogel can be employed to turn on or off the action of a living cell or enzyme that either releases or decomposes a bioactive molecule (see color insert).*

membrane in a skin patch is used for the treatment of agina. The same principle is employed for the transdermal delivery of anti-motion-sickness drugs. Diffusion of steroids through the walls of narrow silicone rubber tubes implanted subcutaneously (Figure 16.5b) is an effective long-term (~5 years) method for birth control, which has the added advantage that the device can be removed at any time. Membrane-based drug release is an extremely effective method for controlled drug delivery, provided the device can be replaced easily after the supply of drug is exhausted.

b. Antibacterial Surfaces. Many bacterial and fungal infections are spread through contact of humans with contaminated surfaces. Moreover, materials that are implanted in the body often become contaminated, and this is a major cause of postoperative infections. The immobilization of antimicrobial agent on metal, ceramic, or polymer surfaces is therefore a subject of widespread interest. Of course, liquid disinfectants are normally used in surgery and in everyday life, but these do not provide a permanent antibacterial film. Instead, attempts are being made to covalently immobilize antimicrobial molecules such as alkylammonium halides to surfaces or to use polymer films as depots for the slow release of such agents. The depot method depends on diffusion of the antimicrobial agents to the surface, and eventually the film becomes depleted of active molecules. Covalent immobilization of the agent to a surface has a strong intuitive appeal. However, it is not clear whether immobilized antimicrobial agents are effective, because their activity may depend on their ability to penetrate a bacterial cell membrane in the form of a free molecule.

c. Responsive Hydrogels. Hydrogels are important materials in biomedicine because their physical properties mimic those of living soft tissues. They are produced by the crosslinking of polymers that would normally be soluble in water. Such materials swell as they absorb water, but, because of the crosslinks, they cannot dissolve. Thus, the degree of crosslinking controls the degree of swelling and the amount of water in the system. The use of hydrogels for the controlled delivery of drugs is described in Chapter 13, and is summarized here in Figure 16.5c. Figure 16.5h illustrates how a responsive hydrogel can be employed to turn on or off the activity of a living cell or enzyme, a process that could allow the timed synthesis and release of biomolecules or the decomposition of toxins.

d. Bioerodible Drug Release Systems. An alternative approach to controlled drug delivery is to disperse drug molecules within solid polymers, microspheres, or micelles made from bioerodible polymers. Erosion of the polymer then allows the drug molecules to escape at a controlled rate. It is important that the polymer decompose by *surface* erosion, because bulk erosion would lead to the catastrophic release of large quantities of the drug in a short period of time. The principle is illustrated in Figure 16.5d. One attraction of this mode of release (shared with membrane diffusion) is the fact that unmodified, licensed drugs may be used. The alternative approach, using a drug that is *covalently linked* to an erodible or even a nonerodible, water-soluble polymer, might be considered less acceptable because such a drug would probably be considered to entirely new and would have to undergo an expensive and time-consuming government approval process.

e. Microspheres, Vesicles, and Micelles.

Although they are not biomaterials in any sense of the word, microspheres, vesicles, and micelles play an important role in experimental biomedicine, particularly in the controlled release of drugs and vaccines and cardiovascular imaging. Microspheres are micrometer-sized spherical particles of polymers that serve as depots for trapped drugs or vaccine molecules (Figures 16.5e and 16.6).

Two mechanisms of drug release have been developed. In one, the water-soluble polymer chains are crosslinked through calcium ionic linkages to carboxylic acid

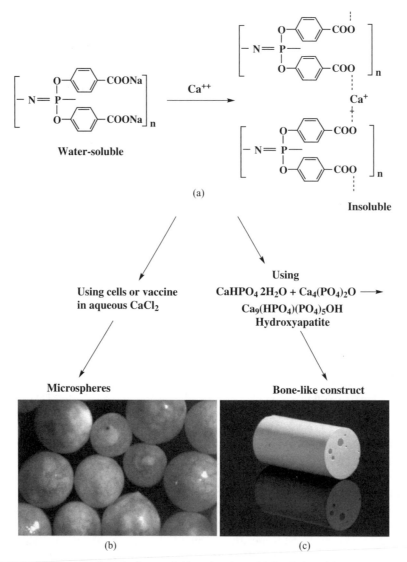

Figure 16.6. *(a) Aqueous phase ionic crosslinking chemistry. (b) A solution of the water-soluble phosphazene polymer with dissolved drug or vaccine is sprayed into an aqueous calcium chloride solution, which causes reversible ionic crosslinking (R. Langer and S. Cohen, MIT). (b) Formation of a synthetic bone material by the formation of hydroxyapatite within a matrix of a calcium crosslinked polymer (P. Brown, K. S. Ten Huisen, and H. R. Allcock, The Pennsylvania State University).*

groups in the side chains (Figure 16.6). The crosslinks are broken in the presence of monovalent cations such as sodium or potassium, and the drug molecules are released as the polymer chains separate from each other and dissolve. This technique has been used to immobilize drug and vaccine molecules in microspheres of alginates and water-soluble polyphosphazenes. It has also been the basis of experimental cardiovascular imaging devices in which gas bubbles trapped in microspheres are injected into the circulatory system and their progress is followed by ultrasound.

The second method is to trap the drug molecules in microspheres of a bioerodible polymer. Subsequent hydrolysis of the polymer (e.g., PLGA) then brings about a release of the drug. Microspheres are also used for the delivery of inhalation drugs to allow active molecules to penetrate deep into the lung system. Microspheres are typically fabricated by coagulation of a polymer from a solvent by a nonsolvent, or by firing a spray of polymer solution into a nonsolvent or into an ionic crosslinking medium. Spraying the solution into an evaporator ("spray drying") to remove the solvent is another method.

Vesicles (Figure 16.5f) are micrometer-sized spheres formed from small molecules that bear a hydrophobic group at one end and a hydrophilic unit at the other. These (amphiphilic) molecules form bilayers in which the hydrophilic units point out toward an aqueous medium while the hydrophobic units are arrayed head-to-head in the middle of the vesicle wall. Drug molecules, vaccines, and even mammalian cells can occupy the aqueous interior and be transported to different parts of the body before being released. Fatty acids are the best-known molecules that form micelles, but many other amphiphilic small molecules behave similarly.

Micelles (Figure 16.5g) are nanometer-sized spheres or elongated "worms" derived from amphiphilic block coplymers. These self-assemble in an aqueous medium with the hydrophilic block pointed outward and the hydrophobic block occupying the interior. Micelles are potentially useful for the delivery to distant parts of the body of hydrophobic drug molecules, which would otherwise be insoluble in blood or muscular tissue. Some mechanism needs to be provided to ensure the release of the drug molecules at the appropriate site. A key requirement for microspheres, vesicles, and micelles is the nontoxicity of the component molecules or their ability to bioerode.

6. Tissue Engineering

The regeneration of a patient's own tissues by the cultivation of cells from the same individual, or from cell lines derived from stem cells, could avoid problems associated with organ transplants or synthetic biomaterials. New blood vessels, bones, skin, liver, pancreas, and other organs could in principle be regenerated in the required shape by using this method either in vivo or ex vivo for subsequent implantation. A key requirement is the need for a scaffold that defines the boundaries beyond which cell spreading will not occur and which provides a "foothold" for the cells. It is also required that the material used for the framework should bioerode to harmless products, preferably at the same rate at which cell colonization is taking place. Complex organs could, in principle, be engineered by having different cell types populating different sections of the framework.

Increased attention is being focused on bone healing through tissue engineering (Figure 16.7). This is a process in which a bioerodible framework of woven fiber or porous solid material provides a platform for colonization by the patient's own osteoblasts (bone-forming cells) to regenerate bone either in vivo or, perhaps, in the future, in vitro. PLGA has been investigated as a framework material, but advanced work with amino acid–substituted polyphosphazenes shows some advantages. The near-neutral pH of the hydrolysis products from these polymers is a considerable attribute. The trend in this area is to use a composite of synthetic calcium hydroxyapatite and the bioerodible polymer to form a rapidly setting cement to repair fractures without the need for reinforcement by metals.

Solid, monolithic polymers are unsuitable as frameworks because cell spreading into the interior would be difficult. Instead, two approaches are under investigation: (1) employing a highly porous solid to allow cells to spread through the pore structure and (2) using woven or nonwoven fabrics, with the space between the fibers providing opportunities for the cells to penetrate into the interior of the construct. Fabrics may be produced from macro- or microfibers, or from nanofiber mats. The latter provide a high-surface-area substrate for cell colonization (Figure 16.7).

Most of the bioerodible polymers discussed elsewhere in this chapter are being studied as platforms for tissue engineering. PLGA has received the most attention in the past, because of its facile availability, but newer polymers such as polyphosphazenes have significant advantages since they appear to be more biocompatible.

One final example of an emerging example of tissue engineering is the use of bioerodible tubes to encourage the reconnection of severed nerves. If the ends of severed nerves become separated by more than one centimeter, they will not rejoin,

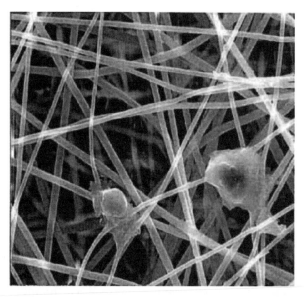

Figure 16.7. Tissue engineering—a process through which a bioerodible framework of a porous polymer or a nanofiber mat is used as a matrix for colonization by mammalian cells; the SEM shows nanofibers of the bioerodible polyphosphazene, $[NP(NHCH(CH_3)COOC_2H_5]_n$, beginning to be colonized by osteoblast cells (C. Laurencin and L. Nair, University of Virginia; J. Bender, and H. R. Allcock, The Pennsylvania State University).

partly because growth factors released by the nerve ends cannot diffuse across the gap. However, if the ends are within one centimeter of each other and confined within a tube, the growth factors will be localized, and the two ends will grow toward each other and eventually rejoin. Early experiments used tubes of glass or silicone rubber, which cannot be removed after nerve healing. However, more recent work with bioerodible polymers provides strong encouragement for the effectiveness of this approach and for total disappearance of the polymer. Nerve tissue is exceedingly sensitive to toxic small molecules and abnormal pH values, so the products from the polymer hydrolysis are a critical issue in this research.

D. FABRICATION AND TESTING OF BIOMEDICAL MATERIALS

1. Fabrication

All the methods used for nonbiomedical materials are employed for the fabrication of biomaterials. These include melt casting; machining and extrusion of metals; powder processing of metals, ceramics, and polymers; and melt extrusion, injection molding, and solvent casting of polymers. Polymer fibers are prepared by melt or solution spinning or solution coagulation, and by centrifugal or electrostatic spinning. Tubes can be fabricated by extrusion or by solvent evaporation from solutions painted or sprayed on a rotating mandrel. Porous polymer constructs are produced by allowing the polymer to absorb a solvent (e.g., a low-boiling-point hydrocarbon or liquid carbon dioxide) and then reducing the pressure to volatilize the solvent and generate an open- or closed-cell foam structure. Polymer surfaces may be modified after fabrication by chemical reactions or by exposure to gas plasmas.

The main difference between biomaterials and nonmedical materials is the crucial need for chemical purity and sterilization. Solvent residues, monomers, plasticizers, or synthesis impurities must be absent from any biomaterial. Sterilization is accomplished by heat (metals and ceramics), gamma or UV irradiation (polymers), or by exposure to chemical antimicrobial agents. Moreover, the materials must be stored in a sterile environment before use.

2. Testing of Biomedical Materials

Physical properties of biostable materials are assessed through the normal evaluation methods discussed elsewhere in this book. Preliminary testing of bioerodible materials involves exposure of films or monoliths of the material to pH 7 aqueous media at human body temperature. Weight loss due to hydrolysis is measured at specific weekly or monthly intervals, and the hydrolysis products in solution are identified by NMR, mass spectrometry, or other techniques. The decline in molecular weight and materials strength can be studied through the use of gel permeation chromatography and micromechanical testing apparatus. Surface erosion may be investigated by means of scanning electron microscopy (SEM) techniques. The pH of the medium surrounding a bioerodible polymer should also be monitored.

In vivo tests for biocompatibility of both biostable and bioerodible polymers are carried out by implantation in test animals such as mice, rats, or rabbits under carefully controlled conditions. After the test period, the materials and the surrounding tissue are examined for evidence of toxicity and capsule or giant cell formation. Following removal of the biomaterial from the animal, the solid is examined by scanning electron microscopy to search for evidence of surface erosion, and through mechanical tests to identify any change in physical properties.

Resistance to microbial colonization requires an initial examination of the surface of a material after exposure to a diverse mixture of microorganisms such as those normally found in soil. More specific tests determine the resistance of the solid to colonization by specific bacteria or fungi. Typically, initial tests involve seeding bacterial or fungal colonies on the surface of a sterilized film of the material, followed by incubation at a specific temperature and with a controlled humidity in a sterile environment for days or weeks. If colonies grow on the surface, the radius of each circular colonization zone is used as a measure of the ease of colonization. Similar techniques are used to assess the compatibility between a surface and mammalian stem cells or mature cells, a process that also requires an assessment of cell adhesion.

The biological experiments may take many months or years before tests can be carried out in human clinical trials. The costs for the whole sequence of evaluations usually exceed several million dollars.

Membranes for controlled drug delivery are tested using the method depicted schematically in Figure 13.8 (in Chapter 13).

E. UNSOLVED PROBLEMS IN BIOMEDICAL MATERIALS SCIENCE

The field of biomaterials is almost unique in the sense that the rate of progress is limited by the inevitable slowness of biological testing and the enormous costs involved in obtaining safety information. This has encouraged a conservative approach by researchers and manufacturers that strongly favors the use of long-existing substances, some of which were used in medicine before modern regulations were formulated. A number of these materials would not win approval today, but are allowed because of precedent. Thus, the introduction of new candidates has been slow. Nevertheless, the need for new biomaterials is acute because progress in many areas of medicine will be limited unless materials with more appropriate property combinations can be used.

Almost all the sections in this chapter present examples of applications where a serious need exists for new biomaterials. In cardiovascular research there is a shortage of elastomers and rigid materials with better antithrombogenic properties, and of new elastomers with flex strengths that at least equal those of polyurethanes. Polymers with the surface character of poly(tetrafluoroethylene) that can be fabricated more easily as elastomers, films, coatings, or tubes are also required. Materials for stents that are bioerodible and have shape memory, or new corrosion-resistant alloys that prevent tissue irritation, are particularly important. Solids with antimicrobial surfaces are needed in nearly all areas of biomedicine, from cardiovascular devices and catheters to surgical instruments. A continuing need exists for new bioerodible tissue engineering materials for the regeneration of skin—particularly important for burn victims and those suffering from skin cancer.

Systems for the *targeted* delivery of drugs to specific regions of the body are a high priority, with an emphasis on antitumor agents. Much of the current experimental work in this area involves research with vesicles or micelles that serve as targeted carriers and depots for highly toxic drug molecules. New drug delivery membranes and bioerodible polymers are also needed.

Materials progress in dentistry has been more rapid than in other areas, partly due to the ease with which materials can be removed from the oral cavity if they prove unsatisfactory. Nevertheless, new impact-resistant ceramic composites for tooth restoration, regenerated teeth by tissue engineering, and antibacterial, antifungal materials for dentures and denture liners are a serious challenge. So too are biostable materials for maxillofacial reconstruction.

In orthopedic research a major challenge is the development of new bioerodible tissue engineering matrices for the rapid regrowth of bone tissue for the repair of fractures or for possible use in the ex vivo growth of new bones. Similar techniques for the regeneration of auditory ossicles in the middle ear are a related challenge. Better orthopedic materials are also needed for regenerated tendons or cartilage or as artificial materials for tendon and cartilage replacement. Tendon replacement materials need to be strong, elastic, biocompatible, and fiber-forming. Also required are nontoxic cements for use in hip replacement surgery, and lightweight, strong prosthetic materials for spikes and for hip joint components.

A related field where new materials will accelerate progress is the area of implanted sensors for monitoring and controlling glucose, insulin, and antibody levels in the blood.

It will be clear that the field of biomaterials provides challenges that exceed those in almost any other field of materials research. Many of these challenges can be met only by collaborative research between experts in chemistry, materials design, engineering, and medicine.

F. SUGGESTIONS FOR FURTHER READING

1. Langer, R.; Tirrell, D. A., Designing materials for biology and medicine, *Nature*, **428**:487–492 (2004).

2. Karp, J. M.; Langer, R., Development and therapeutic applications of advanced biomaterials, *Current Opinion in Biotechnology*, **18**:454–459 (2007).

3. Ma, P. X., Biomimetic materials for tissue engineering, *Advanced Drug Delivery Reviews*, **60**:184–198 (2008).

4. Kopecek, J., Hydrogel biomaterials: A smart future? *Biomaterials*, **28**:5185–5192 (2007).

5. Furth, M. A.; Atala, A.; Van Dyke, M. E., Smart biomaterials design for tissue engineering and regenerative medicine, *Biomaterials*, **28**:5068–5073 (2007).

6. Guelcher, S. A., *An Introduction to Biomaterials*: CRC/Taylor & Francis, Boca Raton, FL, 2006.

7. Vasita, R.; Shanmugam, K.; Katti, D. S., Improved Biomaterials for Tissue Engineering Applications: Surface Modification of Polymers, *Current Topics in Medicinal Chemistry*, **8**:341–353 (2008).

8. Enderle, J. D., *Introduction to Biomedical Engineering*: Elsevier Academic Press, Boston, MA, 2005.

9. Park, J. B., *Biomaterials: Principles and Applications*; CRC Press, Boca Raton, FL, 2003.

10. Ratner, B. D., A paradigm shift: biomaterials that heal, *Polymer International*, **56**:1183–1185 (2007).

11. Wise, D. L. ed., Human Biomaterials Applications, Humana Press, 1996.

12. El-Nokaly, M. A.; Piatt, D. M.; Charpentier, B. A., Polymeric delivery systems: Properties and applications, ACS Symposium Series No. 520. American Chemical Sociey, Washing ton DC, 1993.

13. Brannon-Peppas, L., Polymers in controlled drug delivery, *Medical Plastics and Biomaterials*, **34**, (Nov. 1977).

14. Anusavice, K., Phillips' Science of Dental Materials 11th ed., Elsevier. (2003).

15. Martin, C. R.; Kohli, P., "The emerging field of nanotube biotechnology," *Nature Rev. Drug Discov.* **2**:29–37 (2003).

16. Greish, Y. E.; Bender, J. D.; Lakshmi, S.; Brown, P. W.; Allcock, H. R.; Laurencin, C. T., Composite Formation from Hydroxyapatite with Sodium and Potassium Salts of Polyphosphazenes. *J. Mater. Sci., Mater. in Medicine* **16**(7):613–621 (2005).

17. El-Amin, S. F.; Kwon, M. S.; Starnes, T.; Allcock, H. R.; Laurencin, C. T., The biocompatibility of biodegradable glycine containing polyphosphazenes: acomparative study in bone, *J. Inorg. Organometal, Polymers and Materials*, **16**:387–397 (2006).

18. Kohn J. (ed.) "Tissue Engineering", *M. R. S. Bull.* **21**(11):18–65 (1996). (Eight articles by different authors).

G. STUDY QUESTIONS

1. What types of hydrolysis products would be toxic if a biomaterial degrades in the body? Be specific. Starting from that point, list different polymers, metals, ceramics, or semiconductors that are *not* appropriate for use in internal medicine.

2. Design a series of new membranes that might be useful for (a) controlled drug delivery, (b) oxygenation of blood, and (c) a replacement for the cornea of the eye, and explain why they would be better than existing biomedical materials.

3. Suggest ways in which collagen-based materials such as porcine heart valves might be sterilized and stabilized against deterioration, using methods other than treatment with formaldehyde.

4. Accelerating the regeneration of living bone is a major challenge. Describe different ways in which this might be accomplished, giving clear reasons why you think your suggestions might work.

5. Polymer hydrogels have many physical property resemblances to living soft tissues. How are hydrogels produced, and how can their properties be tuned for different biomedical applications?

6. Assume that the use of metal spikes for bone repair has been shown to lead to the slow release of potentially toxic transition metal ions into the body. Devise ways to continue the use of this metal but prevent the metal ion release.

7. From your knowledge of nonoxide ceramics discussed in Chapter 7, speculate on the possible uses of these materials in biomedicine. What might be their advantages over materials that are currently in use? What would be the disadvantages?

8. Are there medical applications for which a hydrolytically unstable silicate ceramic might prove useful? If so, describe those uses and specify the chemistry involved.

9. Assume that a new material has been synthesized that is designed to last for 50 years when implanted in the human body, but you have only 6 months to obtain the evidence for that stability. How would you obtain the necessary information?

10. Most bioerodible polymers degrade by nonbiological hydrolysis reactions. Are there polymers that might degrade via enzymatic mechanisms? If so, how would you evaluate their rates of decomposition?

17

Materials in Nanoscience and Nanotechnology

A. BACKGROUND AND MOTIVATION

The purpose of this chapter is to summarize the connections that exist between topics discussed earlier in the book and the rapidly expanding field of nanoscience. Thus, this is not a comprehensive survey of nanoscale materials science, but rather a general overview, and the reader is encouraged to use it as a starting point for further study.

Nanoscience and nanotechnology deal with objects that have dimensions in the range from 1 nanometer (10^{-9} meter) to about 100 nm (1 μm) (Table 17.1). Thus, nanostructures lie in the size regime between the Angstrom level (1 Å = 0.1 nm) and the microscale (from 1 μm to roughly 1 mm). The Angstrom level is typically reserved for discussions of atom sizes, bond lengths, and small molecular structures. The microlevel is the range of integrated electronic circuits, living cells, and coarse texture in solids. Table 17.2 summarizes the classification of size features normally used by the materials science community. Nanoscale structures are a source of intense interest because new properties and unique behavior can be found when the size of objects falls below ~100 nm. These new properties result because the ratio of surface area to volume is now so large that surface effects, such as those that give rise to catalysis, adsorption, and adhesion, become dominant characteristics. For example, a nanoparticle 5 nm in diameter has about half of its atoms at the surface. Moreover, electronic and photonic effects enter a new level because electrons are confined within such a small volume that quantum effects now dominate optical and electronic behavior.

Apart from the fundamental interest in these phenomena, nanoscience and nanotechnology are driven by the accelerating initiative to miniaturize devices. The development of smaller and smaller transistors in integrated circuits (see Chapter

TABLE 17.1. Size Regimes

1 micrometer (μm)	= one-millionth of a meter	1×10^{-6} meter
	= one-thousandth of a millimeter	
1 nanometer (nm)	= one-billionth of a meter	1×10^{-9} meter
	= one-thousandth of a micron	
1 angstrom (Å)	= one-tenth of a nanometer	1×10^{-10} meter
1 picometer (pm)	= one-hundredth of an Angstrom	1×10^{-12} meter

TABLE 17.2. Classification of Different Feature Sizes and Visualization

Name	Size Regime	Features Detectable	Visualization[a]
Macro	≥1 mm	Ordinary objects	Unaided eye
Micro	100 nm–1 mm	Microcircuits, living cells, small crystals, diameter of human hair, texture in coarse composites	Optical microscopy, SEM, TEM
Nano	10 Å to 100 nm	Large molecules, nanoparticles, atomic clusters, nanofibers, nanotubes, texture in composite materials	SEM, STM, AFM
Ångstrom level	1–10 Å	Atoms, small molecules	X-ray or electron diffraction

[a]See Chapter 4.

10) provides an obvious example of the advantages of miniaturization in electronic technology. It is now reasoned that further miniaturization using nanoscale structures could bring about another technical revolution that will impact electronics and photonics, the treatment of diseases, and the development of new classes of ultra-high-performance materials that could propel technology in new directions.

The most obvious motivation for miniaturization below the macro- and microscales is the need for more rapid and more extensive information processing. As implied in Chapter 10, the performance of modern integrated *electronic* circuits is limited by the distances between transistors, the resistance of wiring between transistors, and the need to dissipate heat. Thus, smaller electronic circuits lead to faster processing, lower power consumption, and potential cost savings because the amount of silicon used per chip is correspondingly less. However, the miniaturization of *optical* circuits is an equally desirable objective for rapid information processing and transmission. Thus, nanoscale optical circuits might ovrcome many of the problems with electronic circuits. The miniaturization of medical devices, such as substrates for tissue engineering (Chapter 16), or controlled drug delivery through micellar units that can penetrate the smallest capillaries in the human vascular system, is an urgent need. Nanoscale machines that can penetrate into human cells to restore function or alter the replication process are being contemplated. Catalyst particles with nanoscale dimensions are crucial for many chemical and biochemical processes because of the high ratio of surface to volume. Two other areas of science and technology are fertile fields for the study of nanomaterials. The first includes the use of nanoscale powders of ceramics, polymers, and metals to form high-surface-area materials. The second involves the study of nanoscale domains in polymers, ceramics, and metals as a means to improve solid-state properties such as toughness and processing. In this respect, the "critical scale length"—the size below which new properties are generated—becomes a crucial factor in materials design.

Although widespread interest in nanoscience and nanotechnology is fairly recent, nanoscale structures have been known and studied for almost a century. Metallic colloids, for example, consist of nanoscale particles, as do some ceramic powders and precipitated polymers. However, the recognition that the *size* of particles, fibers, or surface features has an enormous influence on their properties is of fairly recent origin, and it underlies the current enthusiasm for all things "nano." Specifically and realistically, the most important aspects of this field are connected with four issues. The first deals with the way in which nanoscale particles interact with their neighbors in solid arrays or with a second phase in composite materials. The second area of keen interest is the miniaturization of electronic and photonic circuits below the current microscale in order to pack an increasing number of features into an integrated array. This aspect brings forth the need to create *nanowires* to connect nanoscale transistors. Third, the development of methods to produce *nanofibers* is perhaps the most advanced area in terms of proximity to commercial success. Nanofibers are of great interest in biological tissue engineering, advanced textiles, and filtration technology. Fourth, the use of micelles, which have diameters in the nanorange, for controlled drug and vaccine delivery is a subject with a strong future. Applications of nanoscience in molecular machines or nano-robots, although prominent in the popular literature, are probably some distance over the horizon.

The continuing imperative toward miniaturization in nearly all areas of materials science presents numerous challenges for chemists, physicists, and engineers. These challenges range from the synthesis of nanoscale structures and their characterization to the manipulation and organization of them in two and three dimensions. The response of living cells to nanoscale objects is also a matter of growing interest.

B. SYNTHESIS AND FABRICATION OF NANOSTRUCTURES

Two different approaches exist for the preparation of nanostructures. These are known as the "top–down" and "bottom–up" methods. Top–down procedures start with micro- or macroscale materials, which are then broken down chemically or physically to nanoparticles. Methods that are used include grinding and abrasion ("ball milling"), chemical etching, electrospinning, and the vaporization of metals using plasmas or high-energy beams.

Bottom–up procedures involve the assembly of very small units (atoms, molecules, or small nanoparticles) to create larger constructs. It often happens that the units employed in the assembly process are formed by top–down procedures. Hence, the two approaches are intimately connected.

1. "Top–Down" Nanostructure Preparations

The grinding or ball milling of ceramics can reduce them to a fine powder with each individual particle having nanoscale dimensions. The main challenge with this method is to avoid contamination of the nanoparticles by the materials used in the abrasion process. A second disadvantage of this process is that the particle sizes are not uniform. This is not a problem if the particles are to be compressed and consolidated into a porous material, but it is a disadvantage if they are to be assembled into a uniform, regular, repeating three-dimensional array.

In general, polymers cannot be reduced to nanoparticles by grinding because of their molecular stucture (covalent bonds must be broken) and their impact resistance. An alternative is to precipitate a polymer from a solution into a nonsolvent. Under appropriate conditions it may be possible to form spherical particles with diameters below 100 nm. However, uniform polymer nanoparticles are also accessible via bottom–up techniques (see next section). Polymer fibers with diameters near or below 100 nm are produced by the process of electrospinning, as described in a later section.

In principle, nanostructures can be obtained through advanced photolithography techniques with the use of X rays or electron beams. Wavelengths below 100 nm would be needed to obtain the necessary resolution, although practical engineering restrictions provide serious hurdles.

2. "Bottom–Up" Synthesis Methods

Perhaps the simplest example of this approach is the compaction of nanoscale powders to form three-dimensional constructs. Uniform-sized particles, when assembled in this way, can yield regular repeating structures that resemble the atomic arrangements found in metals (see Chapter 8), but obviously on a larger scale. Another example is the assembly of molecules on a surface by moving them with an AFM probe.

Chemical synthesis can also yield nanoparticles. Indeed, some of the earliest nanoparticles recognized as such were metallic colloidal particles produced by reducing metal salts in solution by reagents such as formaldehyde. The size of the nanoparticles controls the color of the colloids so produced, with the larger particles giving a reddish color and the smallest providing a blue or green hue. Semiconductor nanoparticles have been produced by allowing a water-soluble cadmium salt to react with sodium sulfide in a polymer matrix which controls diffusion. However, this method often gives particles with a range of different diameters. Reactions carried out in sol–gel media are also capable of yielding nanostructures.

Chemical synthesis methods are also employed to produce polyhedral boranes and carboranes, fullerenes, and carbon nanotubes. For carbon-based structures, the syntheses depend on the volatilization of carbon atoms at high temperatures. These then recombine to form nanoscale cage molecules or, in the presence of metal catalysts, carbon nanotubes.

Self-assembly, another bottom–up technique, occurs when the *shape* of a molecule or nanostructure favors the formation of a regular repeating arrangement. An example is the way in which alkylthiol molecules coordinate to and assemble on a gold surface (see Chapter 15). The sulfur atom on each thiol molecule forms a coordinative attachment to the gold surface, while the alkyl chains assume a zigzag conformation and pack together with their neighbors to form a uniform pattern. This process can be employed to "image" a surface by means of photolithography of a resist attached to a gold surface. Alternatively, soft contact printing may be employed to lay down thiol molecules on discrete regions of the surface.

Self-assembly is also feasible when molecules enter and assemble within the galleries or pores of porous or layered intercalation solids or zeolites (Chapters 7 and 15) or in the tunnels of a clathrate adduct (Chapter 5). Biological structures such as proteins, nucleic acids, or the inorganic skeletal structures of microorganisms may also be used as templates for self-assembly.

Several of these methods for assembling nanoscale building blocks are discussed in more detail in the remainder of this chapter. The following sections cover a few specific examples where progress in nanoscience is readily apparent.

C. EXAMPLES OF NANOSTRUCTURES

1. Nanofibers

The field of nanofibers has proved to be an area with broad advantages in both advanced technology and biomedicine. Nanofibers are produced mainly by the process of electrostatic spinning. The method is illustrated in Figure 6.7 (in Chapter 6).

As a reminder, in this process, a solution of a polymer in a hypodermic syringe is extruded through a needle by the action of a syringe pump. A direct current potential in the range of 15,000–30,000 V is connected between the metal needle and a metallic target. As each droplet of solution emerges slowly from the needle tip, it is attracted to the target by the potential difference, a process that induces thin fibers of solution to be spun out, accelerated, stretched, and deposited on the target. The solvent evaporates from the fiber during this process. The deposited nanofibers form a random fiber mat, which often resembles the appearance of tissue paper.

In addition, composite fibers can be spun by the use of two different polymers in solution or a suspension of nanoparticles in the polymer solution. Hollow nanofibers or fibers with a core that is different from the surface material are accessible by spinning two solutions through a needle that has concentric tubes. Moreover, the surface of the nanofibers can be smooth or rough depending on fabrication conditions.

In practice, very few polymers can be electrospun to dimensions that are less than 100 nm; most examples range within 100–300 nm. Nevertheless, even though these fibers are at the lower end of the microfiber ranges (Tables 17.1 and 17.2), they are usually described by researchers as nanofibers.

Three advantages of nanofibers over microfibers or films are as follows:

1. The overall surface roughness generated by the fiber mat brings about an increase in hydrophobicity, as indicated by a high water droplet contact angle (Chapter 15). For example, hydrophobic polymers have water contact angles near 100°. Nanofiber *mats* of the same polymer have contact angles of 155–160°, a dramatic increase in hydrophobicity brought about by the presence of air between the nanofibers.

2. The small diameter of nanofibers assists bonding to a second phase when the fibers are used for reinforcement in a polymer composite system.

3. Bioerodible polymer nanofiber mats are increasingly important in mammalian tissue regeneration. The small diameter of the fibers and the free space between them encourages the growth of cells in the form of a three-dimensional tissue. Tissue growth is encouraged by the "foothold" provided by the fibers, while the small diameter of each fiber accelerates bioerosion. An example is shown in Chapter 16 (Figure 16.6b).

Preceramic polymers (see Chapter 7) can also be electrospun, and the nanofibers obtained may be pyrolyzed to a nonoxide ceramic with the same nanofiber mat morphology as the precursor fibers. The use of ceramic nanofibers as reinforcement materials for ceramics and metals is a subject of considerable interest, because this provides a way to increase the strength of a host material in the same way that crystal growth in a metal or ceramic does. An example is the preparation of boron–carbon–silicon ceramic nanofibers from an organic-soluble polymeric precursor. Scanning electron micrographs of one such mat are shown in Chapter 7 as Figure 7.20. The different morphologies are obtained by changes in the concentrations of the solutions, the gap between needle tip and the target, and the rate of extrusion through the needle. The formation of nanofibers by electrospinning creates different morphologies depending on the concentration of polymer, the distance from the extrusion site to the target, and the voltage used. Fiber, beads distributed along nanofibers, and nanoscale beads alone are all accessible by changes to the spinning conditions.

2. Nanowires

Metal or conductive polymer nanowires may eventually be employed in electronic circuits. Silver nanowires have been produced by the reduction of solutions of silver salts. Although not fully understood why, it is known that the presence of poly(vinylpyrolidone) in the solution causes growth of silver mainly from the 100 face of the nanocrystals, and this generates nanofibers, some of which are hollow.

Nanorods are short nanowires that may be produced by metal vapor deposition into nanotubes that penetrate thin films of materials such as aluminum oxide (Chapter 13). Removal of the ceramic by etching in acid then releases the nanorods.

Nanowires can also be assembled from conductive or semiconductive nanoparticles by lining them up in contact with each other. Assembly methods that have been suggested include magnetic attraction, organization within nanochannels etched into a surface, or within tunnels that are α-track-etched through a membrane or that penetrate crystalline solids.

Silicon nanowires have been synthesized via chemical vapor deposition (CVD) techniques from silicon tetrachloride gas. One method utilizes a thin layer of gold nanoparticles deposited on a polymer surface. The nanoparticles act as "seeds" for the parallel growth of silicon nanowires from the surface in a CVD apparatus. The gold nanoparticles control the diameter of the fibers that grow from them. Patterning of the nanowire growth on a silicon surface using soft contact printing of the polymer, followed by deposition of gold nanoparticles, allows nanowires to be grown at strategic locations on a semiconductor chip.

3. Nanoscale Particles

a. Ceramic Nanospheres. As described above, inorganic powders with radii below 100 nm may be obtained by grinding ("sculpting") from the bulk phase, by controlled precipitation or crystal growth, and by chemical reactions. Aerosol procedures have also been employed, as have sonochemical acoustic cavitation methods. Ceramic nanoparticles are raw materials used for consolidation to highly porous solids. Symmetrically packed nanospheres can be used as templates for the synthesis of porous materials. For example, a close-packed, three-dimensional assembly of ceramic nanospheres can be infused with a monomer, which polymerizes in the free space between the spheres. Decomposition of the spheres by acid or base then leaves a polymeric honeycomb (Figure 17.1). Honeycombs of this type have been proposed for a variety of uses such as filters, gas storage containers, and reservoirs for the controlled release of drugs.

Fluorescent silica nanoparticles are of interest for labeling biological molecules. In one report a fluorescent dye is first attached to a (sol–gel) silica precursor to form the core of a nanoparticle. This is then surrounded by a shell of sol–gel silica to give nanoparticles in the size range of 20–30 nm. Encapsulation of the dye in this way gives nanoparticles that fluoresce 20 times brighter than the unprotected dye because the surrounding liquid medium cannot interfere with the fluorescence process. Attachment of biological molecules to the outer shell then allows the location of the nanoparticles to be monitored within a living system by their fluorescence.

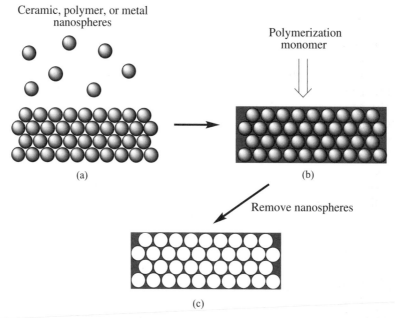

Figure 17.1. *Formation of nanoscale honeycomb structures. (a) Nanospheres of ceramic, polymer, or metal are allowed to form a "crystal"-packing arrangement. (b) The space between nanospheres is filled with a polymerizable liquid (inorganic or organic), which, after polymerization, binds the structure together. (c) The nanospheres are then removed by strong acid or base (ceramic or metal spheres) or pyrolysis (polymer spheres) to leave a honeycomb of polymer or ceramic material.*

b. Polymer Nanospheres. Similar procedures are used to make porous constructs with the use of polymer nanospheres. Nanospheres of polymers are accessible through emulsion polymerization methods in which polymer chain growth takes place at a water–organic solvent interface (see Chapter 6). If a two- or three-dimensional array (crystal) of polymer nanospheres is used to form a template, and a sol–gel ceramic precursor fills in the dead space, then subsequent pyrolysis removes the polymer (as CO_2 and water or depolymerized monomer) to leave a ceramic honeycomb structure (Figure 17.1c).

It has also been reported that drug molecules encased in bioerodible polymeric nanoparticles might be delivered to the patient by inhalation.

c. Metal Nanoparticles. Nanoparticles of metals are typically produced by the treatment of a solution of a metal salt with a reducing agent. The reduction of soluble silver salts by formaldehyde under some conditions gives a silver mirror. Under other conditions the metallic silver precipitates as a colloid of suspended nanoparticles. The color of the colloid depends on the size and shape of the particles. Similar colloids can be prepared from solutions of gold salts. Because the particle size is so small, the nanoparticles are slow to settle to the bottom of the reaction flask. Brownian motion of the solvent molecules keeps them in suspension. The precipitation of metals or metal salts in a molten glass generates similar phenomena, and, as described in Chapter 5, these particles are the basis of photochromism in some sunglasses. Coreduction of two or more metal salts in solution can yield nanoparticles that can be heated to form alloys. In all these examples nanoparticle formation is favored over metal mirror deposition by the presence of water-soluble molecules such as ethylene glycol or poly(vinylpyrolidone) that coat nanoparticles and inhibit coalescence. Gold nanospheres can be assembled into hexagonal close-packed "crystals," and these, too, can be used as templates to generate honycomb structures from materials that fill the space between the spheres.

d. Semiconductor Nanoparticles. As discussed in Chapter 10, materials such as cadmium sulfide precipitate in gels as nanoparticles, and these may be utilized as quantum confinement wells.

e. Fullerenes. Fullerenes ("buckyballs"; buckminster Fullerenes) are polyhedral, electron-delocalized molecules with the formula C_{60} and with nanoscale dimensions. They are formed when carbon is subjected to high temperatures. An early synthesis procedure employed a pulsed laser beam to strike a rotating graphite disk to produce a plasma of vaporized carbon atoms. These are swept from the hot plasma zone by a stream of helium into a cool chamber where they coalesce into clusters. Fullerene C_{60} is the principal product, but C_{70} and other hollow cage clusters are also formed. Fullerene C_{60} has a special stability due to the lower strain inherent in the bond angles. Other methods are available to prepare C_{60}, including the evaporation of graphite electrodes using resistive heating. The C_{60}/C_{70} from the "soot" produced by these methods can be separated first by extraction with benzene and then by column chromatography over alumina. Fullerene C_{60} is a yellowish or brown-colored needle-shaped crystalline solid that gives magenta colored solutions.

The C_{70} crystals are red-brown in color but dissolve in organic solvents to give yellow-red colored solutions. The colors presumably are related to the different particle sizes.

In the solid state the pseudospherical shape of the C_{60} molecules allows them to pack into a solid lattice to yield a face-centered cubic close-packed arrangement, a structure that leaves void space. At room temperature the C_{60} molecules rotate in place within the solid. Metal atoms, such as potassium or rubidium, can be incorporated into the lattice, and these solids become superconductors (Chapter 11) when cooled below ~19 K (K_3C_{60}) and ~28 K (Rb_3C_{60}).

f. Polyhedral Boranes and Carboranes. The discovery of fullerenes was preceded by 40 or 50 years with the synthesis of boron-containing cage structures, some of which show a high level of stability. Polyhedral boranes are prepared by the pyrolysis of small boron hydrides such as diborane, $H_2B—BH_2$. One of the most stable products is the icosahedral $B_{12}H_{12}^{2-}$. These molecules have a delocalized electronic structure within the boron–boron polyhedral framework, with hydrogen atoms attached to each boron. The diameter of each pseudosphere is approximately 8–9 Å. The hydrogen atoms can be replaced by chlorine to provide access to a range of organic and organometallic derivatives. Even more stable are the polyhedral carboranes such as $C_2B_{10}H_{12}$, which are synthesized by the insertion of an acetylene molecule into a nonspherical borane molecule. Polyhedral boranes and carboranes have been studied in great detail, but have not generated as much use-oriented attention as have fullerenes, mainly because the cost of the starting materials is high and scaleup is difficult. However, at least one attempt has been made to use polyhedral carboranes as the wheels of "nanocars."

g. Micelles. As discussed in Chapter 16, micelles are nanometer-sized spherical or wormlike units that are formed when amphiphilic block copolymers are dispersed in water. The hydrophilic block of each macromolecule becomes associated with the aqueous medium to form a "corona", and the hydrophobic block withdraws into the interior to form the "core." The utility of micelles as carriers for hydrophobic drug molecules was mentioned in the Chapter 16, but they are also vehicles for the "solubilization" of hydrophobic dyes and inorganic nanoparticles in aqueous media, a property that is of interest for the dispersion and stabilization of dyes and pigments in liquid media and in solid polymers. Quantum dots of cadmium sulfide have been stabilized in the center of micelles.

4. Nanochannels and Nanotubes

a. Membrane Nanochannels (Nanotunnels). These are pores fabricated in polymer membranes via α-particle irradiation (see Chapter 13). The α-particle tracks penetrate the membrane, break bonds, and provide pathways for subsequent chemical etching. After the tracks are widened by chemical etching, metals or semiconductors are deposited in the tunnels by vapor deposition. Each channel then yields a uniform diameter tube or rod with a length that equals the thickness of the membrane. Dissolution of the polymer in a suitable solvent then yields the free nanotubes or nanorods.

b. Clathrate and Zeolite Nanotunnels. Some compounds with odd shapes crystallize in such a way that nanometer-size channels or tunnels are formed between the molecules (Chapter 5). The space can be occupied by solvent molecules, but these guest molecules can often be removed under vacuum to leave empty tunnels. Two examples are perhydrotriphenylene and a series of spirocyclophosphazenes described in Chapter 5. Typically, the 5–15-Å tunnels are continuous throughout each crystal and are accessible by replacement solvent molecules, by organic polymerization monomers, or even by fully formed polymer molecules. This capacity for guest exchange is the basis of useful separation processes both for small molecules and for polymers.

Zeolites (see Chapter 7) have micro- or nanotunnels that are formed in a different way. Here, a sol–gel-type ceramic synthesis process is used to build a lattice around template molecules that are eventually burned out to leave empty space. The tunnels and chambers are lined with acidic groups that facilitate the re-forming of hydrocarbons. They may also be used to trap a variety of inorganic catalyst molecules, and to bring about molecular separations.

c. Carbon Nanotubes

(1) Description. Carbon nanotubes are cylindrical structures with a network of carbon–carbon bonds arrayed like a coil of chicken wire. Carbon nanotubes exist in a wide variety of different forms that range from single-wall tubes to multiwall structures with 2–50 concentric shells. The tubes may have different diameters and chiral angles (Figure 17.2). They can vary in length from several nanometers or millimeters, or even exist as long multifiber ropes as long as 60 cm.

(2) Synthesis and Fabrication. Carbon nanotubes are synthesized by a variety of methods from elemental carbon by carbon arc discharge, laser ablation of graphite, or chemical vapor deposition. Early arc syntheses yielded a mixture of nanotubes, fullerenes, and amorphous carbon products, but the yield of tubes can be increased dramatically by the laser ablation and CVD techniques, especially when metal nanoparticles such as iron or nickel are used as initiator sites. The diameter of the tubes is often directly related to the diameter of the initiator particles. Single-wall nanotubes have attracted the most intense interest. A typical CVD synthesis procedure uses ethanol or ethylene as the carbon source to give a bundle of nanofibers. The ends of carbon nanotubes are often capped by a curved or cone-shaped terminus. However, the cap is more sensitive to oxidation by concentrated nitric acid, and acid treatment is used to open the ends. Nanofibers may be suspended in a liquid medium, such as sulfuric acid, and spun into an aligned continuous fiber that is coagulated in an ether. It has been demonstrated that metal nanoparticles can be assembled at the opened ends of a severed "rope," and these induce the further growth of nanotubes while maintaining the same diameter. Metal atoms have been induced to enter and fill the space inside nanotubes, and these constructs have been proposed for use as nanowires.

(3) Useful Properties. Carbon nanotubes have a high stiffness and a low density combined with a very large surface area: volume ratio. Thus, one of the earliest proposed uses was as reinforcement agents in polymers to replace carbon fiber

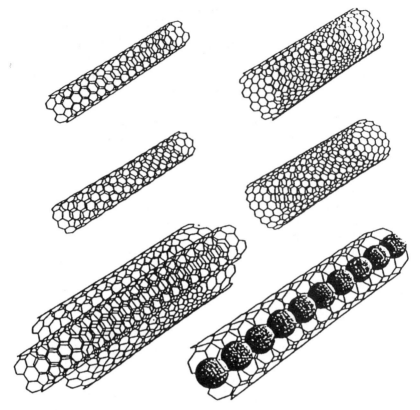

Figure 17.2. *Different forms of carbon nanotubes, including single- and multiple-wall variations and nanotubes containig metal atoms. (From Cox, D. M., in Siegel, R. W.; Hu, E.; Roco, M. C., eds.,* Nanostructure Science and Technology, *Kluwer, 1999, used with permission.)*

(Chapter 9) or carbon powder. However, such uses will depend on the availability of large quantities of nanotubes. One reason why this application is likely to be at the forefront of nanotube technology is that is does not depend on the availability of 100% pure tubes, because impurities such as fullerenes or amorphous carbon particles would not be detrimental to the application.

The discovery of the electronic conductivity of carbon nanotubes may eventually have a major impact on electronic circuits. Carbon nanotubes may be either electronic semiconductors or metal-level conductors depending on tube diameter and chirality. Nanotubes can be manipulated on a surface using AFM techniques, and this offers a mechanism for the assembly of semiconductors into integrated circuits. However, such item-by-item assembly techniques, may be useful only for highly specialized applications. The first significant use for carbon nanotubes has proved to be as electronically conductive reinforcement agents in polymer composites. In addition, much emphasis is being placed on their possible use as field emission materials, which generate electrons when placed in an electric field and are irradiated by laser light.

Carbon nanotubes have been used as templates around which inorganic nanotubes can be fabricated by wet chemical or vapor deposition techniques followed by removal of the template by pyrolysis of the carbon. Structures such as this could, in principle, be employed as reinforcement agents in ceramics.

d. Graphenes. Graphenes are one atom thick sheets of carbon atoms. They can be visualized as the products that would be formed by unrolling carbon nanotubes. Alternatively, they are the products formed if single sheets of carbon can be flaked (exfoliated) from a crystal of graphite. This has in fact been accomplished on a small scale by peeling an adhesive material from a face of graphite. The reason for the growing interest in graphene sheets is that they have the potential to be useful as high carrier mobility semiconductors that could in principle be used as the basis for nanoscale electronic switches and integrated circuits. However, because graphene sheets are soluble in organic solvents a more immediate application may be to print them on suitable surfaces to generate conductive circuits.

5. Nanoscale Features in Electronics and Photonics

The shrinkage of integrated electronic circuits to the nanostructure level involves two challenging questions—first, how to miniaturize semiconductor transistors, and second, how to connect them by nanowires. Current photolithography technology utilizes structures that are 90 nm or 65 nm wide, and 45-nm features are about to be introduced. Transistor formation can, in principle, be further miniaturized by photolithographic procedures. For instance, the use of smaller and smaller imaging wavelengths, such as short-wave UV, electrons, or X rays, may allow silicon wafers to be imaged with feature sizes below 45 nm. However, the surface roughness of a polished silicon wafer becomes a limiting factor at the nanolevel, as does lithographic instrumentation design, including the stepper. For this reason, "writing" on a resist surface by a very narrow beam of electrons rather than the use of photolithography, or the use of soft contact printing (Chapter 15), are alternatives being investigated. In soft contact printing, nanocircuits imaged on a "stamp" are used to print semiconductor circuits using doped, soluble organic polymeric semiconductors as an ink. Circuit wiring may be printed in the same way using polymeric conductors. However, a major problem to be overcome is the stability of polymeric nanowires to electrical burnout, even with the use of very low voltages and currents.

In principle, *photonic* circuits may also be produced by contact printing or by lithography. The ultraminiaturization of optical circuits may prove to be more feasible than the further miniaturization of electronic circuits because the limitations of power density and heat buildup associated with electronic circuits are of less significance in photonic devices.

6. Nanomachines

A definition of a nanomachine is a small nanoscale construct that performs some function such as "swimming" through a solvent by movement of a tail, or rolling across a surface on molecular wheels. The ultimate objective of such devices is to perform some useful activity such as assembling molecules in a predetermined pattern on a surface, or transporting a molecule through a cell wall to a specified location that may not be accessible by diffusion alone. Demonstrations of simple devices of this kind are becoming increasingly common, but translating them into practical applications is probably some distance away.

D. MAJOR CHALLENGES IN NANOSCIENCE AND TECHNOLOGY

The future of nanoscience and technology depends on solutions to four general challenges:

1. A need exists to determine in the broadest possible sense the degree to which theoretical ideas about nanostructures and their properties can be verified by experiments, and the degree to which nanostructures provide a marked improvement over the existing properties of micro- and macromaterials.

2. A major challenge is to synthesize, fabricate, and manufacture in a reproducible manner components with sizes between 1 and 100 nm. This will require the integrated utilization of chemistry, physics, engineering, and perhaps biology.

3. Progress in this field requires the development of better analytical tools to enable scientists and engineers to "see" nanostructures, and to guide their fabrication and assembly, and to monitor their long-term stability.

4. Finally, because the size of nanostructures is in the range where unpredictable biological responses are expected, considerable research is needed to examine the health consequences of inhalation and ingestion of nanoscale materials.

Overall, the future of the field of nanomaterials appears to be extremely promising from a scientific viewpoint, but the rate of progress in technology will depend on the speed with which the four challenges described above can be met. Some of these challenges may take many years before they are resolved. Hence a steady evolution in the technology of the field may be anticipated during the coming decades, rather than a series of dramatic overnight breakthroughs.

E. SUGGESTIONS FOR FURTHER READING

1. Siegel, R. W.; Hu, E.; Roco, M. C., eds., *Nanostructure Science and Technology*, Kluwer Academic Publishers, Boston, 1999.

2. Kamperman, M.; Garcia, C. B. W.; Du, P.; Ow, H.; Wiesner, U., "Ordered mesophase ceramics stable up to 1500° from diblock coplymer mesophases," *J. Am. Chem. Soc.* **126**:14708–14709 (2004).

3. Velev, O. D.; Kaler, E. W., "Structured porous materials via colloidal crystal templating: From inorganic oxides to metals," *Adv. Mater.* **12**:531–534 (2000).

4. Hochbaum, A. I.; Fan, R.; He, R.; Yang, P., "Controlled growth of Si nanowire arrays for device integration," *Nano Lett.* **5**:457–460 (2005).

5. Wang, Y.; Kim, M. J.; Shan, H.; Kittrel, C.; Fan, H.; Ericson, L. M.; Hwang, W.-F.; Arepalli, S.; Hauge, R. H.; Smalley, R. F., "Continued growth of single-walled carbon nanotubes," *Nano Lett.* **5**:997–1002 (2005).

6. Wang, Y.; Maspoch, D.; Zou, S.; Schatz, G. C.; Smalley, R. E.; Mirkin, C. A., "Controlling the shape, orientation, and linkage of carbon nanotube features with nano affinity templates," *Proc. Natl. Acad. Sci.* **103**(7):2026–2031 (Feb. 14, 2006).

7. Shirai, Y.; Osgood, A. J.; Zhao, Y.; Kelly, K. F.; Tour, J. M., Directional control in thermally driven single-molecule nanocars," *Nano Lett.* **5**:2330–2334 (2005).

8. Ow, H.; Larson, D. R.; Srivastava, M.; Baird, B. A.; Webb, W. W.; Weisner, U., "Bright and stable core-shell fluorescent silica nanoparticles," *Nano Lett.* **5**:113–117 (2005).

9. Wang, Y.; Hernandez, R. M.; Bartlett, D. J.; Bingham, J. M.; Kline, T. R.; Sen, A.; Mallouk, T. E., "Bipolar electrochemical mechanism for the propulsion of catalytic nanomotors in hydrogen peroxide solutions," *Langmuir* **22**:10451–10456 (2006).

10. Welna, D. T.; Bender, J. D.; Wei, X.; Sneddon, L. G.; Allcock, H. R., "Preparation of boron carbide nanofibers from poly(norbornenyl-decacarborane) single-source precursor via electrostatic spinning," *Adv. Mater.* **17**:859–862 (2005).

11. Olshavsky, M. A.; Allcock, H. R., "Synthesis of CdS nanoparticles in solution in a polyphosphazene matrix," *Chem. Mater.* **9**:1367–1376 (1997).

12. Hatzor, A.; Weiss, P. S., "Molecular rulers for scaling down nanostructures," *Science* **291**:1019 (2001).

13. Martin, C. R., "Nanomaterials—a membrane-based synthetic approach," *Science* **266**:1961–1966 (1994).

14. Laicer, C. S. T.; Chastek, T. Q.; Lodge, T. P.; Taton, T. A., "Gold nanorods seed coaxial, cylinder-phase domains from block-copolymer solutions," *Macromolecules* **38**:9749–9756 (2005).

15. Sun, Y.; Xia, Y., "Shape-controlled synthesis of gold and silver nanoparticles," *Science* **298**:2176–2179 (2002).

16. Smalley, R. E.; Li, Y.; Moore, V. C.; Price, B. K.; Colorado, R., Jr., Schmidt, H. K.; Hauge, R. H.; Barron, A. R.; Tour, J. M., "Single Wall Carbon Nanotube Amplification: En Route to a Type-Specific Growth Mechanism," *J. Am. Chem. Soc.* **128**:15824–15829 (2006).

17. Lörstcher, E.; Ciszek, J. W.; Tour, J. M.; Riel, H., "Reversible and Controllable Switching of a Single-Molecule Junction," *Small* **2**:973–977 (2006).

18. Morin, J.-F.; Sasaki, T.; Shirai, Y.; Guerrero, J. M.; Tour, J. M., "Synthetic Routes toward Carborane-Wheeled Nanocars," *J. Org. Chem.* **72**:9481–9490 (2007).

19. Rolland, J. P.; Maynor, B. W.; Euliss, L. E.; Exner, A. E.; Denison, G. M.; DeSimone, J. M. "Direct fabrication and harvesting of mondisperse, shape-specific nanobiomaterials," *J. Am. Chem. Soc.* **127**:10096–10100 (2005).

20. Kelly, J. Y.; DeSimone, J. M. "Shape-specific monodisperse nano-molding of protein particles," *J. Am. Chem. Soc.* **130**:5438–5439 (2008).

F. STUDY QUESTIONS

1. Discuss the reasons why nanomaterials are of such great interest. What properties change most as the size of an object is reduced from macroscale, to microscale, to nanoscale, and why?

2. The ultimate foreseeable level of miniaturization of electronic and photonic circuits is at the molecular (i.e., Angstom) level, rather than the nanolevel. How do you think this might be accomplished?

3. Compare the "top–down" and "bottom–up" methods for producing nanoparticles, and decribe the advantages and disadvantages of each.

4. How might the sol–gel process be adapted to produce silicate or titanium dioxide nanoparticles?

5. Highly convoluted, high-surface-area features on a semiconductor surface are of great interest for solar cell development. Suggest ways in which these

features can be constructed and how they would improve the performance of solar cells.

6. Why are bioerodible nanofibers of special interest for mammalian tissue regeneration? What differences would you expect compared to using micro- or macrofibers?

7. Nanomachines that contain cog wheels, or fishtail propulsion features, have been reported. How might such devices be fabricated and assembled? What might they be used for?

8. What uses can you think of for nanoscale honeycomb structures? How are they produced?

9. Discuss some of the obstacles that stand in the way of the large-scale use of nanowires, nanofibers, and nanospheres. Give your opinion on how these obstacles may be overcome.

Glossary

adhesive strength The strength of the interface between a solid substrate and a second phase or layer. Adhesive strength is measured by the force needed to peel the adhered layer from the substrate.

amphiphilic Containing both water-attracting and water-repelling units. Amphiphilic surfaces can become either water attracting or water-repellent depending on conditions.

amphoteric Capable of behaving like an acid or a base under different circumstances.

biocompatible A material that elicits no serious biological response when it is in contact with biological tissues.

bioerodible A material that decomposes, usually by hydrolysis, to nontoxic small molecules when implanted within a living organism.

blow molding A method for forming hollow objects. A tube or bubble of a semi-molten material (usually a polymer or a molten glass) is expanded inside a mold by means of internal gas pressure. The tube expands to coat the inside of the mold, and is then cooled to solidify it. Bottles and flasks are fabricated in this way.

ceramic A material comprised of inorganic elements connected by ionic, coordination, or covalent bonds. Typically a ceramic is a high-melting material with good thermal stability. Although the term is widely used to include inorganic glasses as well as crystalline and composite materials, specialists may restrict use of the word to opaque semicrystalline, composite-type solids. Some "organic" ceramics are also known that are based on highly crosslinked organic or organometallic polymers that are stable at elevated temperatures. In this book the term is used in its broadest sense.

Introduction to Materials Chemistry, by Harry R. Allcock
Copyright © 2008 by John Wiley & Sons, Inc.

chain flexibility or rigidity   The ability of a polymer backbone to undergo twisting (torsional) motions activated by thermal energy or by physical stress.

compressive strength   The resistance of a solid to dimensional distortion when subjected to an applied compressive force.

creep   A distortion of the shape of a solid, often at high temperatures, and frequently a response of the semimolten material to gravity. Creep occurs with uncrosslinked polymers, metals at high temperatures, and some glasses.

CVD   Chemical vapor deposition. See **vapor deposition**.

delocalized electrons   Electrons that can move widely along a molecule or throughout a solid. Such electrons are responsible for many examples of electronic conductivity.

dendrimer   A highly branched molecule with branch points present at regular intervals from a central core. After a few "generations" of branch sites, the molecule will have a pseudospherical geometry.

dimensional stability   The ability of a solid to retain its size and shape over an extended period of time, especially when subjected to a force such as gravity or when heated to high temperatures.

domain   A region within a solid that is physically separated from the surrounding material.

doping   A process for the introduction of "impurities" into a material. The term was originally used to describe the introduction of trace impurity atoms into inorganic semiconductors, but is now more widely used to include the addition of compounds to organic polymers to make them conductive, or of salts to polymers or ceramics to produce ionic conducting solids.

elastomer   A material that is elastic; in other words, a substance that readily changes its shape in response to physical stress or strain. Many elastomers re-form their original shape when the stress or strain is removed, often because entropy factors return them to an internal state of maximum disorder. Crosslinks help to generate elasticity. A rubber band is a good example of an elastomer.

electronic conductivity   The movement of an electric current through a solid in which the current is carried by moving electrons. Often associated with electronic conductivity is "hole conduction" in which positively charged "holes" (regions of positive charge left behind by a moving electron) are the current carriers, moving in the opposite direction. An electronic conductor has a conductivity of roughly $\geq 10 \, \text{S/cm}$. A semiconductor conducts in the range of 10^{-2}–$10^{-6} \, \text{S/cm}$, while an insulator typically has a conductivity of $<10^{-10} \, \text{S/cm}$.

ex vivo   Using living tissues, but in experiments carried outside a living organism.

filler   A particulate or fibrous material added to a polymer or ceramic to increase its strength.

film casting   This is a method for producing films from polymers. A solution of the polymer in a volatile solvent is spread on a flat surface. Evaporation of the solvent leaves a film, which can be peeled from the substrate.

flex strength   The capacity of a solid to resist breaking when subjected to bending motions. Typically, a material with poor flex strength will break after a few flexures. Elastomers often have high flex strength because they can sustain hundreds or thousands of flexural motions without fission or permanent distortion.

gel   A materials phase intermediate between a liquid and a solid. Gels are usually formed when a lightly crosslinked polymer absorbs a liquid but cannot dissolve because of the crosslinks. Gels can be hydrogels (with water as the swelling agent) or organogels (with an organic solvent as the swelling agent).

glass   An amorphous material, often transparent, which is sometimes described as a supercooled liquid. All types of materials can form glasses by cooling from the melt in such a way that crystallization is inhibited.

glass transition temperature (T_g)   A temperature below which the material is a rigid solid and above which it is a more flexible or elastomeric material. Inorganic glasses, semicrystalline inorganic solids, and polymers all have glass transition temperatures. Thermofabrication of these materials must usually be carried out at a temperature above the T_g.

gum   A highly viscous liquid. A characteristic of many gums is that they are able to flow slowly under the influence of gravity. Crosslinking of the molecules in the material converts a gum to a dimensionally stable solid or elastomer.

holes   A hole is a site in a material that lacks an electron that would normally be needed to maintain electrical neutrality. Thus, holes are positively charged, and can move through a delocalized network in the same way as an electron, but in the opposite direction.

hybrid materials   Solids that contain both organic and inorganic components.

hydrogen bonds   Weak bonds formed between hydrogen atoms in, for example, O—H or N—H groups and oxygen atoms in water, ice, ethers, esters, ketones, and other compounds. Hydrogen bonds are broken as the temperature is raised.

hydrophilic   Able to attract water.

hydrophobic   Able to repel water.

impact strength   A measure of the resistance of a material to shattering when subjected to a sharp impact. Glasses and ionic crystalline solids generally have low impact strength, but elastomers can absorb the energy of impact without fracturing.

in vitro   Outside a living organism.

in vivo   Inside a living organism (see also **ex vivo**).

injection molding   This is a fabrication technique in which a molten or semimolten fluid is forced by pressure into a mold, where it solidifies in a predetermined shape. This is a widely used method for materials fabrication.

inorganic material   A solid based on elements other than carbon. In other words, an inorganic material is derived from elements or compounds obtained from minerals. However, graphite, a form of carbon, is present in some mineral deposits, and can be considered to be inorganic.

ionic conductivity   The passage of an electric current through a solid or liquid caused by the movement of ions rather than electrons. Anions and cations move through the medium toward the oppositely charged electrodes.

liquid crystalline   Liquid crystallinity occurs when a solid undergoes a transition from a crystalline phase to a semiordered but not totally molten state when heated.

lithography   A process for transferring a pattern from a "master" to one or many receiver surfaces. Lithography is widely used for printing newspapers and books and, in a specialized way, for producing integrated electronic circuits (see **photolithography**).

lone-pair electrons   A spin-paired electron duo located in a normally nonbonding orbital. Lone-pair orbitals are common features in the oxygen atoms found in ethers, ester, and the nitrogen atoms of amines. Hydrogen bonding occurs when a lone pair forms a weak linkage to the hydrogen atom of an O—H, COOH, or N—H group. Lone pairs are usually present when an atom exists in a valence state that is two lower than its position in the periodic table would suggest. For example, oxygen, in group 6 (16), could in principle form six bonds to other elements. In practice, it prefers to form two bonds and locates the other four electrons in two lone-pair orbitals.

melamine   Tri-amino-s-triazine, $[N = C(NH_2)]_3$. A condensation monomer often treated with formaldehyde to give methylol, namely, $-NHCH_2OH$, derivatives.

melting point   For a covalent small molecule, the melting point is the temperature at which the crystalline structure breaks down completely to form a liquid. The same definition applies to ionic solids and to metals. However, polymers and many ceramics behave differently. At low temperature, the material is a glass or a composite of glass and crystalline material. As the temperature is raised, the material undergoes a phase change called the *glass transition* that converts the glassy part into a semiliquid or an elastomer but does not affect the crystalline domains. Only at higher temperatures do the crystallites melt to give a homogeneous liquid. This is the melting point (T_m).

microcrystalline   A composite material in which small crystals are embedded in an amorphous matrix.

microgel   A microgel is formed by the crosslinking of a polymer or a sol–gel system to give small particles of gel.

microlithography   See photolithography.

oligomer   A molecule with a size between that of a small molecule and a polymer. This term covers a wide range of materials from those with only about 10 repeating units to those with several hundred.

organic material   "Organic" is a frequently misused term. In its narrowest sense, the term means compounds or materials based on the element carbon, such as hydrocarbons or chlorocarbons. However, it is also used to include organosilicon, organophosphorus, and other compounds that contain a heteroelement.

organometallic   Strictly speaking, an organometallic compound is one that contains a metallic element as well as a carbon-based unit. However, in practice, the term is often used to describe organosilicon species and other compounds that contain metalloid elements such as boron, germanium, or arsenic.

photolithography   A process by which a pattern on a "negative" is projected through lenses onto a surface that is coated with a photosensitive thin film. Ultraviolet light is employed in the projection system, which can be configured to reduce the size of the image to yield image features in the micrometer or nanometer range for the production of integrated electronic circuits. Enlargement or same-size reproduction is used in the printing industry to produce newspaper or book printing plates (see Chapter 10).

photostable A material that does not decompose when exposed to visible or ultraviolet light. Photostability usually depends on the radiation wavelength, so the wavelength should always be stated in any discussion. Stability to higher-energy radiation, such as X rays, gamma (γ) rays, and electron beams, is normally described as *radiation stability*, although this term is sometimes used to cover all wavelengths from the γ ray to the microwave region.

poling This is a process that uses a direct-current (DC) high voltage to align polar molecules so that they all point in the same direction.

polymer A very large molecule, typically with a molecular weight higher than 10,000 or containing 1000 or more repeating units. Many different shapes or "architectures" of polymers exist—ranging from linear to branched, star-shaped, dendritic, block copolymers, comb structures, and so on (see Chapter 6).

pyrolysis A process of converting or decomposing materials at elevated temperatures. Pyrolysis is sometimes used to "burn out" the organic parts of an organic–inorganic material. It is also employed to convert one type of material into another. For example, preceramic polymers are converted to nonoxide ceramics by pyrolysis. Because oxidation is a constant possibility when solids are heated in air, pyrolysis often requires heating the material in the absence of oxygen.

reptation The snake-like movement of a polymer molecule moving through a matrix of other polymer molecules.

semiconductor A material with an electrical conductivity higher than that of an insulator but lower than that of a typical metal. Semiconductivity normally falls within the conductance range of 10^{-4}–$1\,S/cm$ (see Chapter 10).

silk–screen printing A process for the mass production of designs (as in printed circuit boards) by painting a design on a fabric with a substance that blocks the pores in specific areas. The screen is then placed in contact with a surface, and a liquid is squeezed through the remaining porous areas to reproduce the pattern on the underlying substrate. This process is used by artists to produce multiple copies of a picture, but it is also widely used to produce low-resolution electronic circuit patterns.

strength Strength is the ability of a material to resist distortion when subjected to compressive or tensile stress or strain, or to resist breakage on impact. Because strength is one of the principal attributes of most materials, the measurement of the various types of strength is a major interest of engineers.

superconductor A material with no resistance to the flow of an electric current (see Chapter 11).

surface turnover Although the surface of a material may appear to be static, many solids are subject to atomic or molecular motions that allow the surface to reorganize. In other words, those units originally on the surface become buried and those below the surface emerge. This can create a problem if a property such as adhesion, biocompatibility, or corrosion protection is compromised.

spin casting A process for the preparation of thin fims usually from polymers. A solution of the polymer in a volatile solvent is allowed to drip onto the center of a rapidly revolving disk or plate. Centrifugal force spread the solution toward the perimeter of the substrate as the solvent evaporates.

tensile strength A measure of the ability of a material to resist breakage or elongation when opposite ends of the object are pulled apart. Tensile strength is important in fiber technology, aircraft design, and many other applications.

thermal stability This term is used to describe the ability of a material to resist decomposition when heated. The thermal behavior usually depends on conditions such as the rate of temperature increase, the time the material is held at a certain temperature, and the behavior when the material is heated in air, vacuum, or an inert gas. *Thermal stability* usually refers to chemical stability, whereas phase changes as the temperature is raised are described as *thermal transitions*. Thermogravimetric analysis (TGA) is widely used to study thermal stability (see Chapter 4).

thermal transitions A thermal transition occurs when a solid undergoes a change of properties when heated, such as glass to elastomer, crystalline to amorphous, or solid to liquid. Thermal transition are detected by observation of the flexibility of a material as the temperature is raised or, more accurately, by techniques such as differential thermal analysis (see Chapter 4).

thermooxidative stability The ability of a material to resist decomposition at high temperatures in the presence of air or oxygen. The presence of oxygen often decreases the thermal stability of carbon-containing materials because of facile carbon–carbon bond cleavage reactions that lead eventually to the formation of carbon dioxide.

thermoplastic A material that softens sufficiently to be fabricated when heated. The word is normally used for polymers.

thermosetting A polymeric material that permanently hardens when heated, usually due to crosslinking.

torsional mobility A term usually applied to covalent chemical bonds. A bond with high torsional mobility can undergo twisting through $360°$ without encountering a serious energy barrier. On the other hand, a bond with a high torsional barrier requires the input of a large amount of energy (thermal or radiation energy) before it can surmount the barrier and access another conformation.

torsional strength The ability of a solid to survive multiple cycles of torsion before failing.

transparency The property of many solids to transmit radiation without absorbance or scattering in a target region of the electromagnetic spectrum.

ultrastructure This term refers to substances that are of very high molecular weight, usually branched, dendritic, or crosslinked, where the object you can see with the unaided eye is one large molecule.

vapor deposition A method for coating a surface with molecules or molecular fragments that have been generated using heat, plasma, or radiation-induced vaporization. Also known as *chemical vapor deposition* (CVD).

Index